ULTRASONIC WAVES

Edited by **Auteliano Antunes dos Santos Júnior**

Ultrasonic Waves
Edited by Auteliano Antunes dos Santos Júnior

Published by InTech
Janeza Trdine 9, 51000 Rijeka, Croatia

Publishing Process Manager Bojan Rafaj

Technical Editor Teodora Smiljanic

Cover Designer InTech Design Team

Additional hard copies can be obtained from orders@intechopen.com

Ultrasonic Waves, Edited by Auteliano Antunes dos Santos Júnior
p. cm.
ISBN 978-953-51-0201-4

For EU product safety concerns:
IN TECH d.o.o., Prolaz Marije Krucifikse Kozulić 3, 51000 Rijeka, Croatia,
info@intechopen.com or visit our website at intechopen.com.

Meet the editor

Auteliano Antunes dos Santos Junior is an Associate Professor and Head of the Mechanical Design Department at the University of Campinas – Unicamp. He obtained his bachelor degree in Mechanical Engineering from the University of the State of São Paulo in 1987, his MS in Mechanical Design in 1992, and PhD in Mechanical Engineering in 1996, both from Unicamp. From 1998 to 1999, he worked as a Visiting Scholar at the Texas A&M University, USA researching acoustoelasticity. His main subjects of research are stress measurement using ultrasound (acoustoelasticity) and ESPI (electronic speckle pattern interferometry), tribology and brake design modeling of elasto-plastic contact between rolling bodies and multi-body dynamic analysis of railroad freight cars.
He developed partnerships for research and service projects with companies like VALE, Petrobras, Faiveley Transport, Thyssenkrup Metalúrgica Campo Limpo, Fras-le and Union Park Limited, as well as with research foundations like São Paulo Research Foundation (FAPESP), National Council for Scientific and Technological Development (CNPq) and Air Force Office for Scientific Research – USA (AFOSR).

Contents

Preface

The idea of using ultrasonic waves in medical and engineering fields came from military applications of sound waves in sonar systems. Sokolov, Firestone, Simons, and several other researchers are among the ones who studied this high frequency kind of wave in the first half of twentieth century.

Ultrasonic waves are defined as those with frequencies higher than 20 kHz. The constant development of data acquisition systems and probes seems to work, removing the upper limit of the wave frequency that can be acquired. The phenomenal improvement of this technique, allied with its simplicity and the low cost of the required equipment transformed the initially focused technique to one of the most accepted ways to inspect biological structures, fluid flow, and mineral bodies.

Today, medical applications of ultrasonic techniques (UT) allow physicians to see virtually all parts and details of organic bodies, permitting them to detect abnormalities. The same happens with the inspection of non-organic bodies, for which non-destructive and semi destructive ultrasonic methods have been standardized as the principal way to evaluate safety and quality. Fluid flow and its contents and uniformity are among the important applications, mainly in oil industry and environmental studies.

This book is part of the effort to spread the knowledge about ultrasonic applications and the theory involved. In it, a group of outstanding researchers present insights into their respectivebranches of study related to ultrasonic waves. The book covers topics from different areas of study without losing the focus on innovative developments.

Chapter 1 presents modeling of the generation and propagation of ultrasonic waves in cylindrical waveguides, based on Pochhammer-Chree equations and modal analysis. The authors show the theoretical background used in the development of a code named PCDISP, and application of the methodology to study the wave propagation in a hollow cylinder.

Chapter 2 describes the application of ultrasonic pulses to create images to be used in medical diagnosis. 3D projection techniques are discussed and 2D ultrasonic arrays are presented as an evolution of the technique to be utilized with soft tissue. Tomographic reconstruction is one of the benefits reached with rotation of pairs of probes, and 3D reconstruction of internal structures is possible.

Chapter 3 shows the study of the radiated beam from phased array transducers (PA) in solids. The authors present the principles involved in using PA transducers and the way to effectively set the delay time required to avoid grating and side lobes. Frequency and time delay evaluations are presented. The technique is used to evaluate a welded T-joint with adequate results.

Chapter 4 discusses the influence of Goldberg's number on the validity domain for quasi-linear approach used with finite amplitude acoustic waves. The parameter measures the relative importance of nonlinear and dissipative phenomena that appears when ultrasonic waves propagate inside biological media.

Chapter 5 is an investigation into the acoustic properties of gas hydrate-bearing sediments. Two kinds of ultrasonic methods, namely the flat-plate transducers and the one based on a new kind of bender elements, have been successfully used in measuring such properties. The study of gas hydrates is becoming highly important due to their application as a possible source of energy.

Chapter 6 introduces a new probe design to generate intense ultrasonic waves in air. The fixture is based on a flexural vibrating plate and is used to remove unnecessary gas. The author shows successful experiments with a low-hydrophilicity gas in water particles of two sizes, proving the efficiency of the technique.

In Chapter 7, ultrasonic Doppler technique is applied to measure velocity and concentration in two-phase flows. The authors propose a new method, expanding the characteristics of the conventional PDUV, which only gives the velocity as result. Their approach was tested in an experimental set up specially built for this purpose, and the results proved that the method is effective for small concentrations of the phase under analysis.

The application of ultrasonic thrusters as an alternative small-scale propulsion system for underwater robotic devices is the problem explored in Chapter 8. The authors present a comparison among the main technologies for thrusters and describe the characteristics of UT thrusters. They also present a detailed outline of the construction of a prototype and an analysis of the behavior of UT transducers in such devices.

Chapter 9 addresses real time positioning of robots using spread spectrum ultrasonic waves. The authors built a system that is able to evaluate positioning, receive the signal from sensors, calculate correlation, detect peak, and estimate the distance from a known position. All tasks were accomplished in less than 6 microseconds, with resolution of 1.5 cm.

Application of ultrasonic waves to inspect rocks is the main theme studied by the authors in Chapter 10. They present a method to find discontinuities in marble rocks based on the method used for evaluation of concrete. The simplicity, low cost, and adequate results are pointed out and confirmed by experiments.

The main subject of Chapter 11 is the design of an ultrasonic system for Tomography based on an array of transducers. The authors developed a method able to identify the size and position of gas bubbles in measurement columns. Thresholding techniques were used to improve the resulting images, making the methodology useful to measure liquid/gas concentrations.

The corrosion products that cover the pits in stainless steel are responsible for the growth of these pits. Chapter 12 addressed the ultrasonic method to remove such corrosion products, resulting in extending life of underwater structures. The authors studied the effect of power of the UT transducer, as well as its frequency and distance where it is placed. From those parameters, the best setting of the system was achieved.

Concluding the book, Chapter 13 presents the new trends in material characterization using ultrasound. The authors concentrate their analysis on concrete samples and aerospace composites. They show that it is possible to characterize mortar specimens with inhomogeneities (small flakey inclusions). Besides that, they show that ultrasonic microscopy and nonlinear acoustics can be used to inspect aerospace materials. The chapter also shows the limits of the techniques and presents suggestions for improving them.

The editor would like to congratulate the authors for their contribution to this book. It's a privilege to work with researchers filled with such enthusiasm for scientific work. The quality of the book is a natural result of the competency and dedication of each of the participants. The editor trusts that the reader will find this book to be a valuable and up to date source of knowledge.

Prof. Auteliano Antunes dos Santos Júnior
School of Mechanical Engineering
University of Campinas - Unicamp
Brazil

1

Modelling the Generation and Propagation of Ultrasonic Signals in Cylindrical Waveguides

Fernando Seco and Antonio R. Jiménez
Centro de Automática y Robótica (CAR)
Ctra. de Campo Real, Madrid
Consejo Superior de Investigaciones Científicas (CSIC)-UPM
Spain

1. Introduction

Elongated cylindrical structures like rods, pipes, cable strands or fibers, support the propagation of mechanical waves at ultrasonic frequencies along their axes. This waveguide behaviour is used in a number of scientific and engineering applications: the Non Destructive Evaluation (NDE) of the structural health of civil engineering elements for safety purposes (Rose, 2000), in linear displacement sensors (Seco et al., 2009) for high accuracy absolute linear position estimation, in the evaluation of material properties of metal wires, optical fibers or composites (Nayfeh & Nagy, 1996), and as fluid sensors in pipes transporting liquids (Ma et al., 2007). These applications demand exact quantitative models of the processes of wave generation, propagation and reception of the ultrasonic signals in the waveguides.

The mathematical treatment of mechanical wave propagation in cylindrical structures was provided by J. Pochhammer and C. Chree at the end of the XIX century, but its complexity prevented researchers from obtaining quantitative results until the advent of computers. D. Gazis (Gazis, 1959) reported the first exact solutions of the Pochhammer-Chree frequency equation, as well as a complete description of propagation modes and displacement and stress distributions for an isotropic elastic tube, found with an IBM 704 computer. Since then, the literature on the topic has grown steadily, and references are too numerous for this book chapter. We will only mention a few landmark developments: the study of multilayered waveguides beginning with a composite (two-layer) cylinder by H. D. McNiven in 1963; the extension of Gazis' results to anisotropic waveguides, initiated by I. Mirsky in 1965; the consideration of fluids and media with losses surrounding, or contained in the waveguides, beginning with V. A. Del Grosso in 1968; and finally, the demonstration of ultrasonic guided waves generated with electromagnetic transducers by W. Mohr and P. Holler in 1976, and piezoelectrically by M. Silk and K. Bainton in 1979, for the nondestructive testing of pipes.

1.1 Modelling the response of the waveguide to external excitation

Of particular importance for transducer design is the determination of the mechanical response of a waveguide when subjected to an external excitation. Several approaches exist to consider this problem.

Integral transform methods (Graff, 1991) convert the differential equations that physically model the excitation forces and the behaviour of the waveguide into a set of algebraic equations, which are more easily solvable. However, in order to find the actual distribution of the elastic field excited in the waveguide, inverse contour integration in the complex plane has to be performed, which is usually complicated. Due to the complexity of the Pochhammer-Chree equations, this procedure is only practical with simplified versions of the wave equation, which in general are not accurate enough for ultrasonic frequencies. See for example, Folk's solution for the transient response of a semi-infinite rod to a step pressure applied to its end (Folk et al., 1958).

The **Semi-Analytical Finite Element (SAFE)** method is a modification of Finite Element Methods (FEM) in which the elastic field is expanded as a superposition of harmonic waves in the azimuthal-axial (θ-z) plane, while discretized mechanical equations are used in the radial (r) direction of the waveguide. This reduction of the number of dimensions permits a much higher efficiency in the computation of the elastic fields (Hayashi et al., 2003). Waveguides surrounded by infinite media (like a pipe submerged in soil) can be handled by SAFE techniques with proper discretized elements (Jia et al., 2011), as well as waveguides with arbitrary profiles: for example, a railroad rail in (Damljanovic & Weaver, 2004). Although finite element methods are powerful and flexible, they have the shortcoming of great requirements on computer memory and processing time when large structures or high frequencies of operation are considered, and the difficulty encountered in the parameterization of transducer designs (for example, the determination of the transfer function of the transducer-waveguide coupling).

Spectral methods are another numerical technique which approximate the differential elastic equations of the waveguide (Helmholtz equations) by differentiation operators, turning the problem of finding the wavenumber-frequency roots into a matrix eigenvalues determination (Doyle, 1997). This numerical method, which is computationally simple and reportedly does not suffer from the problems associated with large diameter waveguides at high frequencies, has been recently applied to model multi-layered cylindrical waveguides (Karpfinger et al., 2008).

Modal analysis is an analytical method based on the expansion of the forcing terms acting in the waveguide into the set of its proper modes (Auld, 1973). In (Ditri & Rose, 1992), modal analysis is employed to model the loading of a waveguide by a transducer array. This treatment is extended to more general transducers and antisymmetric modes by (Li & Rose, 2001). Modal analysis is a mathematically exact technique that leads to a closed form integral equation for the elastic fields in the waveguide, and which incorporates in a natural way the issue of mode selectivity, offering insight on the physics of waveguide behaviour. For these reasons, modal analysis will be the approach used in this work.

1.2 Intention and scope of the research

With this book chapter we contribute a numerical simulation treatment of the ultrasonic behaviour of cylindrical waveguides, based on the Pochhammer-Chree (PC) theory, and covering the aspects of assembly of the description matrix of the waveguide, tracing of the frequency-wavenumber curves, computation of mode shapes, use of modal analysis to determine the response of the waveguide to external excitations, and the dispersive propagation of signals.

The work described here has resulted in a software package, named PCDISP, written in the Matlab environment (*Matlab*, 2004), and freely available[1] to be adapted to particular

[1] PCDISP webpage: `http://www.car.upm-csic.es/lopsi/people/fernando.seco/pcdisp`

circumstances. The main features of the PCDISP software will be introduced in this chapter alongside with the theoretical concepts upon which it is based.

The purpose of PCDISP is freeing the researchers from the numerically delicate, time consuming issues arising in the solution of the PC equations, such as the creation of the waveguide matrix, the numerical instabilities encountered when the thickness of the waveguide or the operating frequency are high, the determination of proper modes and the tracing of the dispersive wavenumber-frequency curves. In this way, the researcher can concentrate in the study of the waveguide/transducer interaction as such.

As far as we are aware of, only two other software suites specifically designed for modelling elastic wave propagation in cylindrical waveguides exist. Disperse (Pavlakovic & Lowe, 1999) is a commercial package, based on matrix techniques, capable of analyzing cylindrical or plate waveguides made of perfectly elastic or damped solids, as well as fluids. GUIGUW (Bocchini et al., 2011) is a Matlab-based software which utilizes a SAFE-based approach to model ultrasonic propagation in cylindrical, plate, and arbitrary cross section waveguides. However, none of these computer solutions permit to model the waveguide response to external excitations.

The organization of this chapter is detailed next. Section 2 briefly reminds the mathematical background of the PC theory. Section 3 properly describes the main features of our methodology and how it is implemented in the PCDISP package. Two common transducer setups for the generation of ultrasonic waves are studied in section 4 with the help of PCDISP. Finally, we will offer some conclusions and point to lines in which this research could be further extended.

2. Background and nomenclature

In this section we present a summarized theoretical background on wave propagation in cylindrical waveguides, treating such aspects as relevant for our purposes; standard references can be consulted for further information (Graff, 1991; Meeker & Meitzler, 1972; Rose, 1999).

A **waveguide** is a physical structure which supports the propagation of mechanical waves along its elongated direction z, and modifies the behaviour of such waves with respect to free propagation in the bulk material. There are two fundamental characteristics of waveguide propagation. The first is the discretization of waves into **propagating modes**, of which only a finite number are permitted for a given frequency, and whose properties are determined by the shape of the cross section and boundary conditions of the waveguide. The second is the existence of **dispersion**, which is the nonlinear relationship between wavenumber and frequency. As a consequence, signals with a significant bandwidth are distorted as they travel along the waveguide, because their spectral components propagate at different phase speeds.

The solutions of the wave equation in a cylindrical material are readily found by the use of potentials and the technique of separation of variables, arriving at the following general form for the displacement vector ($\widehat{u}$) and stress tensor ($\widehat{\sigma}$):

$$\widehat{u}(r,\theta,z) = \widetilde{u}(r,\theta)e^{jkz} = u(r)e^{jn\theta}e^{jkz} \qquad \widehat{\sigma}(r,\theta,z) = \widetilde{\sigma}(r,\theta)e^{jkz} = \sigma(r)e^{jn\theta}e^{jkz}, \tag{1}$$

where the cylindrical system is used (with coordinates (r,θ,z), and unit vectors (e_r, e_θ, e_z)), harmonic time variation $e^{-j\omega t}$ is assumed, and ω is the angular frequency, k the wavenumber, and integer n is a separation constant called the circumferential order, which determines the symmetry of the solutions in the azimuthal direction.

The radial dependent part of the displacement vector and stress tensor is expressed in matrix form as (Gazis, 1959):

$$u(r) = \begin{bmatrix} u_r(r) \\ u_\theta(r) \\ u_z(r) \end{bmatrix} = D^u(r) \cdot \begin{bmatrix} \mathrm{L}_+ \\ \mathrm{L}_- \\ \mathrm{SV}_+ \\ \mathrm{SV}_- \\ \mathrm{SH}_+ \\ \mathrm{SH}_- \end{bmatrix}, \qquad \sigma(r) = \begin{bmatrix} \sigma_{rr}(r) \\ \sigma_{\theta\theta}(r) \\ \sigma_{zz}(r) \\ \sigma_{\theta z}(r) \\ \sigma_{rz}(r) \\ \sigma_{r\theta}(r) \end{bmatrix} = D^\sigma(r) \cdot \begin{bmatrix} \mathrm{L}_+ \\ \mathrm{L}_- \\ \mathrm{SV}_+ \\ \mathrm{SV}_- \\ \mathrm{SH}_+ \\ \mathrm{SH}_- \end{bmatrix}. \qquad (2)$$

The amplitude coefficient vector $A = [\mathrm{L}_+\ \mathrm{L}_-\ \mathrm{SV}_+\ \mathrm{SV}_-\ \mathrm{SH}_+\ \mathrm{SH}_-]^T$ consists of longitudinal (L), shear vertical (SV), and horizontal (SH) deformation components, and the $+$ and $-$ terms stand for perturbations propagating in the direction of increasing and decreasing radius, respectively.

The coefficients of matrices D^u and D^σ are of the general form $D_{ij}(r; n, k, \omega, c)$, with c being the elastic compliance tensor of the solid. These matrices are given explicitly in tables 1 and 2, for the case of a mechanically isotropic material. They have been obtained from the equations of motion with a symbolic computation program (*Maple*, 2007), and match those found in (Gazis, 1959), except for some typographical errors in the paper, also propagated to later works as (Graff, 1991). Matrices D^u and D^σ are implemented in the PCDISP package in function `pcmat`.

In tables 1 and 2, $\alpha^2 = \omega^2/c_{\mathrm{vol}}^2 - k^2$ and $\beta^2 = \omega^2/c_{\mathrm{rot}}^2 - k^2$, where c_{vol} and c_{rot} are respectively the volumetric and rotational speeds of the solid (Rose, 1999). Functions $Z_n(x)$ and $W_n(x)$ are two independent solutions of Bessel's differential equation, with, in general, complex arguments $x = \alpha r, \beta r$. Of the possible choices for $Z_n(x)$ and $W_n(x)$, the numerical stability of the frequency equation determinant is increased when Bessel's ordinary functions $J_n(x)$ and $Y_n(x)$ are employed for real arguments, and the modified Bessel functions $I_n(x)$ and $K_n(x)$ for purely imaginary arguments. With this choice, the contributions to the elastic field from the standing waves propagating towards increasing (+) and decreasing (-) radius are separated (see section 3.1.4 for more on the stability of the frequency equation determinant). To cope with the fact that the recurrence relationships between Bessel's ordinary functions are different from those of the modified functions (Abramowitz & Stegun, 1964), signs λ_1 and λ_2 are introduced, following the scheme of table 3. For complex wavenumbers, PCDISP uses the ordinary Bessel functions $J_n(x)$, $Y_n(x)$ with complex values, and $\lambda_1 = \lambda_2 = 1$.

The solutions of the wave equation are classified in family modes according to their **symmetry properties**, which depend on the circumferential index n of equation 1. Modes for which $n = 0$ have no dependence on the azimuthal coordinate θ and are labelled axisymmetric. They are further divided into **torsional** modes $\mathrm{T}(0, m)$ (which only involve the azimuthal component, and can be thought of as waves which twist the waveguide), and **longitudinal** modes $\mathrm{L}(0, m)$ (with both radial and axial components). Antisymmetric modes ($n \geq 1$) are labelled **flexural** $\mathrm{F}(n, m)$, and involve all three components of the displacement vector. In general, multiple propagating modes exist for a given circumferential order and frequency, so a second index m is used to order them. Table 4 summarizes this information.

$D^u_{11} = nW_n(\alpha r) - \alpha r W_{n+1}(\alpha r)$
$D^u_{12} = nZ_n(\alpha r) - \lambda_1 \alpha r Z_{n+1}(\alpha r)$
$D^u_{13} = krW_{n+1}(\beta r)$
$D^u_{14} = krZ_{n+1}(\beta r)$
$D^u_{15} = nW_n(\beta r)$
$D^u_{16} = nZ_n(\beta r)$
$D^u_{21} = jnW_n(\alpha r)$
$D^u_{22} = jnZ_n(\alpha r)$
$D^u_{23} = -jkrW_{n+1}(\beta r)$
$D^u_{24} = -jkrZ_{n+1}(\beta r)$
$D^u_{25} = jnW_n(\beta r) - j\beta r W_{n+1}(\beta r)$
$D^u_{26} = jnZ_n(\beta r) - j\lambda_2 \beta r Z_{n+1}(\beta r)$
$D^u_{31} = jkrW_n(\alpha r)$
$D^u_{32} = jkrZ_n(\alpha r)$
$D^u_{33} = j\lambda_2 \beta r W_n(\beta r)$
$D^u_{34} = j\beta r Z_n(\beta r)$
$D^u_{35} = 0$
$D^u_{36} = 0$

Table 1. Coefficients of the displacement matrix D^u of equation 2 (all D^u_{ij} coefficients must be multiplied by $1/r$).

3. Methodology for the simulation of the waveguide behaviour

In this section we describe the methodology used to study waveguide generation and propagation of ultrasonic signals, discuss the numerical issues encountered, and the approach used in the PCDISP package. The full process consists in four stages: assembly of the waveguide description matrix, tracing of the dispersion curves, modal analysis of the excited modes, and modelling of the signal propagation along the waveguide.

While dealing with these topics, we will introduce the relevant PCDISP routines that should be used for the computations. Table 5 shows the components of the PCDISP software, arranged by their functionality. Throughout this chapter, we will use monospace fonts (like `pcmat`) to refer to programs of the package.

3.1 Assembling the waveguide description matrix

The waveguide description matrix contains the necessary information to study the mechanical behaviour of the waveguide. It is built by matching the displacement vector and stress tensor between adjacent layers, and applying the external boundary conditions. In PCDISP, the physical data of the waveguide is provided in routine `pcwaveguide`, and the description matrix itself is built in `pcmatdet`.

$D^{\sigma}_{11} = ((k^2 - \beta^2)r^2 + 2(n-1))W_n(\alpha r) + 2\alpha r W_{n+1}(\alpha r)$
$D^{\sigma}_{12} = ((k^2 - \beta^2)r^2 + 2(n-1))Z_n(\alpha r) + 2\lambda_1 \alpha r Z_{n+1}(\alpha r)$
$D^{\sigma}_{13} = 2\lambda_2 \beta k r^2 W_n(\beta r) - 2(n+1)krW_{n+1}(\beta r)$
$D^{\sigma}_{14} = 2\beta k r^2 Z_n(\beta r) - 2(n+1)krZ_{n+1}(\beta r)$
$D^{\sigma}_{15} = 2n(n-1)W_n(\beta r) - 2n\beta r W_{n+1}(\beta r)$
$D^{\sigma}_{16} = 2n(n-1)Z_n(\beta r) - 2n\lambda_2 \beta r Z_{n+1}(\beta r)$
$D^{\sigma}_{21} = ((2\alpha^2 - \beta^2 + k^2)r^2 - 2n(n-1))W_n(\alpha r) - 2\alpha r W_{n+1}(\alpha r)$
$D^{\sigma}_{22} = ((2\alpha^2 - \beta^2 + k^2)r^2 - 2n(n-1))Z_n(\alpha r) - 2\lambda_1 \alpha r Z_{n+1}(\alpha r)$
$D^{\sigma}_{23} = 2(n+1)krW_{n+1}(\beta r)$
$D^{\sigma}_{24} = 2(n+1)krZ_{n+1}(\beta r)$
$D^{\sigma}_{25} = -2n(n-1)W_n(\beta r) + 2n\beta r W_{n+1}(\beta r)$
$D^{\sigma}_{26} = -2n(n-1)Z_n(\beta r) + 2n\lambda_2 \beta r Z_{n+1}(\beta r)$
$D^{\sigma}_{31} = (2\alpha^2 - \beta^2 - k^2)r^2 W_n(\alpha r)$
$D^{\sigma}_{32} = (2\alpha^2 - \beta^2 - k^2)r^2 Z_n(\alpha r)$
$D^{\sigma}_{33} = -2\lambda_2 \beta k r^2 W_n(\beta r)$
$D^{\sigma}_{34} = -2\beta k r^2 Z_n(\beta r)$
$D^{\sigma}_{35} = 0$
$D^{\sigma}_{36} = 0$
$D^{\sigma}_{41} = -2nkrW_n(\alpha r)$
$D^{\sigma}_{42} = -2nkrZ_n(\alpha r)$
$D^{\sigma}_{43} = k^2 r^2 W_{n+1}(\beta r) - \lambda_2 n \beta r W_n(\beta r)$
$D^{\sigma}_{44} = k^2 r^2 Z_{n+1}(\beta r) - n\beta r Z_n(\beta r)$
$D^{\sigma}_{45} = -nkrW_n(\beta r) + \beta k r^2 W_{n+1}(\beta r)$
$D^{\sigma}_{46} = -nkrZ_n(\beta r) + \lambda_2 \beta k r^2 Z_{n+1}(\beta r)$
$D^{\sigma}_{51} = 2jnkrW_n(\alpha r) - 2jk\alpha r^2 W_{n+1}(\alpha r)$
$D^{\sigma}_{52} = 2jnkrZ_n(\alpha r) - 2j\lambda_1 k \alpha r^2 Z_{n+1}(\alpha r)$
$D^{\sigma}_{53} = j\lambda_2 n \beta r W_n(\beta r) - j(\beta^2 - k^2)r^2 W_{n+1}(\beta r)$
$D^{\sigma}_{54} = jn\beta r Z_n(\beta r) - j(\beta^2 - k^2)r^2 Z_{n+1}(\beta r)$
$D^{\sigma}_{55} = jnkrW_n(\beta r)$
$D^{\sigma}_{56} = jnkrZ_n(\beta r)$
$D^{\sigma}_{61} = 2jn(n-1)W_n(\alpha r) - 2jn\alpha r W_{n+1}(\alpha r)$
$D^{\sigma}_{62} = 2jn(n-1)Z_n(\alpha r) - 2jn\lambda_1 \alpha r Z_{n+1}(\alpha r)$
$D^{\sigma}_{63} = -j\lambda_2 \beta k r^2 W_n(\beta r) + 2jkr(n+1)W_{n+1}(\beta r)$
$D^{\sigma}_{64} = -j\beta k r^2 Z_n(\beta r) + 2jkr(n+1)Z_{n+1}(\beta r)$
$D^{\sigma}_{65} = j(2n(n-1) - \beta^2 r^2)W_n(\beta r) + 2j\beta r W_{n+1}(\beta r)$
$D^{\sigma}_{66} = j(2n(n-1) - \beta^2 r^2)Z_n(\beta r) + 2j\lambda_2 \beta r Z_{n+1}(\beta r)$

Table 2. Coefficients of the stress matrix D^{σ} of equation 2 (all D^{σ}_{ij} coefficients must be multiplied by G/r^2, where G is the shear modulus of the material).

Wavenumber	Frequency range	Coefficients	Bessel functions
Real	$\omega/k < c_{\text{vol}}, c_{\text{rot}}$	$\alpha^2, \beta^2 < 0$ $\lambda_1, \lambda_2 = -1$	$Z_n(\alpha r) = I_n(\alpha r)\ W_n(\alpha r) = K_n(\alpha r)$ $Z_n(\beta r) = I_n(\beta r)\ W_n(\beta r) = K_n(\beta r)$
Real	$c_{\text{rot}} < \omega/k < c_{\text{vol}}$	$\alpha^2 < 0, \beta^2 > 0$ $\lambda_1 = -1, \lambda_2 = 1$	$Z_n(\alpha r) = I_n(\alpha r)\ W_n(\alpha r) = K_n(\alpha r)$ $Z_n(\beta r) = J_n(\beta r)\ W_n(\beta r) = Y_n(\beta r)$
Real	$\omega/k > c_{\text{vol}}, c_{\text{rot}}$	$\alpha^2, \beta^2 > 0$	
Imaginary	any	$\lambda_1, \lambda_2 = 1$	$Z_n(\alpha r) = J_n(\alpha r)\ W_n(\alpha r) = Y_n(\alpha r)$
Complex	any	α^2, β^2 complex $\lambda_1, \lambda_2 = 1$	$Z_n(\beta r) = J_n(\beta r)\ W_n(\beta r) = Y_n(\beta r)$

Table 3. Choice of Bessel functions in the solution of the Pochhammer-Chree's equations.

Modes	Coefficients	Displacement	Stress
Torsional $\text{T}(0, m)$	$\text{SH}_\pm$	u_θ	$\sigma_{\theta z}, \sigma_{r\theta}$
Longitudinal $\text{L}(0, m)$	$\text{L}_\pm, \text{SV}_\pm$	u_r, u_z	$\sigma_{rr}, \sigma_{\theta\theta}, \sigma_{zz}\ \sigma_{rz}$
Flexural $\text{F}(n, m)$	$\text{L}_\pm, \text{SV}_\pm, \text{SH}_\pm$	u_r, u_z, u_θ	$\sigma_{rr}, \sigma_{\theta\theta}, \sigma_{zz}, \sigma_{rz}, \sigma_{\theta z}, \sigma_{r\theta}$

Table 4. Notation and non-null components of the amplitude coefficients, displacement vector, and stress tensor, for the three family modes of a cylindrical waveguide.

3.1.1 Single layer waveguides

Consider an isotropic tube of inner radius r_{int} and outer radius r_{ext} in vacuum or air. The boundary conditions specify that the traction part of the stress tensor is null in both surfaces of the tube (Gazis, 1959), so:

$$\sigma_t = \sigma \cdot e_r = [\sigma_{rr}, \sigma_{r\theta}, \sigma_{rz}]^T = 0 \quad \text{for } r = r_{\text{int}}, r_{\text{ext}}, \tag{3}$$

which leads to the following matrix determinant equation:

$$\det D(\omega, k) = \det \begin{bmatrix} D^{\sigma t}(r_{\text{int}}) \\ D^{\sigma t}(r_{\text{ext}}) \end{bmatrix} = 0, \quad \text{where} \quad D^{\sigma t} = D^{\sigma}_{ij} \quad \text{with} \quad i = 1, 5, 6. \tag{4}$$

Equation 4 is called the frequency or characteristic equation of the waveguide, and its roots (ω, k) determine the proper modes supported by it. Once these roots are known, the vector of amplitude coefficients A is determined (up to a multiplicative constant) by solving the following homogeneous system of equations:

$$D(\omega, k) \cdot A = 0. \tag{5}$$

Since matrix $D(\omega, k)$ is singular at the mode's frequency-wavenumber roots, a robust method for computing the amplitude, like the singular value decomposition (SVD), is recommended (Press et al., 1992). With A determined, the distribution of $u(r)$ and $\sigma(r)$ is computed by routine `pcmatdet`.

The original Pochhammer-Chree formulation was developed for the simple case of a one-layer isotropic waveguide in vacuum; this, however, represents just a fraction of the waveguides of practical importance. Waveguides may be constituted by several layers, might be built with

Core routines	
`pcwaveguide`	Physical description of the waveguide
`pcmat`	Computes matrices $D^u(r)$ and $D^\sigma(r)$ (tables 1 and 2)
`pcmatdet`	Assembles and solves the waveguide description matrix
`pcviewmatdet`	View the entries of the matrix determinant
Plotters and solvers of the frequency equation	
`pcplotmatdet1D`	One-dimensional plot of the freq. eq. determinant vs. k, f, or c_{ph}
`pcplotmatdet2D`	Two-dimensional plot of the freq. eq. determinant in k-f space
`pcsolvebisection`	Bisection method to find roots of the freq. eq. vs k, f, or c_{ph}
`pcdisp`	Plots the phase and group speeds vs. frequency
`pckfcurves`	Traces k-f curves for real, imaginary, and complex k
`pcsolverandom`	Random solutions of the freq. eq. for complex k
Field computing and wave propagation	
`pcwaveform`	Finds the displacement vector $u(r)$ and the stress tensor $\sigma(r)$
`pcorthogonalcheck`	Checks the orthogonality of modes in the waveguide
`pcsignalpropagation`	Simulates the propagation of a signal along the waveguide
Modal analysis	
`pcextsurfacestress`	External traction stresses σ_e acting on the waveguide
`pcextvolumforce`	External volumetric forces f_e acting on the waveguide
`pcplotexcitation`	Plots the excitation volumetric force and surface stress
`pcmodalanalysis`	Finds the amplitudes of modes excited in the waveguide

Table 5. Components of the PCDISP software.

anisotropic materials, be embedded in the ground, or transport (or be surrounded by, or both) fluids. We will consider next the extensions of the PC theory which permit to model these situations.

3.1.2 Multilayered waveguides

Some waveguides are formed by several layers: for example, a metallic rod with external insulation, or a tube embedded in rock. The Pochhammer-Chree approach was first used to analyze a two-layer waveguide in (McNiven et al., 1963; Whittier & Jones, 1967), and later extended to laminated waveguides (formed by an arbitrary number of layers) in (Nelson et al., 1971). The modern technique to simulate multilayered waveguides is called the **Multiple Layer Matrix (MLM)** (Lowe, 1995), and is adapted from the transfer matrix and global matrix techniques developed by W.T. Thomson and L. Knopoff in the period 1950-1964 to study wave propagation in stratified media in seismology.

Following the MLM approach, we assemble a system of linear equations for the complete waveguide, which includes the equations of the elastic waves for each individual layer (obtained with `pcmat`), the equations which match the displacement and traction stresses at the interface between adjacent layers, and the boundary conditions. Consider the example of a multilayered waveguide shown in figure 1, where a solid cylinder of radius r_1 is enclosed

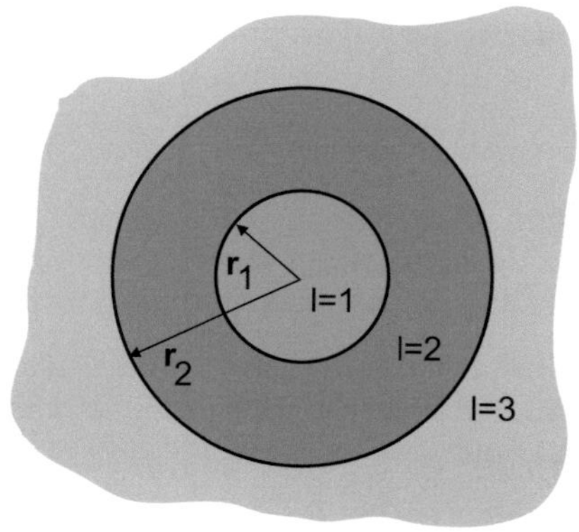

Fig. 1. Example of a three-layer cylindrical waveguide.

by a tube of inner radius r_1 and outer radius r_2, in turn surrounded by an infinite medium. The corresponding system of equations is:

$$\begin{bmatrix} D^u_{1-}(r_1) & -D^u_{2\pm}(r_1) & \\ D^{\sigma t}_{1-}(r_1) & -D^{\sigma t}_{2\pm}(r_1) & \\ & D^u_{2\pm}(r_2) & -D^u_{3+}(r_2) \\ & D^{\sigma t}_{2\pm}(r_2) & -D^{\sigma t}_{3+}(r_2) \end{bmatrix} \cdot \begin{bmatrix} A_{1-} \\ A_{2\pm} \\ A_{3+} \end{bmatrix} = 0. \qquad (6)$$

Note that the radiation conditions are used to simplify the system matrix, leading to discard the + terms in region 1, since no waves can emanate from $r = 0$; similarly, the - terms in region 3 are not considered, as no energy comes from $r = \infty$; however, both outgoing and incoming terms are allowed in the middle region. In each case the unneeded columns of matrix $D^{\sigma t}$ are removed. Next, this equation is solved by the SVD method to find the mode's amplitude vector, in the same way as the single layer waveguide of section 3.1.1.

3.1.3 Anisotropic and fluid-loaded waveguides

Materials with mechanical **anisotropy** are routinely employed to build waveguides, and their behaviour is modelled by taking into account the adequate compliance tensor for the material (Pollard, 1977). The important case of hexagonal symmetry is found in waveguides with transverse isotropy (with respect to the propagation axis z), like beryllium, or in fiber reinforced composite cylinders. Although in this case there are five elastic constants (up from two for an isotropic material), the mechanical fields are still separable with the treatment of Gazis for isotropic waveguides, with different coefficients for the D^u and D^σ matrices, obviously (Mirsky, 1965). In the case of purely orthotropic symmetry (three orthogonal planes of symmetry), the solution of the wave equation is *not* separable; still, closed solutions can be achieved in the form of a Frobenius power series (obtained in (Mirsky, 1964) for the axisymmetric case, and extended later in (Markus & Mead, 1995) to the asymmetric problem). A recent state of the art in the theory of mechanical waves in anisotropic cylindrical waveguides is found in (Grigorenko, 2005).

Materials with **elastic losses** (damping) are treated by the Kelvin-Voigt viscoelastic model (Lowe, 1995), which replaces the elements of the compliance tensor c of the material by operators which contain time derivatives, therefore modifying Hooke's law:

$$\sigma = c \cdot \epsilon \quad \Rightarrow \quad \sigma = c' \cdot \epsilon + c'' \cdot \frac{d\epsilon}{dt} = (c' - j\omega c'')\epsilon,$$

where σ and ϵ are the stress and strain tensors, and component c'' of the compliance tensor models the viscoelastic losses. The solutions of the frequency equation for a waveguide with viscoelastic losses, are, by default, complex wavenumbers.

A case of particular practical importance is that of waveguides including a fluid layer (for example, a pipe carrying a fluid, submerged in a fluid, or both). The theoretical treatment depends on the viscosity of the fluid.

The influence of **inviscid fluids** (which do not support shear stresses) on cylindrical wave propagation is treated theoretically and experimentally in (Sinha et al., 1992). If the waveguide is submerged in a liquid, propagating modes with complex wavenumber appear, which radiate (leak) energy into the surrounding fluid. That does not happen for waveguides containing fluids and surrounded by vacuum, although the propagating modes themselves are modified from the unloaded situation.

A treatment for Newtonian **viscous fluids** which is compatible with the PC based formulation of wave propagation in cylinders is introduced in (Nagy & Nayfeh, 1996). In a form similar to the Kelvin-Voigt model, the viscous liquid is modelled as an isotropic solid whose compliance tensor includes complex elements:

$$c_{11} = \lambda + \frac{4}{3}c_{44} \qquad c_{12} = \lambda - \frac{2}{3}c_{44} \qquad c_{44} = -j\omega\eta, \tag{7}$$

where λ is the compressibility of the fluid, and η its viscosity. This simple model has shown a good accuracy in predicting propagation in waveguides with viscous fluids (Aristegui et al., 2001). Indeed, the changes in the propagation of ultrasonic waves in a pipe caused by the presence of a fluid in its interior can be used to measure the longitudinal wave speed and the viscosity of the fluid (Ma et al., 2007).

3.1.4 The case of large frequency × thickness product

Solutions of the frequency equation become numerically unstable when the product frequency times thickness of the waveguide is high. Physically, this phenomenon arises because the standing waves established in the radial direction of the waveguide are formed by a combination of terms which increase exponentially with the radius r and others that decrease exponentially with it. Since it is the sum of both terms which must match the boundary conditions at $r = r_{\text{int}}$ and $r = r_{\text{ext}}$, the dynamic range of positive and negative exponentials in the frequency equation will eventually overflow the numerical capacity of the machine if the radius or frequency increase. With the 64 bits double precision arithmetic of the IEEE 754 standard, we have found that a direct implementation of the frequency equation determinant fails when the product $f \cdot (r_{\text{ext}} - r_{\text{int}})$ is higher than approximately 30 MHz·mm. This threshold is easily reached in the NDE of piping with ultrasonic waves, where frequencies of a few megahertz are common (Rose, 2000).

From table 3, we can see that the problem arises when a mode's phase velocity falls below the volumetric (c_{vol}) or rotational (c_{rot}) speeds of the solid, making parameters α or β, respectively, become imaginary. This changes the radial dependence of the mode amplitude from the bounded Bessel functions J and Y, to the exponentially varying I and K (you can

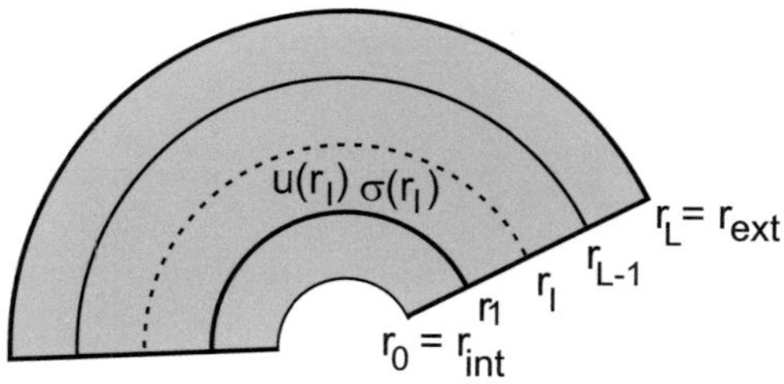

Fig. 2. Partition of the waveguide's cross section for solving the large fd problem.

use `pcviewmatdet` to see the individual entries of the waveguide description matrix). The behaviour is different depending on the frequency range:

- For $c_{ph} < c_{rot}$, both α and β are imaginary, and there exist two solutions of the frequency equation, corresponding to two Rayleigh modes propagating close to the outer and inner surfaces of the waveguide, and with amplitudes decaying exponentially from them. As the frequency increases, these two modes become decoupled, the opposite waveguide surface can ultimately be ignored, and the solutions are numerically stable.
- For $c_{rot} < c_{ph} < c_{vol}$, α is imaginary and β real. While the $SV_\pm$ and $SH_\pm$ terms remain bounded, the terms L_+ and L_- decrease and increase, respectively, with the radius, as described above. The condition number of the waveguide matrix grows with the frequency, and its solution will eventually become unstable.
- For $c_{ph} > c_{vol}$, both α and β are real, and the solution is stable (this is also the case with purely imaginary wavenumber).

Since the detection of the large fd problem in 1965, several solutions have been proposed to increase the numerical stability in plane waveguides (Lowe, 1995). We have not been able to locate similar studies for cylindrical waveguides, so, for the PCDISP software, we have developed an algorithm adapted from the transfer matrix and global matrix approaches and discussed here. We consider this method as a new contribution to the literature.

The cross section of the pipe is divided into L layers of equal thickness, where the l-th layer is given by $r_{l-1} < r < r_l$, and $r_0 = r_{int}$ and $r_L = r_{ext}$ are the inner and outer radii of the pipe (see figure 2). In the l-th layer, the displacement vector and traction part of the stress tensor are given by:

$$\begin{bmatrix} u(r) \\ \sigma(r) \end{bmatrix}_l = \begin{bmatrix} D^u(r) \\ D^{\sigma t}(r) \end{bmatrix} \cdot A_l, \tag{8}$$

where the vector of amplitude coefficients $A_l = [\mathrm{L}_+^l\ \mathrm{L}_-^l\ \mathrm{SV}_+^l\ \mathrm{SV}_-^l\ \mathrm{SH}_+^l\ \mathrm{SH}_-^l]^T$ is permitted to be different for each layer.

We use the shorthand notation:

$$D_l = \begin{bmatrix} D^u(r_l) \\ D^{\sigma t}(r_l) \end{bmatrix},$$

and scale this matrix by columns for each layer as:

$$D_l^s = D_l \cdot G_l, \tag{9}$$

where G_l is a diagonal matrix whose entries are taken as $1/\max \mathrm{col}|D_l|$.

The elastic field $[u\ \sigma]^T$ is propagated from the inner to the outer part of a layer by the following equation:

$$\begin{bmatrix} u(r_l) \\ \sigma(r_l) \end{bmatrix} - P_l \cdot \begin{bmatrix} u(r_{l-1}) \\ \sigma(r_{l-1}) \end{bmatrix} = 0, \tag{10}$$

where P_l is the propagator matrix of layer l, and is given by:

$$P_l = D_l^s \cdot (G_l^{-1} \cdot G_{l-1}) \cdot (D_{l-1}^s)^{-1}. \tag{11}$$

Applying equation 10 to all the layers of the waveguide, we can assemble a global matrix:

$$\begin{bmatrix} P_1 & -I_6 & & & & & \\ & P_2 & -I_6 & & & & \\ & & & \ddots & & & \\ & & & & P_{L-1} & -I_6 & \\ & & & & & P_L & -I_6 \end{bmatrix} \begin{bmatrix} u_0 \\ \sigma_0 \\ u_1 \\ \sigma_1 \\ \vdots \\ u_{L-1} \\ \sigma_{L-1} \\ u_L \\ \sigma_L \end{bmatrix} = 0, \tag{12}$$

where I_6 is the identity matrix of size 6×6.

The boundary conditions to match the wave fields to the surrounding medium (or to other layers in multilayered waveguides) are introduced at radii r_{int} and r_{ext}. In the case of stress free boundaries, the terms σ_0 and σ_L are zero, and their corresponding columns are simply removed from the global matrix.

Once equation 12 has been solved, and we have determined the displacements and traction stresses at each layer boundary $[u_l, \sigma_l]$, the vector of amplitude coefficients for each layer is found by solving:

$$\begin{bmatrix} D_{l-1}^s \\ D_l^s \cdot (G_l^{-1} \cdot G_{l-1}) \end{bmatrix} \cdot A_l^s = \begin{bmatrix} u_{l-1} \\ \sigma_{l-1} \\ u_l \\ \sigma_l \end{bmatrix}, \tag{13}$$

for $l = 1, \dots L$. The unscaled amplitude vector is simply: $A_l = G_l \cdot A_l^s$.

The algorithm described increases the fd stability limit by a factor proportional to the number of layers L, at the expense of larger matrices and longer computational time. In PCIDSP, the algorithm described is written into routine `pcmatdet`, and is activated automatically if the waveguide defined in `pcwaveguide` consists of $N_L > 1$ layers of the same material.

3.2 Computation of the dispersion curves

The roots of the waveguide's frequency equation represent the mechanical modes which satisfy the boundary conditions. The procedure for computing such solutions is described in this section.

Material	Aluminium	Standard	DN 25, SCH 80
Inner radius (r_{int})	12.15 mm	Outer radius (r_{ext})	16.70 mm
Poisson's ratio (ν)	0.35	Bar velocity (c_0)	5000 m/s
Density (ρ)	2700 kg/m^3	Shear modulus (G)	25.5 GPa

Table 6. Physical data for the aluminium tube used for demonstration of the PCDISP software in this book chapter.

3.2.1 Nature and ordering of the solutions of the frequency equation

For a given frequency f, the roots of the frequency equation are in general complex wavenumbers $k = k_r + jk_i$. Wavenumbers with $k_i = 0$ correspond to the **propagating** or **proper** modes of the waveguide; those with $k_i \neq 0$ represent **evanescent** modes which attenuate along the axial distance z. Due to the symmetry of the coefficients of the frequency equation, purely real or imaginary solutions appear in pairs ($\pm k_r$ and $\pm jk_i$), while complex solutions do so in quartets ($\pm k_r \pm jk_i$). The principle of the conservation of energy dictates which solutions are valid in a waveguide problem. For example, waves which propagate towards the $z+$ axis of an infinite waveguide, will necessarily have $\text{Im}\{k\} \geq 0$. Although signal propagation does not occur for imaginary or complex wavenumbers, those solutions are needed to fulfill the condition of completeness which will be stated in section 3.3. Furthermore, stationary waves formed by combination of the two wavenumbers $k_r + jk_i$ and $-k_r + jk_i$ (provided that $k_i > 0$), can exist locally, storing but not dissipating energy (Meeker & Meitzler, 1972). In ultrasonic applications, these waves are important in the region of generation of ultrasonic waves, at waveguide discontinuities (like defects), and at the waveguide ends.

Solutions of the frequency equation can be traced like continuous curves in the k-f space, where $k = k_r + jk_i$. As an example, we show the dispersion curves of the longitudinal modes L(0,m) of a sample waveguide with the data shown in table 6 (this waveguide will be further used in the examples of section 4). The complete spectrum, up to 3 MHz, has been computed with the `pckfcurves` routine of PCDISP, and is shown in figure 3. As it can be seen, the wavenumber curve of a given mode changes from a real value (shown in blue colour) to purely imaginary (coloured red, and projected into the negative wavenumber axis), or complex (plotted in green, with the real part on the positive k axis, and the imaginary part on the negative k axis), in a complicated fashion.

After finding all possible roots of the frequency equation, branches L(0,m) have to be ordered in such a way that each mode is assigned a unique, continuous wavenumber between the maximum frequency f_{max} and zero frequency. Part (b) of figure 3 shows a reduced frequency range of the spectrum in part (a), and illustrates the convention for labelling modes. PCDISP uses the following rules about the behaviour of dispersion curves (Meeker & Meitzler, 1972):

- Only the first torsional T(0,1), longitudinal L(0,1), and first flexural F(1,1), modes propagate down to zero frequency with real wavenumber.
- Higher order modes switch from real to imaginary wavenumber at the **cutoff frequencies** when the wavenumber becomes null ($k = 0, f = f_{\text{cutoff}}$), and the phase speed infinite.
- Miminum points ($d\omega/dk = 0$, $d^2\omega/dk^2 > 0$) of either real or imaginary wavenumber branches are also cutoff frequencies (with finite phase speed), from which complex wavenumber branches start.

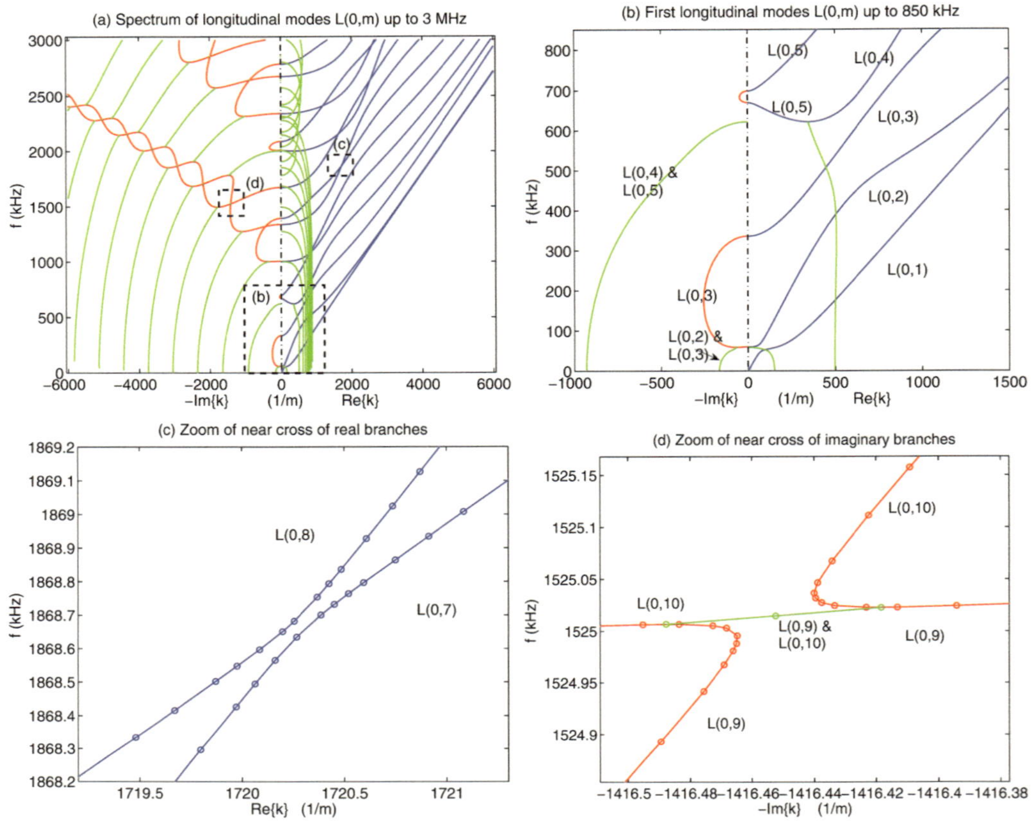

Fig. 3. Wavenumber-frequency plot for the L(0,m) modes of the aluminum pipe of table 6, for frequencies up to 3 MHz (part a). Part b shows the low-frequency spectrum, and the labelling scheme for the modes. Parts c and d show near crosses of the branches of two different modes; they have to be largely magnified to be visible.

- Complex branches terminate either at zero frequency or at the maximum points ($d\omega/dk = 0$, $d^2\omega/dk^2 < 0$) of imaginary branches.
- The branches are ordered such that k_i is positive for modes propagating in the $z+$ direction, and that the sign of the group speed does not change along the curve (although it becomes null at the cutoff frequencies, and at the purely imaginary branches).

3.2.2 Algorithm for curve tracing in PCDISP

In order to generate continuous curves $k = k(f)$, from $f = 0$ to $f = f_{\max}$, for each mode, `pckfcurves` first traces the real wavenumber branches, which start at solutions found with `pcsolvebisection` at the $f = f_{\max}$ and $k = k_{\max}$ axes, and finish at the $k = 0$ axis. If one is only interested in propagating modes, this finishes the procedure, and the `pcdisp` routine can be used to plot the phase and group speeds of propagating modes in the specified frequency range. Otherwise, the next step consists in tracing branches with purely imaginary wavenumber, which start at points $(0, f_{\text{cutoff}})$ on the vertical axis of zero wavenumber, and finish when they reach the $f = f_{\max}$, $k = jk_{\max}$ axes, or at another cutoff frequency in the $k = 0$ axis.

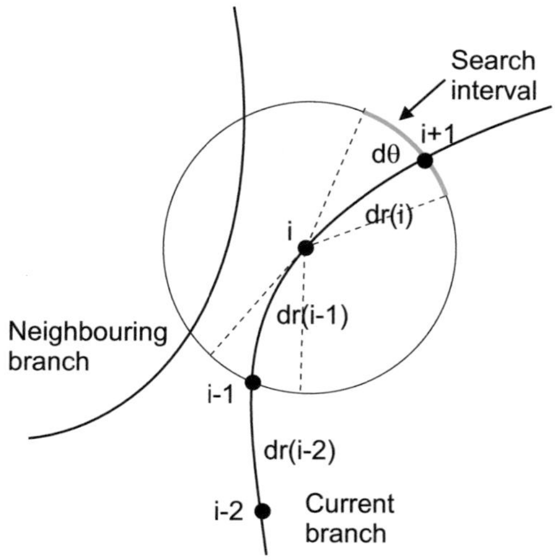

Fig. 4. Illustrates the root tracing algorithm used by PCDISP.

Parts (c) and (d) of figure 3 show the reason why a robust algorithm for curve tracing of the dispersion branches is needed, in order to avoid the apparent crossings between branches, such as modes L(0,7) and L(0,8), with real wavenumbers, and modes L(0,9) and L(0,10), with imaginary wavenumbers. The curve tracing method used in `pckfcurves` is shown in figure 4. The dispersion curve being traced is extrapolated from the three last computed points $\{i-2, i-1, i\}$ to define an angular interval of width $d\theta$ in a circle of radius $dr(i)$ centered in the last correctly determined point (i) (dr is the step size of the algorithm, in normalized coordinates of the k-f space). If a sign change is found in this interval, the algorithm proceeds with a bisection method to accurately estimate the position of point $i+1$ (this is the normal situation). Otherwise, the dispersion curve might have undergone a sudden change of curvature, or another mode might have come very close to the one being traced, provoking multiple sign changes. In this case, the step $dr(i)$ is decreased, or, if needed, points $i-1$, $i-2$, etc, are recomputed with a smaller step dr. Summarizing, the tracing algorithm of PCDISP keeps track of the curvature of the branch and the proximity of neighbouring branches, adjusting the interval step between consecutive points accordingly.

It must be pointed out that the frequency equation has spurious solutions at the lines with slope equal to the volumetric and rotational speeds of the solid ($\omega/k = c_{\text{vol}}$ and $\omega/k = c_{\text{rot}}$), which have to be removed by the root finding algorithm. In the case of multilayered waveguides, the same phenomenon happens for the speeds of each layer.

The dispersion curves are completed by tracing the complex wavenumber branches $k = k_r + jk_i$, starting from the extrema points of the real/imaginary wavenumber branches. Tracing the complex wavenumber branches needs more computational effort, since it is required to solve simultaneously for the real and imaginary parts of the wavenumber; we have obtained satisfactory results with Muller's method (Press et al., 1992) using also an adaptive step.

Finally, the rules enumerated at the end of section 3.2.1 are used to convert the initially obtained branches into continuous dispersion curves $k = k(f)$ for each mode. An example of this procedure is shown in part (b) of figure 3. First, note that all longitudinal modes except L(0,1) exhibit cutoff. Mode L(0,2) is cut off at ($k = 0, f = 59.7$ kHz), where it switches to a branch with imaginary wavenumber which goes on until ($k = 67j$ m^{-1}, 58.5 kHz),

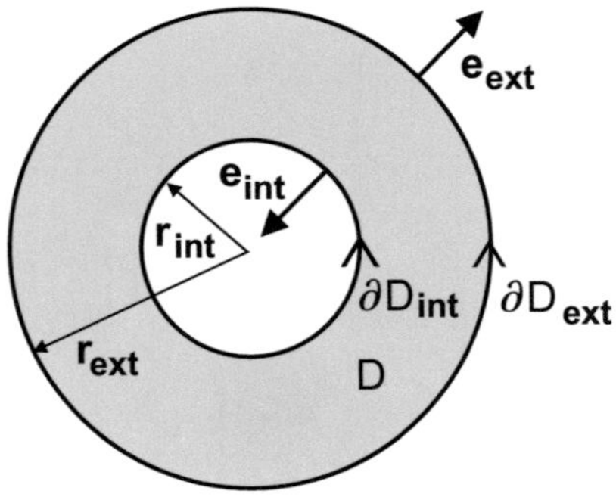

Fig. 5. Cross section of the waveguide and definition of the regions of integration for modal analysis.

the minimum point of the imaginary branch. At that point, it is joined by the branch corresponding to mode L(0,3), and both go down to zero frequency with negative complex conjugate wavenumbers $k_r + jk_i$ and $-k_r + jk_i$.

Similarly, mode L(0,4) is cut off at the minimum point of the real wavenumber branch ($k = 345\ \mathrm{m}^{-1}$, $f = 621.5$ kHz). Mode L(0,5) is cut off at ($k = 0, f = 698.6$ kHz), changes to imaginary wavenumber until it reaches point ($k = 0, f = 669.4$ kHz), and then again to real, *but negative*, wavenumber down to ($k = 345\ \mathrm{m}^{-1}, f = 621.5$ kHz), where it joins mode L(0,4). Mode L(0,5) has a negative wavenumber (and consequently, negative phase speed ω/k), in order to maintain a positive group velocity $d\omega/dk$, since this is a propagating mode in the $z+$ direction of the waveguide, according to the last rule of section 3.2.1. Below 621.5 kHz, the dispersion curves of modes L(0,4) and L(0,5) descend to zero frequency with negative complex conjugate wavenumbers, in the same way as modes L(0,2) and L(0,3).

3.3 Modal analysis

Modal analysis is a mathematical technique which permits to compute the dynamic response of a waveguide subject to arbitrary external forces, as an expansion of the excited wave over the set of normal modes of the waveguide, as defined in section 2 (Auld, 1973). Modal analysis is based upon two properties of normal modes: orthogonality, the existence of a scalar product which is null for any two different modes; and completeness, the capacity of the set of normal modes to span arbitrary waveforms in the waveguide.

For two different modes of the waveguide (1) and (2) of the form given in equation 1, the **orthogonality relationship** (Auld, 1973) establishes that:

$$\widehat{\nabla} \cdot (\widehat{u}_1 \cdot \widehat{\sigma}_2^* - \widehat{u}_2^* \cdot \widehat{\sigma}_1) = 0, \tag{14}$$

which is applicable to linear elastic materials and also to piezoelectric or magnetostrictive linear materials, assuming no elastic or dielectric losses. Later on this result will be generalized to include external forces and stresses.

For separable vector fields in the z coordinate, the tridimensional divergence operator can be written as:

$$\widehat{\nabla} \cdot \{\} = \widetilde{\nabla} \cdot \{\} + \frac{\partial}{\partial z}\{\} \cdot e_z,$$

so equation 14 becomes:

$$\widetilde{\nabla} \cdot (\widetilde{u}_1 \cdot \widetilde{\sigma}_2^* - \widetilde{u}_2^* \cdot \widetilde{\sigma}_1) e^{j(k_1 - k_2^*)z} + j(k_1 - k_2^*) e^{j(k_1 - k_2^*)z} (\widetilde{u}_1 \cdot \widetilde{\sigma}_2^* - \widetilde{u}_2^* \cdot \widetilde{\sigma}_1) \cdot e_z = 0. \quad (15)$$

Discarding the common factor $e^{j(k_1 - k_2^*)z}$, integrating over the cross section D of the waveguide, and applying the divergence theorem, we find that:

$$\oint_{\partial D} (\widetilde{u}_1 \cdot \widetilde{\sigma}_2^* - \widetilde{u}_2^* \cdot \widetilde{\sigma}_1) \cdot e_n \, dl + j(k_1 - k_2^*) \iint_D (\widetilde{u}_1 \cdot \widetilde{\sigma}_2^* - \widetilde{u}_2^* \cdot \widetilde{\sigma}_1) \cdot e_z \, dS = 0. \quad (16)$$

In equation 16, $\partial D = \partial D_{\text{int}} \cup \partial D_{\text{ext}}$ represents the inner and outer surfaces of the waveguide, and the normal unit vector e_n is taken on each surface pointing out of the waveguide's interior, as shown in figure 5.

Because for proper modes the surface traction stress is null ($\widehat{\sigma}_t = \widehat{\sigma} \cdot e_n = 0$ in ∂D), the first integral of equation 16 is zero. Then a suitable scalar product of modes (1) and (2) is:

$$P_{12} = -\frac{j\omega}{4} \iint_D (\widetilde{u}_1 \cdot \widetilde{\sigma}_2^* - \widetilde{u}_2^* \cdot \widetilde{\sigma}_1) \cdot e_z \, dS = -\frac{j\pi\omega}{2} \int_{r_{\text{int}}}^{r_{\text{ext}}} (u_1 \cdot \sigma_2^* - u_2^* \cdot \sigma_1) \cdot e_z \cdot r \, dr. \quad (17)$$

In the right part of equation 17 we have assumed that the circumferential order n of modes (1) and (2) is the same; otherwise, P_{12} is zero automatically due to the integration over the θ coordinate. The factor $-j\omega/4$ is introduced so that the quantity P_{11} equals to the integral of the acoustic Poynting vector in the cross section of the waveguide, i.e., the power transported by the mode. For nonpropagating modes with $k_i \neq 0$, P_{11} is zero.

With this notation, equation 16 reduces to:

$$(k_1 - k_2^*) P_{12} = 0, \quad (18)$$

which implies that $P_{12} = 0$ unless $k_1 = k_2^*$. In PCDISP, mode orthogonality can be verified with routine `pcorthogonalcheck`.

The second condition for modal analysis is **completeness**, which is based on the premise that an arbitrary perturbation in the waveguide can be expanded in the set of normal modes:

$$\widehat{u}_1(r,\theta,z) = \sum_p a_p(z) \widetilde{u}_p(r,\theta) \qquad \widehat{\sigma}_1(r,\theta,z) = \sum_p a_p(z) \widetilde{\sigma}_p(r,\theta). \quad (19)$$

where the modes are indexed by p, and no distinction has been made between propagating and evanescent modes. The problem lies in computing the set of coefficients $a_p(z)$, when the waveguide is under an arbitrary excitation composed of:

1. A vector force field $\widehat{f}_e(r,\theta,z)$ acting on the bulk material of the tube (region D). In PCDISP, this vector force field is defined in `pcextvolumforce`.
2. A traction stress $\widehat{\sigma}_e(r,\theta,z)$ applied to the tube surfaces (region ∂D). In PCDISP, this surface stress is defined in `pcextsurfacestress`.

In the case of existence of external fields, the orthogonality relationship, equation 14, must be generalized to (Auld, 1973):

$$\widehat{\nabla} \cdot (\widehat{u}_1 \cdot \widehat{\sigma}_2^* - \widehat{u}_2^* \cdot \widehat{\sigma}_1) = -\widehat{u}_1 \cdot \widehat{f}_2^* + \widehat{u}_2^* \cdot \widehat{f}_1, \quad (20)$$

where $\widehat{f}_{1,2}(r,\theta,z)$ represent the forcing terms.

If we take subscript (1) for the wave existing in the waveguide (equation 19) and (2) as the q-th proper mode of the waveguide:

$$\widehat{u}_2(r,\theta,z) = \widetilde{u}_q(r,\theta)e^{jk_qz} \qquad \widehat{\sigma}_2(r,\theta,z) = \widetilde{\sigma}_q(r,\theta)e^{jk_qz}, \tag{21}$$

we can insert both expressions into equation 20, and, letting $\widehat{f}_2 = 0$ since it corresponds to a normal mode in the waveguide, we obtain that:

$$\widehat{\nabla} \cdot \left[\sum_p a_p(z)(\widetilde{u}_p \cdot \widetilde{\sigma}_q^* - \widetilde{u}_q^* \cdot \widetilde{\sigma}_p)e^{-jk_qz} \right] = \widetilde{u}_q^* \cdot \widehat{f}_1 e^{-jk_qz}. \tag{22}$$

Operating with the divergence operator:

$$\sum_p a_p(z)\widetilde{\nabla} \cdot (\widetilde{u}_p \cdot \widetilde{\sigma}_q^* - \widetilde{u}_q^* \cdot \widetilde{\sigma}_p) + \sum_p \left[-jk_q a_p(z) + \frac{da_p(z)}{dz} \right] (\widetilde{u}_p \cdot \widetilde{\sigma}_q^* - \widetilde{u}_q^* \cdot \widetilde{\sigma}_p) \cdot e_z = \widetilde{u}_q^* \cdot \widehat{f}_1, \tag{23}$$

which, when integrating across the transversal section of the tube, with the divergence theorem, reduces to:

$$\underbrace{\sum_p a_p(z) \oint_{\partial D} (\widetilde{u}_p \cdot \widetilde{\sigma}_q^* - \widetilde{u}_q^* \cdot \widetilde{\sigma}_p) \cdot e_n \, dl +}_{(A)}$$

$$\underbrace{\sum_p \left[-jk_q + \frac{d}{dz} \right] a_p(z) \iint_D (\widetilde{u}_p \cdot \widetilde{\sigma}_q^* - \widetilde{u}_q^* \cdot \widetilde{\sigma}_p) \cdot e_z \, dS}_{(B)} = \underbrace{\iint_D \widetilde{u}_q^* \cdot \widehat{f}_1 \, dS}_{(C)}. \tag{24}$$

The first term simplifies as $\widehat{\sigma}_q \cdot e_n = 0$ in ∂D, and we can combine the sum over the index p to find:

$$(A) = -\oint_{\partial D} (\widetilde{u}_q^* \cdot \widehat{\sigma}_1) \cdot e_n \, dl.$$

As for the second term, if we assume that p is a propagating mode, we recall the property of orthogonality of the propagating modes of the waveguide, and the definition of acoustic power P_p in equation 17, to eliminate all the terms of the sum except the one for which $p = q$:

$$(B) = -\frac{4P_p}{j\omega} \left(-jk_p + \frac{d}{dz} \right) a_p(z).$$

Inserting this result in the preceding equation:

$$-\frac{4P_p}{j\omega} \left(-jk_p + \frac{d}{dz} \right) a_p(z) = \oint_{\partial D} (\widetilde{u}_p^* \cdot \widehat{\sigma}_1) \cdot e_n \, dl + \iint_D \widetilde{u}_p^* \cdot \widehat{f}_1 \, dS, \tag{25}$$

where the contributions to the mode amplitude due to the volumetric forces ($\widehat{f}_1 = \widehat{f}_e$) and the surface tractions ($\widehat{\sigma}_1 = \widehat{\sigma}_e$) appear clearly separated. With the following definitions:

$$f_p^s(z) = -j\omega \oint_{\partial D} [\widetilde{u}_p^*(r,\theta) \cdot \widehat{\sigma}_e(r,\theta,z)] \cdot e_n \, dl = -j\omega \oint_{\partial D} e^{-jn_p\theta} \cdot [u_p^*(r) \cdot \widehat{\sigma}_e(r,\theta,z)] \cdot e_n \, dl, \tag{26}$$

and

$$f_p^v(z) = -j\omega \iint_D [\widetilde{u}_p^*(r,\theta) \cdot \widehat{f}_e(r,\theta,z)]\, dS = -j\omega \iint_D e^{-jn_p\theta} \cdot [u_p^*(r) \cdot \widehat{f}_e(r,\theta,z)]\, dS, \qquad (27)$$

equation 25 is changed into an ordinary differential equation, solvable by a standard change of variables, resulting in:

$$a_p(z) = \frac{e^{jk_pz}}{4P_p} \int_{R_g} e^{-jk_pz'} [f_p^s(z') + f_p^v(z')]\, dz', \qquad (28)$$

where the integration takes place in the region R_g where the generating terms f^s and f^v are not null, and z is the point where the ultrasonic signal is observed, in the direction of increasing z from region R_g.

If p is a non-propagating mode, our computation method is changed slightly, since $P_p = 0$. However, $P_{p,p*} \neq 0$, and we can set $q = p^*$, $k_q = k_p^*$, and modify equations 26-28 accordingly.

As a summary, we have established the equations that permit to find the amplitude of the proper modes excited in the waveguide by an arbitrary set of external driving forces. These equations are used by routine `pcmodalanalysis` of PCDISP .

3.4 Propagation of waveforms in the waveguide

The modal analysis equations discussed in section 3.3 permit to obtain the frequency response of a transducer exciting the waveguide. In many applications we want to predict what ultrasonic waveforms will be obtained at a certain distance z from the excitation source, when the transducer is excited by a finite length time signal, i.e., to model the transient behaviour of the system.

The method to study the propagation of signal waveforms is relatively straightforward (Doyle, 1997). Let $u(0,t)$ be the input signal in the transducer (placed in region R_g), and $U(0,\omega) = \mathcal{F}[u(0,t)]$ its Fourier transform. When this signal excites the waveguide, we determine the corresponding volumetric forces and surface stresses, and evaluate terms $a_p(z)$ from equation 28 for each significant frequency component of U and all normal modes of the waveguide. Note that the terms $a_p(z)$ incorporate both the frequency response of the transducer itself (inside the integral term) and the effect caused by signal propagation ($\exp(jk_p(\omega)z)$). Thus, the frequency components at a distance z from the generating region are given by:

$$\begin{aligned} U(z,\omega) &= U(0,\omega) \cdot a_p(z,\omega) \\ U(z,-\omega) &= U^*(z,\omega), \end{aligned} \qquad (29)$$

where the values of the Fourier transform for negative frequencies are taken as complex conjugate of the positive ones, in order to obtain a real time signal. The waveform at z is recovered by the inverse Fourier transform:

$$u(z,t) = \mathcal{F}^{-1}[U(z,\omega)]. \qquad (30)$$

The imaginary and complex wavenumber parts of the spectrum are required in equation 29 if the exciting signal has significant frequency content below cutoff of the propagating mode, and the measurement point is not far away from the transducer, as was described in section 3.2.1.

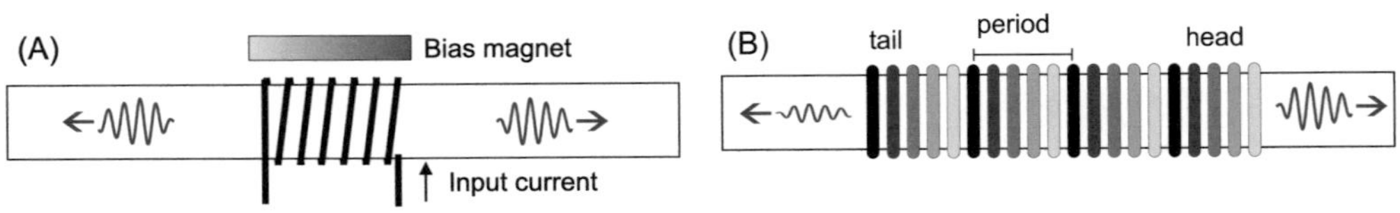

Fig. 6. Transducers used for generation of ultrasonic waves in cylindrical waveguides: (a) Electromagnetic Acoustic Transducer (EMAT) for Lorentz force excitation; (b) Time-Delay Periodic Ring Array (TDPRA) for piezoelectric excitation.

The dispersive effect characteristic of waveguide propagation is frequently undesired in applications, since it implies a distortion of the original signal. One practical way of minimizing dispersion is to employ signals with narrow spectral content (typically by windowing a sine pulse train) with a central frequency in the region where the curve $c_{ph}(\omega) = \omega/k(\omega)$ is relatively flat (Lowe et al., 1998). Where this is not feasible, compensation methods based on inverting the nonlinear $k = k(\omega)$ dependence have been developed (Wilcox, 2003).

The propagation of ultrasonic signals in the waveguide is simulated in PCDISP with routine `pcsignalpropagation`. When customising this routine, the user must be careful to zero pad the excitation signal $u(0,t)$ at the end such that the resulting time window has enough duration to allow propagation of the signal for the distance between transducer and receiving point. Likewise, a sampling frequency high enough to cover all the spectral content of the signal should be used; in practical applications, oversampling the signal above the Nyquist rate is advantageous since it enhances the signal to noise ratio.

4. Demonstration of the methodology

In this section we will illustrate the use of the methodology described in this chapter and the developed software, to model the performance of a given transducer. The complete procedure is summarized in table 7, along with the needed routines of the PCDISP package.

Two transducer setups commonly found in ultrasonic guided wave applications will be analyzed (see figure 6). The first one is an electromagnetic acoustic transducer (EMAT) used to generate ultrasound in metallic waveguides without physical contact between the transducer and the sample; the second, an array of piezoelectric rings which generates ultrasound by mechanically loading the external tube surface. We begin by considering the mechanical behaviour of the sample waveguide.

4.1 Dispersive curves of longitudinal modes

We will continue to use the waveguide described in table 6. The complete signal spectrum of the longitudinal modes was already shown in figure 3; in figure 7 we plot the phase and group speeds in the range of 0 to 800 kHz. An important requisite for guided waves applications of ultrasound is the selection and exploitation of a single propagating mode, in a region where dispersive effects are minimum, since, as a general principle, an external force will excite all propagating modes existing within its bandwidth (Lowe et al., 1998). We will consider two possibilities: excitation of mode L(0,2) at frequency $f_1 = 250$ kHz, and use of mode L(0,3) at frequency $f_2 = 565$ kHz (see figure 7 b). The dispersion curves of these modes are relatively flat at these frequencies, and their group speeds are higher than those of other coexisting modes.

Part A: Input data of the waveguide (`pcwaveguide`)
Assembling of the waveguide description matrix (`pcmatdet`)
Compute the dispersion curves $k = k(\omega)$ (`pckfcurves`)

Part B: Frequency / transient response of the waveguide (`pcmodalanalysis` and `pcsignalpropagation`)
Fourier transform of excitation signal: $U(0,\omega) = \text{FFT}[u(0,t)]$
► Loop over propagating modes p
  ► Loop over frequencies ω of the signal's spectrum
    Compute amplitude vector A_p by SVD solution of $D(\omega, k_p) \cdot A_p = 0$
    Computation of ultrasonic fields of p-th mode: $\widetilde{u}_p, \widetilde{\sigma}_p$ (`pcmatdet`)
    ► Integration in the region of generation $z' \in R_g$
      Compute $f_p^s(z')$ by integrating $\widehat{\sigma}_e(r,\theta,z)$ (`pcextsurfacestress`) in ∂D
      Compute $f_p^v(z')$ by integrating $\widehat{f}_e(r,\theta,z)$ (`pcextvolumforce`) in D
    ◄ End of integration in region R_g
    Computation of mode amplitude $a_p(z)$ (equation 28)
    Determine gain for frequency ω: $U_p(z,\omega) = U(0,\omega) \cdot a_p(z)$
  ◄ End loop over frequency
  Set negative frequencies of the FFT: $U_p(z,-\omega) = U_p^*(z,\omega)$
  Inverse Fourier transform: $u_p(z,t) = \text{IFFT}[U_p(z,\omega)]$ (eq. 30)
◄ End loop over propagating modes
Sum over propagating modes $u(z,t) = \sum_p u_p(z,t)$

Table 7. General method used for modal analysis computations, and required PCDISP routines.

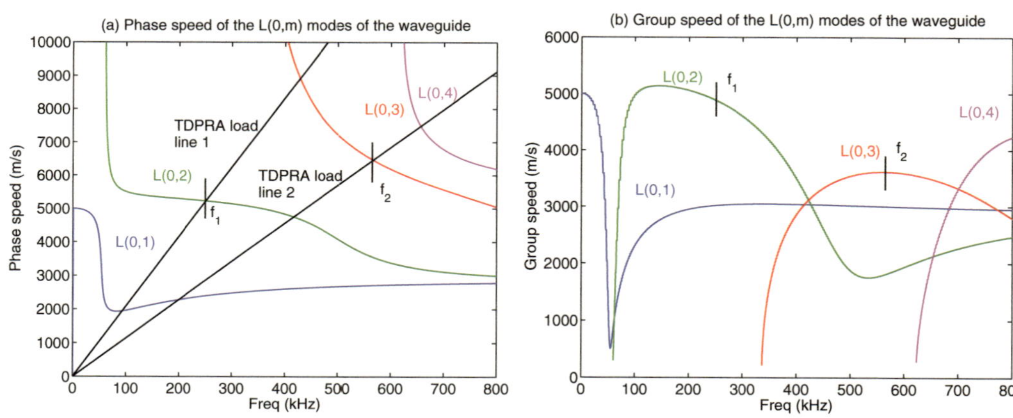

Fig. 7. (a) Phase and (b) group speeds of the first axisymmetric modes of the waveguide of table 6.

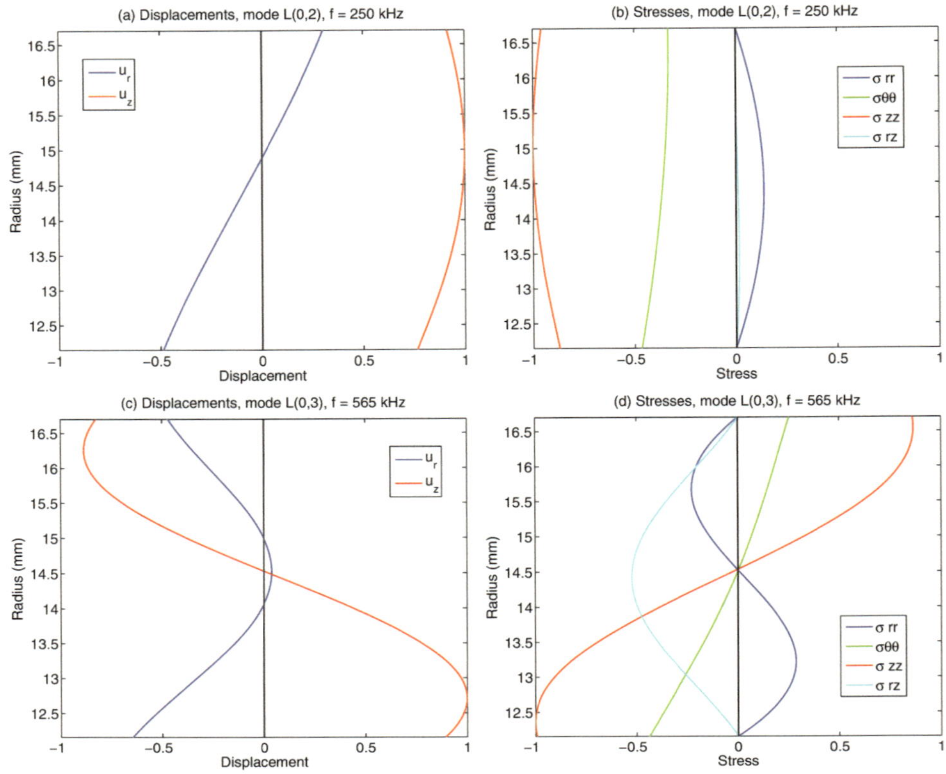

Fig. 8. Displacement and stress profiles for mode L(0,2) at f = 250 kHz (upper row), and mode L(0,3) at f = 565 kHz (bottom row), computed with `pcwaveform`.

The displacement and stress profiles of the selected modes are shown in figure 8. Their determination is important in NDE applications since the sensitivity of a propagating mode to a waveguide defect depends on the matching between the defect's shape and the mode profile (Ditri, 1994).

In order to minimize the influence of dispersion in the propagation of signals, we use as excitation signal a tone burst consisting of $n_{\text{cyc}} = 16$ cycles of a central frequency modulated by a raised cosine window. This waveform does a good job in exciting a single frequency of the waveguide with a finite length signal and minimum sidelobes (Oppenheim et al., 1999).

4.2 Generation of ultrasonic waves with an electromagnetic acoustic transducer

Electromagnetic acoustic transducers (EMATs) are used to excite ultrasonic waves in metallic waveguides by non contact means (Cawley et al., 2004). Basically, an EMAT consists of a generating coil, which creates a dynamic field $H(t)$ at ultrasonic frequencies, and a bias magnet providing a constant magnetic field H_0, as shown in figure 6 (a). The physical phenomenon that couples the magnetic field with the elastic field in our non-ferromagnetic aluminium waveguide is the Lorentz force resulting from the interaction between the eddy currents $J(t)$ induced in the tube by the dynamic field and the bias field H_0, which create a volumetric force given by:

$$f^{\text{em}} = \mu_0 J \times H_0. \tag{31}$$

Previously to computing the Lorentz force, we must determine the distribution of the electromagnetic field in the waveguide. Although an exact solution exists, it is complicated (Dodd & Deeds, 1968), so, for the purposes of this example, we will consider a simplified model in which the penetration depth of the EM field in the metal ($\delta = (2/\omega\mu_0\sigma_e)^{1/2}$, with μ_0 being the magnetic permeability of vacuum and $\sigma_e = 38$ MS/m the electrical conductivity of aluminum) is small compared with the thickness $r_{\text{ext}} - r_{\text{int}}$ of the tube (in our case $\delta < (r_{\text{ext}} - r_{\text{int}})/10$ for $f > 30$ kHz). Then the magnetic field in the tube is mainly axial and can be written as:

$$H(r,z) = H_{\text{zext}}(z)\exp[-(1+j)(r_{\text{ext}} - r)/\delta]e_z \qquad r_{\text{int}} \le r \le r_{\text{ext}}, \tag{32}$$

where $H_{\text{zext}}(z)$ is the axial field at the outer surface ($r = r_{\text{ext}}$) of the tube. We will further assume that $H_{\text{zext}}(z)$ can be obtained by the elementary formula:

$$H_{\text{zext}}(z) = \frac{N_s I_s}{2L_s}\left[\frac{z}{\sqrt{R_s^2 + z^2}} + \frac{L_s - z}{\sqrt{R_s^2 + (L_s - z)^2}}\right], \tag{33}$$

where L_s and R_s are the solenoid's length and radius, N_s the number of turns, and I_s the current through one turn.

Computing the eddy current with Ampère's law, $J(r,z) = \nabla \times H(r,z)$, and since the bias field is $H_0 = H_0 e_z$, we obtain:

$$f^{\text{em}}(r,z) = -\mu_0\frac{1+j}{\delta}H_{\text{zext}}(z)H_0\exp[-(1+j)(r_{\text{ext}} - r)/\delta]e_r. \tag{34}$$

And using equation 27, we can compute the volumetric forcing term for mode p in the waveguide as :

$$f_p^v(z) = -2\pi j\omega\int_{r_{\text{int}}}^{r_{\text{ext}}} u_{pr}^*(r)f_r^{\text{em}}(r)\,r\,dr. \tag{35}$$

If we consider an EMAT with a solenoid length $L_s = 30$ mm and $R_s = r_{\text{ext}} = 16.70$ mm we obtain the dependence of transducer gain with frequency shown in figure 9 (a), where the radial component of displacement at the surface, $u_r(r_{\text{ext}})$, is plotted on a log scale on the vertical axis. As we can see, the EMAT exhibits high gain in the low frequency region, which unfortunately coincides with the zone where the dispersive behaviour of mode L(0,1) is maximum, making it difficult for guided waves applications. In parts (b) and (c) of the same figure we show the transient response of the waveguide when excited with the pulse train described in section 4.1, for central frequencies of 250 kHz and 565 kHz, respectively, when the radial component of surface displacement (u_r) is measured at a point $z = 1.5$ m from the EMAT. For 250 kHz, the faster propagating L(0,2) mode is excited with lower amplitude than the mode L(0,1), while for 565 kHz all modes are excited with approximately equal amplitudes, appearing also very close in the time domain.

Summarizing, the basic EMAT described in this section shows poor mode selectivity control, exciting all modes within the bandwidth of the source signal with relatively equal amplitudes, which makes it a poor choice for this waveguide.

4.3 Generation of ultrasonic wave with piezoelectric surface loading

In this section we consider a Time-Delay Periodic Ring Array (TDPRA), an ultrasonic transducer with very good mode selectivity and also capable of achieving one directional emission, emitting much more ultrasonic energy from its enhanced side than in the direction

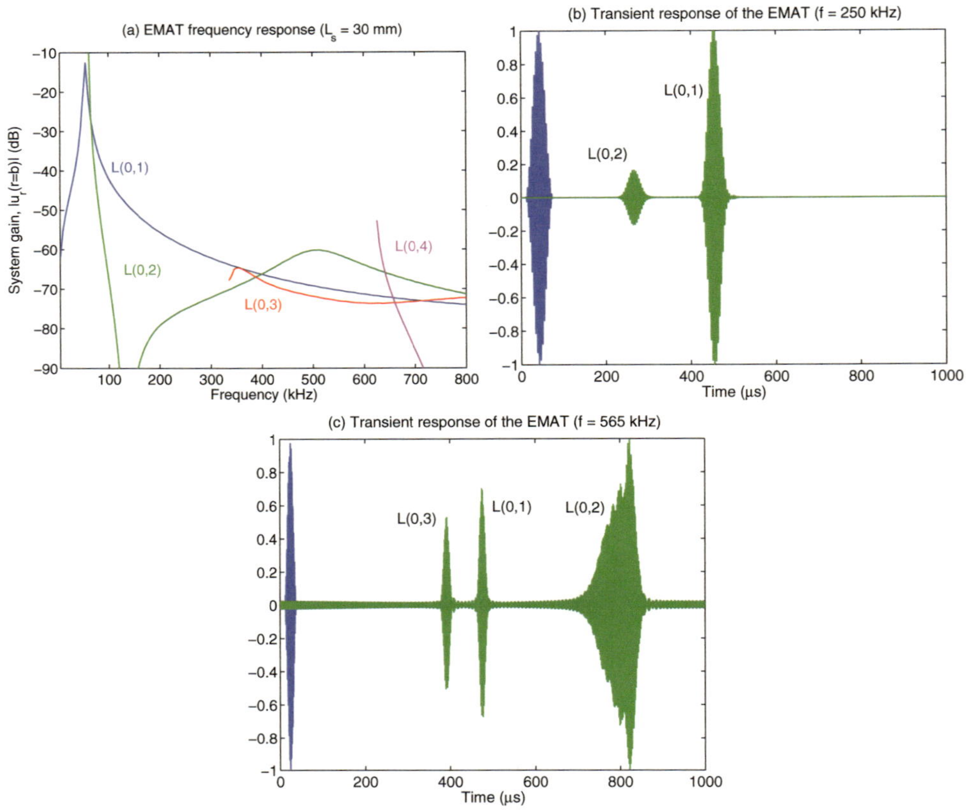

Fig. 9. (a) Computed frequency response of the EMAT for solenoid length $L_s = 30$ mm; transient waveforms at (b) 250 kHz and (c) 565 kHz. The signals in the transient plots have been normalized to unit amplitude.

of its weakened side (Zhu, 2001). A TDPRA, shown in figure 6 (b), consists in a number of piezoelectric rings capable of exerting a pressure loading on the outer surface of the tube. The rings are organized into N_p identical periods of N_r rings each, with the length of a period matched to the wavelength λ of the mode to be excited. The rings of a period are connected to the same excitation source, but with a relative delay between them proportional to their position within the period of the TDPRA; this scheme is repeated throughout the TDPRA. This reinforces the wavelength matching of a single period and also creates constructive interference in the enhanced direction (the "head" of the array) and destructive in the other (the "tail").

To model numerically the TDPRA, we assume that each ring has width z_w, with a separation z_s between adjacent rings and vibrates in its thickness mode, exerting a pressure loading, $\sigma_{rr} = -p$ over the outer surface of the waveguide, constant for all frequencies. This leads to an axisymmetric loading (no θ dependence). In this case, the term corresponding to the surface loading (equation 26) is:

$$f_p^s(z) = 2\pi j\omega r_{\text{ext}} u_r^*(r_{\text{ext}}) p(z), \tag{36}$$

while the volumetric term is null. The total pressure $p(z)$ is a sum over the N_p periods of N_r rings each:

$$p(z) = p_0 \cdot \sum_{i=0}^{N_p N_r - 1} \sqcap\left(\frac{z_{ci}}{z_w}\right) e^{j2\pi \mathrm{mod}(i,N_r)/N_r}, \qquad (37)$$

where $z_{ci} = (z_w + z_s)i + z_w/2$ is the position of the center of each ring, and $\sqcap(x) = 1$ for $|x| < 1/2$, $\sqcap(x) = 0$ for $|x| > 1/2$, is the rectangular function.

The parameters of the TDPRA must be tuned to the frequency to be excited. The load line of the TDPRA, given by $c_{\mathrm{ph}} = N_r(z_w + z_s)f$, is shown in figure 7 (a), along with the phase speed curves of the aluminum tube given in section 4.1, for two different designs: the first one with $N_p = 4$, $N_r = 8$, $z_w = 2.2$ mm and $z_s = 0.4$ mm, intended to excite mode L(0,2) at 250 kHz, and the second with $N_p = 5$, $N_r = 6$, $z_w = 1.5$ mm, $z_s = 0.4$ mm for excitation of mode L(0,3) at 565 kHz. The intersection points of these lines with the phase speed curves correspond to the frequencies for which the TDPRA achieves maximum efficiency in mode coupling.

The transducer gain of the first TDPRA is shown in figure 10 (a). For 250 kHz, the mode L(0,2) is effectively excited, and the excitation frequency can be fine tuned to make it coincide with

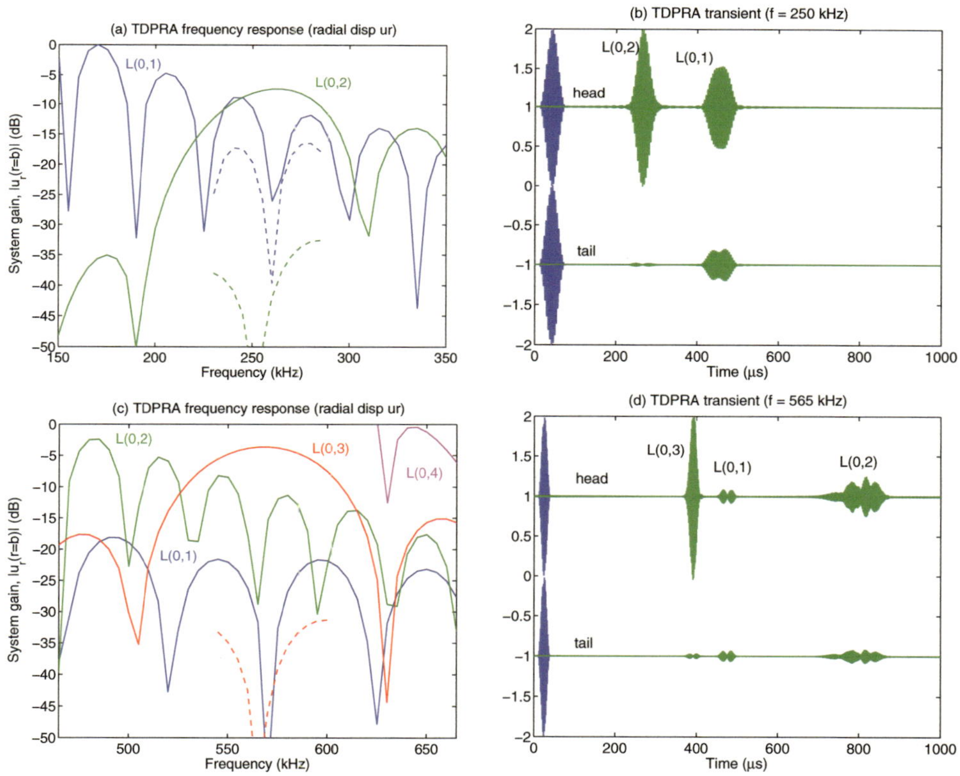

Fig. 10. Plots of the (a) frequency and (b) transient response of the first design of the TDPRA at $f = 250$ kHz; plots of the (c) frequency and (d) transient response of the second design of the TDPRA at $f = 565$ kHz. In the gain plots, the dashed lines correspond to the opposite ("tail") direction. The signals in the transient plots have been normalized to unit amplitude.

a minimum of the amplitude of the L(0,1) mode, improving the dynamic range between the two modes. The simulation of the propagated wave (at a distance $z = 1.5$ m from the TDPRA) gives the expected results (part (b)). The results obtained with the second TDPRA design are shown in parts (c) and (d) of figure 10. Mode L(0,3) dominates in this case at 565 kHz, with a higher relative amplitude over the L(0,1) and L(0,2) modes. In this case, however, as the group velocities are similar, the received signals appear closer in the time view of part (d). In all cases, ultrasonic generation in the "head" side has higher amplitude than in the "tail" side.

As a conclusion, the TDPRA is an efficient transducer for generating ultrasonic signals in cylindrical waveguides, and its feature of phase and wavelength matching permits to excite modes selectively, fine tune the system gain to a desired frequency, and direct the generated signal in only one direction. Although in this communication we have concerned ourselves only with axisymmetric transducers, PCDISP can also be used to study excitation of nonsymmetric modes by piezoelectric arrays (see reference (Li & Rose, 2001) for an example).

5. Conclusions

In this chapter we have presented a methodology to model the dynamic response of a waveguide of cylindrical symmetry when subject to an arbitrary set of external forces acting at ultrasonic frequencies, based on the combination of the mechanical Pochhammer-Chree equations and modal analysis techniques. Furthermore, a software package (named PCDISP), created in the Matlab environment, is offered freely with the intention of saving other researchers from the time needed for implementation of the PC theory equations, permitting them to focus on their particular problems.

Throughout this communication, we have paid special attention to the numerical issues of stability of the matrix determinant for large frequency thickness products, provided algorithms for robust root solving and tracing of the dispersion curves, and modelled the dispersive effect of the waveguide on signal propagation. The methods described in this chapter are valid for waveguides formed by any number of layers as long as they have cylindrical symmetry. The PCDISP software can be further extended to consider materials with anisotropy (transversely isotropic and orthotropic), as well as materials with elastic damping and waveguides surrounded by, or containing, fluids. Guidelines for such extensions are given in the text.

The performance of modal analysis is illustrated by studying two common transducers employed in guided wave ultrasonic applications: an electromagnetic-acoustic transducer and a time-delay piezoelectric ring array. We believe that transducer analysis with quantitative results is achieved comparatively easier and faster than with other competing techniques like spectral or finite element methods, obtaining significant time savings in the design stage of ultrasonic transducers.

6. Acknowledgments

The financial support for this work was provided by the Spanish Ministerio de Ciencia e Innovación through project Lemur (TIN2009-14114-C04-03).

7. References

Abramowitz, M. & Stegun, I. A. (1964). *Handbook of Mathematical Functions with Formulas, Graphs, and Mathematical Tables*, 9 edn, Dover.

Aristegui, C., Lowe, M. J. S. & Cawley, P. (2001). Guided waves in fluid-filled pipes surrounded by different fluids, *Ultrasonics* 39(5): 367–375.

Auld, B. A. (1973). *Acoustic Fields and Waves in Solids*, Wiley Interscience.

Bocchini, P., Marzani, A. & Viola, E. (2011). Graphical user interface for guided acoustic waves, *Journal of Computing in Civil Engineering* 25(3): 202–210.

Cawley, P., Lowe, M. & Wilcox, P. (2004). An EMAT array for the rapid inspection of large structures using guided waves, *Journal of Nondestructive Testing* 9(2): 1–8.

Damljanovic, V. & Weaver, R. L. (2004). Propagating and evanescent elastic waves in cylindrical waveguides of arbitrary cross section, *Journal of the Acoustical Society of America* 115(4): 1572–1581.

Ditri, J. J. (1994). Utilization of guided elastic waves for the characterization of circumferential cracks in hollow cylinders, *Journal of the Acoustical Society of America* 96(6): 3769–3775.

Ditri, J. J. & Rose, J. L. (1992). Excitation of guided elastic wave modes in hollow cylinders by applied surface tractions, *Journal of Applied Physics* 72(7): 2589–2597.

Dodd, C. V. & Deeds, W. E. (1968). Analytical solutions to eddy-current probe-coil problems, *Journal of Applied Physics* 39(6): 2829–2838.

Doyle, J. F. (1997). *Wave Propagation in Structures: Spectral Analysis Using Fast Discrete Fourier Transforms*, 2nd edn, Springer.

Folk, R., Fox, G., Shook, C. A. & Curtis, C. W. (1958). Elastic strain produced by sudden application of pressure to one end of a cylindrical bar. I. theory, *Journal of the Acoustical Society of America* 30(6): 552–558.

Gazis, D. C. (1959). Three-dimensional investigation of the propagation of waves in hollow circular cylinders. I. Analytical foundation. II. Numerical results., *Journal of the Acoustical Society of America* 31(5): 568–578.

Graff, K. F. (1991). *Wave Motion in Elastic Solids*, Dover.

Grigorenko, A. Y. (2005). Numerical analysis of stationary dynamic processes in anisotropic inhomogeneous cylinders, *International Applied Mechanics* 41(8): 831–866.

Hayashi, T., Kawashima, K., Sun, Z. & Rose, J. L. (2003). Analysis of flexural mode focusing by a semianalytical finite element method, *Journal of the Acoustical Society of America* 113(3): 1241–1248.

Jia, H., Jing, M. & Rose, J. (2011). Guided wave propagation in single and double layer hollow cylinders embedded in infinite media, *Journal of the Acoustical Society of America* 129(2): 691–700.

Karpfinger, F., Gurevich, B. & Bakulin, A. (2008). Modeling of wave dispersion along cylindrical structures using the spectral method, *Journal of the Acoustical Society of America* 124(2): 859–865.

Li, J. & Rose, J. L. (2001). Excitation and propagation of non-axisymmetric guided waves in a hollow cylinder, *Journal of the Acoustical Society of America* 109(2): 457–464.

Lowe, M. J. S. (1995). Matrix techniques for modeling ultrasonic waves in multilayered media, *IEEE Trans. on Ultrasonics, Ferroelectrics and Frequency Control* 42(4): 525–542.

Lowe, M. J. S., Alley, D. N. & Cawley, P. (1998). Defect detection in pipes using guided waves, *Ultrasonics* 36(1-5): 147–154.

Ma, J., Lowe, M. J. S. & Simonetti, F. (2007). Measurement of the properties of fluids inside pipes using guided longitudinal waves, *IEEE Trans. on Ultrasonics, Ferroel. and Freq. Control* 54(3): 647–658.

Maple (2007). Release 11 edn, Waterloo Maple, Inc.

Markus, S. & Mead, D. J. (1995). Axisymmetric and asymmetric wave motion in orthotropic cylinders, *Journal of Sound and Vibration* 181(1): 127–147.

Matlab (2004). Release 14 edn, The Mathworks, Inc.

McNiven, H., Sackman, J. & Shah, A. (1963). Dispersion of axially symmetric waves in composite, elastic rods, *Journal of the Acoustical Society of America* 35(10): 1602–1609.

Meeker, T. R. & Meitzler, A. H. (1972). Guided wave propagation in elongated cylinders and plates, *in* W. P. Mason & R. N. Thurston (eds), *Physical Acoustics, Principles and Methods*, Vol. 1A, Academic Press.

Mirsky, I. (1964). Axisymmetric vibrations of orthotropic cylinders, *Journal of the Acoustical Society of America* 36(11): 2106–2112.

Mirsky, I. (1965). Wave propagation in transversely isotropic circular cylinders: Part I: theory; part II: numerical results, *Journal of the Acoustical Society of America* 37(6): 1016–1026.

Nagy, P. B. & Nayfeh, A. H. (1996). Viscosity-induced attenuation of longitudinal guided waves in fluid-loaded rods, *J. Acoust. Soc. Am.* 100(3): 1501–1508.

Nayfeh, A. H. & Nagy, P. B. (1996). General study of axisymmetric waves in layered anisotropic fibers and their composites, *J. Acoust. Soc. Am.* 99(2): 931–941.

Nelson, R., Dong, S. & Kalra, R. (1971). Vibrations and waves in laminated orthotropic circular cylinders, *Journal of Sound and Vibration* 18(3): 429–444.

Oppenheim, A. V., Schafer, R. W. & Buck, J. R. (1999). *Discrete-Time Signal Processing*, 2 edn, Prentice Hall.

Pavlakovic, B. & Lowe, M. (1999). A general purpose approach to calculating the longitudinal and flexural modes of multi-layered, embedded, transversely isotropic cylinders, *Annual Review of Progress in Quantitative Nondestructive Evaluation*, pp. 239–246.

Pollard, H. F. (1977). *Sound Waves in Solids*, Pion Limited.

Press, W. H., Teukolsky, S. A., Vetterling, W. T. & Flannery, B. P. (1992). *Numerical Recipes in C. The Art of Scientific Computing*, 4 edn, Cambridge University Press.

Rose, J. L. (1999). *Ultrasonic Waves in Solid Media*, 1 edn, Cambridge University Press.

Rose, J. L. (2000). Guided wave nuances for ultrasonic nondestructive evaluation, *IEEE Trans. on Ultrasonics, Ferroelectrics and Frequency Control* 47(3): 575–583.

Seco, F., Martín, J. M. & Jiménez, A. R. (2009). Improving the accuracy of magnetostrictive linear position sensors, *IEEE Trans. on Instrumentation and Measurement* 58(3): 722–729.

Sinha, B. K., Plona, T. J., Kostek, S. & Chang, S.-K. (1992). Axisymmetric wave propagation in fluid-loaded cylindrical shells. i: Theory, *J. Acoust. Soc. Am.* 92(2): 1132–1143.

Whittier, J. & Jones, J. (1967). Axially symmetric wave propagation in a two-layered cylinder, *Int. J. Solids Structures* 3(4): 657–675.

Wilcox, P. D. (2003). A rapid signal processing technique to remove the effect of dispersion from guided wave signals, *IEEE Trans. on Ultrasonics, Ferroelectrics and Frequency Control* 50(4): 419–427.

Zhu, W. (2001). A finite element analysis of the time-delay periodic ring arrays for guided wave generation and reception in hollow cylinders, *IEEE Trans. on Ultrasonics, Ferroelectrics and Frequency Control* 48(5): 1462–1470.

2

Ultrasonic Projection

Krzysztof J. Opieliński
Institute of Telecommunications, Teleinformatics and Acoustics,
Wroclaw University of Technology
Poland

1. Introduction

Ultrasonic technique of imaging serves an increasingly important role in medical diagnostics. In most of applications, echographic methods are used (ultrasonography, ultrasonic microscopy). Using such methods, an image presenting changes of a reflection coefficient in the interior of analysed structure is being constructed. This chapter presents possibilities of utilising information included in ultrasonic pulses, which penetrate an object in order to create images presenting the projection of analysed structure (Opielinski & Gudra, 2004b, 2004c, 2005, 2006, 2008, 2010a, 2010b, 2010c; Opielinski et al., 2009, 2010a, 2010b) in the form of a distribution of mean values of a measured acoustic parameter, for one or numerous planes, perpendicular to the direction of ultrasonic waves incidence (analogical as in X-ray radiography). Due to the possibility of obtaining images in pseudo-real time, the device using this method was named the ultrasonic transmission camera (UTC) (Ermert et al., 2000). Only some centres in the world work on this issue and there are a low number of laboratory research setups, which enable pseudo-real time visualisation of biological structures using UTC: Stanford Research Institute in USA (Green et al., 1974; Green et al., 1976), Gesellschaft für Strahlen- und Umweltforschung at Neuherberg in Germany (Brettel et al., 1981; Brettel et al., 1987), Siemens Corporate Technology in Germany (Ermert et al., 2000; Granz & Oppelt, 1987; Keitmann et al., 2002), Wroclaw University of Technology in Poland (Opielinski & Gudra, 2000; Opielinski & Gudra, 2005, Opielinski et al., 2010a, 2010b), University of California in San Diego and University of Washington Medical Center in Seattle (Lehmann et al., 1999). In most of studies, the projection parameter is the signal amplitude, not its transition time, which seems to more attractive due to the simplicity and precision of measurements. The majority of 2-D ultrasonic multi-element matrices are designed for miniature 3-D volumetric medical endoscopic imaging as intracavital probes provided unique opportunities for guiding surgeries or minimally invasive therapeutic procedures (Eames & Hossack, 2008; Karaman et al., 2009; Wygant et al., 2006a; Wygant et al., 2006b). It can be concluded based on worldwide literature review that most of the 2-D ultrasonic matrices are assigned to work of echo method (Drinkwater & Wilcox, 2006). Moreover, the commercial devices (e.g. Submersible Ultrasonic Scanning Camera made of Matec Micro Electronics, AcoustoCam produced by Imperium Inc.) work using reflection method and are designed for non-destructive inspection (NDI), most of all. It allows manufacturers to instantly visualize a variety of material subsurface faults (voids, delaminating, cracks and corrosion).

This chapter, with the use of a computer simulation and real measurements, includes the complex analysis of the precision of images obtained using visualisation of mean values of sound speed and a frequency derivative of the ultrasonic wave amplitude attenuation coefficient (by the means of measurements of the transition time and the frequency down shift projection values of transmitted ultrasonic pulses) from the point of view of possibilities of using UTC for visualisation of biological structures, especially for soft tissue examinations (*in vivo* female breast). The principles of ultrasonic projection methods are described at the beginning (Section 2), and then the projections of acoustic parameters are clearly defined (Section 3). Section 4 contains theoretical analysis of ultrasonic projection method accuracy. By the means of elaborated software, there was done a simulation of the ultrasonic projection data for several three-dimensional objects immersed in water, what is presented in Section 5, as well as obtained projection images of these objects. The simple measurement setup for examining biological structures by the means of ultrasonic projection, in the set of single-element ultrasonic sending and receiving probes and projection measurement results are presented in Section 6. Next, the construction, parameters and operating way of three different types of ultrasonic 2-D flat multi-element matrices (standard, passive and active one) elaborated by the author and his team, are described (Section 7). The models of ultrasonic transmission camera were constructed in the result, what is presented in Section 8. The conclusion (Section 9) contains a summary and the plane for the future to improve the quality of ultrasonic projection images and increase scanning resolution.

2. Ultrasonic projection methods

Analogically as in case of X-ray pictures (RTG - roentgenography), for visualisation of biological structures, it is possible to use the projection method with utilisation of ultrasonic waves. In case of generating of ultrasonic plane waves (parallel beam rays), we shall obtain an image in the parallel projection and in case of generating of ultrasonic spherical waves (divergent beam rays), it shall be an image in the central projection; it is also possible to use a source of cylindrical waves (beam rays are divergent on one plane and parallel in perpendicular plane) – central-parallel projection (Fig.1) (Opielinski & Gudra, 2004c).

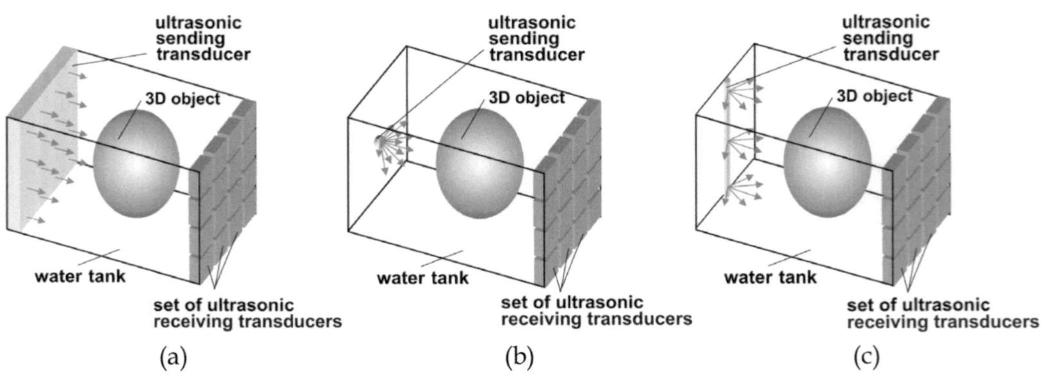

Fig. 1. The way of visualization of a biological structure by means of the ultrasonic projection: a) parallel, b) central (divergent), c) central–parallel (divergent–parallel)

Phenomena related with propagation of ultrasonic waves in biological structures (diffusion, diffraction, interference, refraction and reflection) induce small distortions of a projection image, if the local values of acoustic impedance in an examined structure are not significantly diversified (Opielinski & Gudra, 2000). The greatest advantage of ultrasonic diagnostics is its non-invasiveness, by dint of which, it is possible to attain a multiple projection of an examined biological structure at numerous different directions *in vivo*, what enables a three-dimensional reconstruction of heterogeneity borders in its interior (Opielinski & Gudra, 2004a).

By the means of using the transmission method, it is also possible to obtain a twice higher level of amplitude of ultrasonic receiving pulses in comparison with the echo method. In this case, the subject of imaging can simultaneously be several acoustic parameters, digitally determined on the basis of information directly contained in ultrasonic pulses, which penetrate a biological structure (e.g. amplitude, transition time, mid-frequency down shift, spectrum of receiving pulse). Such method enables obtaining various projection images, every of which characterises slightly different traits of a structure (e.g. distribution of mean (projective) values of attenuation coefficient and propagation velocity of ultrasonic waves, frequency derivative of attenuation coefficient, nonlinear acoustic parameter B/A (Greenleaf & Sehgal, 1992; Kak & Slaney, 1988). What is more, if projective measurements of a distribution of a specified acoustic parameter of an examined projection plane of a biological medium, are recorded from numerous directions around the medium, it is possible to reconstruct a distribution of local values of this parameter in 3-D space by determining the inverse Radon transform (Kak & Slaney, 1988; Opielinski & Gudra, 2010b) (ultrasonic transmission tomography). Such complex tomographic characteristics may have a key importance, for example at detecting and diagnosing cancerous changes in soft tissues (e.g. in female breast).

3. Projection of acoustic parameters

According to the definition of function projection, projective values of an acoustic parameter, measured at projections, are an integral of their local values, in the path of an ultrasonic beam, transmitted from a source to a detector. Moreover, point sizes of the transmitter and the receiver, the rectilinear path between them and an infinitely narrow ultrasonic wave beam are assumed. It is easy to prove that transition time of an ultrasonic wave is an integer of an inverse of local values of sound velocity through the propagation path L (Opielinski & Gudra, 2010b):

$$t_p = \int_L dt_p = \int_L \frac{dt_p}{dl} dl = \int_L \frac{1}{c(x,y,T)} dl \tag{1}$$

where $c(x,y,T)$ denotes the sound speed local value at point (x,y) of an object's cross-section on the propagation path, at set temperature T, $(dl)^2 = x^2 + y^2$. By a direct measurement of values of transition time of an ultrasonic wave, it is not difficult to image a distribution of projective values of sound velocity $c_p = L/t_p$ in determined plane of projection of a biological medium, immersed in water, assuming the linear propagation path of wave.

By the means of measuring the amplitude of an ultrasonic pulse after a transition through the structure of a biological medium, it is possible to obtain information about the projective value of an amplitude attenuation coefficient $a_p = ln(A_N/A_L)/L$, measured at the set frequency of a transmitting pulse f_N and temperature T (Opielinski & Gudra, 2010b):

$$\ln\frac{A_N}{A_L} = \int_L \frac{1}{dl}\ln\frac{A(l_i)}{A(l_{i+1})}dl = \int_L \alpha(x,y,f_N,T)dl \quad (2)$$

where A_N – amplitude of an ultrasonic pulse before transition (near the source), A_L – amplitude of an ultrasonic pulse after transition through an object (dipped in water) on the path of L, $A(l_i)$ and $A(l_{i+1})$ – amplitudes of an ultrasonic pulse after transition through an object on segments l_i and l_{i+1}, respectively, of the path L, and distance $dl = l_{i+1} - l_i$. Due to difficulties related with measuring the amplitude A_N, it is possible to directly determine the difference of projective values of the amplitude attenuation coefficients $(a_p - a_w) = \ln(A_w/A_L)/L$, where a_w – attenuation coefficient in water, A_w – amplitude of an pulse after passing the path L in water without any biological medium.

Assuming a linear change of attenuation with frequency (as a certain approximation for soft tissues (Kak & Slaney, 1988)), it is possible to obtain an amplitude attenuation coefficient, independent from frequency $a_o(x,y,T)$:

$$\alpha(x,y,T,f) = (\alpha_o(x,y,T)) \cdot f \quad (3)$$

On this basis, it is possible to use in ultrasonic projection measurements the method of measuring down shifting of mid-frequency of a receiving pulse. After transition of an ultrasonic signal through an object, there is a slight change of its mid-frequency f_r, which can be measured by FFT or by a zero-crossing counter (Opielinski & Gudra, 2010b). Then, the projective value of a frequency derivative of an amplitude attenuation coefficient $a_o(x,y,T)$ can be determined in the following form (Kak & Slaney, 1988):

$$\frac{f_N - f_r}{2\sigma^2} = \int_L \alpha_o(x,y,T)dl \quad (4)$$

where variance σ^2 is the measure of the power spectrum bandwidth of receiving signal after a transition through water. More complex models, e.g. with considering the relationship $a = a_o f^n$, where $n \neq 1$ (in case of soft tissues $1 \leq n \leq 2$), can be found in references (Narayana & Ophir, 1983).

The development of computer technologies enables now also projective measurements of acoustic parameters of biological media, which require time-consuming calculations. One of such parameters is non-linear acoustic parameter B/A, which characterises non-linear response of a measured tissue structure on propagation of an ultrasonic wave (Opielinski & Gudra, 2010b). Parameter B/A can be determined by the means of measuring transition times for different static pressures or by measurements of higher harmonics as a distance function (Greenleaf & Sehgal, 1992; Zhang et al., 1997).

4. Theoretical analysis of ultrasonic projection method accuracy

The theoretical analysis, which enables estimating of the accuracy of the projective visualisation of heterogeneities in a tissue structure was conducted for a projection of ultrasonic wave propagation velocity (mean sound velocity on a rectilinear section of the beam path from the source to the detector), obtained after wave's transition through the model of a heterogeneous sphere in its axis (Fig.2) (Opielinski & Gudra, 2004b, 2006).

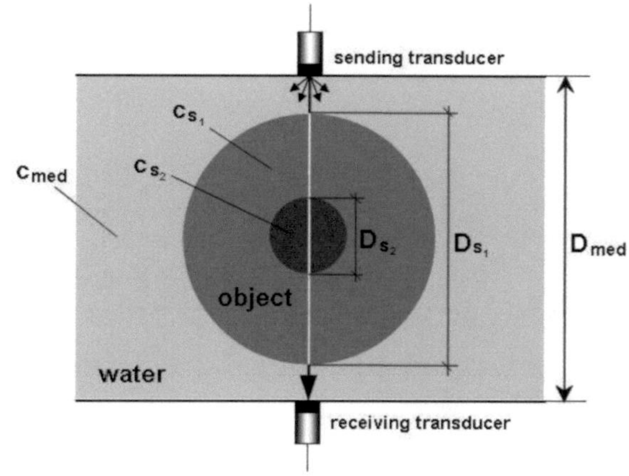

Fig. 2. Theoretical model of a heterogeneous sphere (because of ultrasound propagation speed)

It was assumed that a sphere of the diameter D_{s1} and sound speed c_{s1} = 1500 m/s contains a spherical heterogeneity of the diameter D_{s2} and sound speed c_{s2}. It was also assumed that the sound speed in water where the sphere is immersed c_{med} = 1485 m/s (for ~21°C) and the distance between surfaces of the sending and receiving transducers is D_{med} = 20 cm (simulation of female breast examination). The mean value (projection) of sound speed c_p between surfaces of the transmitter and the receiver was determined using the formula (Opielinski & Gudra, 2004b, 2006):

$$c_p = \frac{D_{med}\, c_{med}\, c_{s1}\, c_{s2}}{c_{s1}\, c_{s2}\left(D_{med} - D_{s1}\right) + c_{med}\, c_{s2}\left(D_{s1} - D_{s2}\right) + c_{med}\, c_{s1}\, D_{s2}} \tag{5}$$

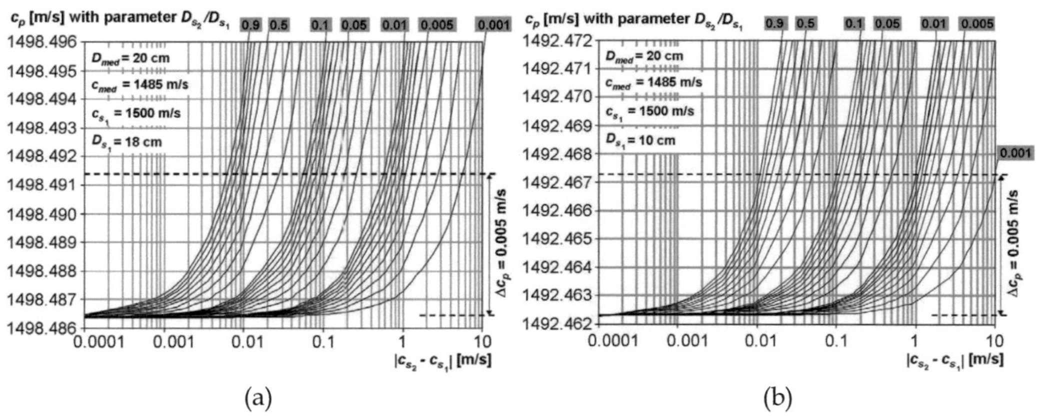

Fig. 3. Dependence of c_p on $|c_{s2} - c_{s1}|$ (for $c_{s2} > c_{s1}$) with parameter D_{s2}/D_{s1} (formula (5)), for different ratios of D_{s1}/D_{med} = 0.9 (a), 0.5 (b)

Figures 3a and 3b (for $c_{s2} > c_{s1}$) and Figures 4a and 4b (for $c_{s1} > c_{s2}$) present dependences between mean (projective) sound speed value, determined using formula (5) and the absolute value of the speed difference in heterogeneity and its spherical surrounding (Fig.2) for 29 various ratios of their diameters (0.001, 0.002, ..., 0.01, 0.02, ..., 0.1, 0.2, ..., 0.9) and for two relations between sphere diameter and distance between transducers (0.9 and 0.5) (Opielinski & Gudra, 2006).

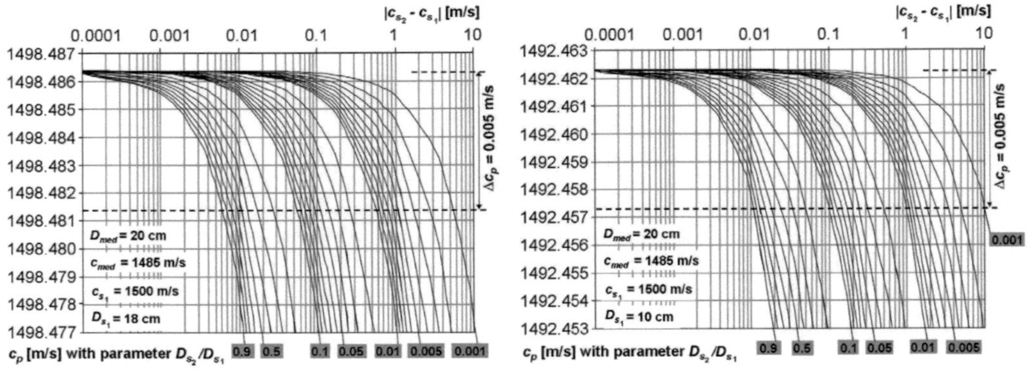

(a) (b)

Fig. 4. Dependence of c_p on $|c_{s2} - c_{s1}|$ (for $c_{s1} > c_{s2}$) with parameter D_{s2}/D_{s1} (formula (5)), for different ratios of D_{s1}/D_{med} = 0.9 (a), 0.5 (b)

If we assume that the resolution of digital measurement of transition time at sampling frequency f_s, is $1/f_s$, then depending on the distance between surfaces of sending and receiving transducers and depending on the value of measured mean speed, the resolution of measurement of mean propagation speed of an ultrasonic wave (projection) can be determined using the formula (Fig.5) (Opielinski & Gudra, 2004b):

$$\Delta c_p = \frac{D_{med}}{D_{med}/c_p - 1/f_s} - c_p \qquad (6)$$

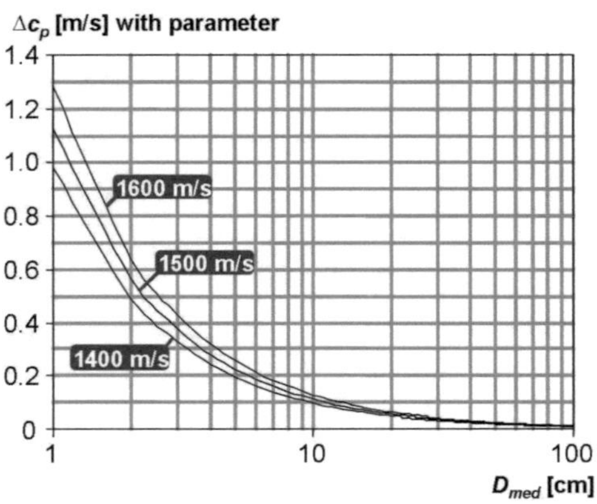

Fig. 5. Dependence of Δc_p on D_{med} (formula (6)) for different values of c_p in a measurement area

For the diameter of measurement area D_{med} = 20 cm, f_s = 200 MHz (T_s = 5 ns), c_p = 1500 m/s, resolution $\Delta c_p \approx 0.06$ m/s. Using burst type ultrasonic pulses of the frequency of 5 MHz, 20 samples accrues to a half of wave period.

It means that at digital measuring of transition time, using the method of determining zero-crossing by the means of linear interpolation between samples of negative and positive amplitude values, it is possible to achieve the resolution of transition time measurement better than $\Delta t_p \approx T_p/2 = 2.5$ ns ($\Delta c_p \approx 0.03$ m/s); at a significant interval between a signal and noise (e.g. using algorithms of noise reduction), uncertainty of such measurements can be even better than about $\pm\Delta c_p/6 \approx \pm 0.005$ m/s (formula (6)).

Calculations show (Fig.3, Fig.4) that by the means of projection of ultrasonic wave propagation speed it is possible to detect differences of its local values in a biological structure. In case of a difference of speed in heterogeneity and its surrounding $|c_{s2} - c_{s1}| \approx 0$ (homogeneous sphere), all values of projection of speed c_p tend to the determined value, marked by a dotted line in Fig.3a,b and Fig.4a,b. In order to conduct an interpretation of charts obtained from calculations, it is necessary to determine the resolution of measuring transition time of a wave through a structure. If we assume that using good class devices for digital measurements of transition time with interpolation it is possible to detect changes of sound speed of minimum about 0.005 m/s (dotted line in Fig.3a,b and Fig.4a,b) it can mean e.g. that for biological media of size of about 20 cm (in case of adjusting a distance between sending and receiving transducers to external dimensions of that media), by the means of ultrasound speed projection it is possible to detect heterogeneity of a local value of this speed in a structure (diversification) at the minimal level of about (Fig.3a):

a. 1 m/s for heterogeneity size $D_{s2} \approx 1$ mm,
b. 0.6 m/s for heterogeneity size $D_{s2} \approx 2$ mm,
c. 0.2 m/s for heterogeneity size $D_{s2} \approx 5$ mm,
d. 0.1 m/s for heterogeneity size $D_{s2} \approx 10$ mm,

considering that there still is an obligatory limitation due to the diffraction of an ultrasonic wave (a wave penetrates heterogeneities of sizes larger than a half of wave's length). Resolution of a detection of differences of local sound speed values in heterogeneity and its surrounding strongly depends on the ratio of their dimensions D_{s2}/D_{s1} and the ratio of the size of an analysed object in the axis of ultrasonic transducers – sending and receiving – to the distance between the transducers D_{s1}/D_{med} (the less ratios, the worse resolution – compare: Figures: 3a with 3b, and 4a with 4b).

A similar analysis was conducted for the projection of a frequency derivative of the ultrasonic wave amplitude attenuation coefficient, obtained after a transition of a wave through the model of a heterogeneous sphere in its axis (Fig.2). In case of projective measurements of the value of a frequency derivative of the amplitude attenuation coefficient of an ultrasonic wave by the means of detecting the frequency of a pulse of an ultrasonic wave after the transition, it is possible to assume a minimal uncertainty of the measurement of the projective value of frequency of about 2 kHz, what at assuming the ultrasonic wave frequency of 2 MHz, results in the uncertainty of determining the projective value of a frequency derivative of attenuation coefficient of about 0.001 dB/(cm·MHz). It means that in the measured projective values of receiving signal frequencies, it is possible to distinguish an influence of a heterogeneity for its dimensions proper correlated with a difference of a local value of the amplitude coefficient of attenuation of an ultrasonic wave in the structure of this heterogeneity and in the structure of its surroundings Δa. If the presence of a heterogeneity changes measured projective values of a receiving pulse frequency, this heterogeneity is able to be detected. Calculations prove that at assuming the total uncertainty of projective measurements of values of the receiving pulse frequency of 2 kHz, it shall be possible to detect in a projective image heterogeneities, which differ from surrounding tissue by the minimum value of sound attenuation about: 0.1 dB/cm of the size > 9 mm, 0.2 dB/cm of the size > 4 mm, 0.5 dB/cm of the size > 2 mm, 1 dB/cm of the size > 900 μm, 2 dB/cm of the size > 400 μm, 5 dB/cm of the size > 200 μm. Calculated contrast resolution increases at higher spectrum width of the receiving pulse and a change of sound attenuation in water around a sphere has a negligible influence on this resolution.

In case of heterogeneities of diversified sizes, it can happen that in projective images for several projection planes, heterogeneities can be visible, while in others can be not.

5. Computer simulation of ultrasonic projection data

By the means of elaborated software, there was done a simulation of a distribution of local values of propagation velocity of an ultrasonic wave inside several three-dimensional objects immersed in water (Fig.6), and next, for a selected projection plane, there was done a calculation (using the Radon transform (Kak & Slaney, 1988)) of distributions of mean values (projections) in a parallel-ray geometry, at the set distance D_{med} = 20 cm between surfaces of the transmitter and the receiver (Opielinski & Gudra, 2004c, 2006). In the space surrounding each of the objects, the speed of ultrasonic wave propagation was assumed at c_{med} = 1485 m/s, what corresponds to the speed of sound in water of the temperature of about 21°C. In case of ultrasonic projection, using water as a coupling medium is obligatory due to a matching to the acoustic impedance of an examined biological structure. The Section 5 presents simulations of projective data for exemplary virtual objects *C* and *D*. Object *C* is a sphere of the diameter of 14 cm, containing 7 homogeneous balls of 0.5 cm diameters each, situated on the major axis, at 2 cm intervals between their centres, apart from two terminal ones - 1.85 cm intervals. Object *D* is an ellipsoid of the semi-major axis of 7 cm and the 1 cm semi-minor axis, containing 7 homogeneous balls of 0.5 cm diameters each, located on the diameter in the same way as in the object *C*. Local values of ultrasonic wave propagation speed are 1500 m/s for each point of the interior of objects *C* and *D*, apart from small balls – 1499 m/s.

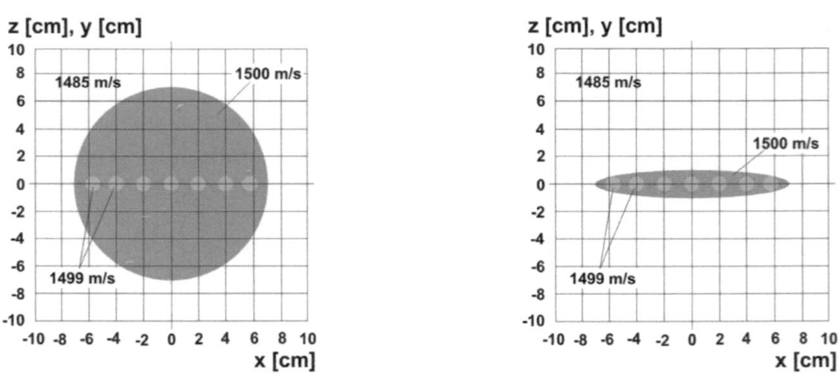

(a) (b)

Fig. 6. Shape of created 3-D virtual objects in cross-section: a) object *C* – small balls in sphere, b) object *D* – small balls in ellipsoid

Figure 7 presents images of ultrasonic projection for objects *C* and *D* obtained on the basis of computer simulated measurements of mean values of sound speed in an orthogonal projection (Opielinski & Gudra, 2004c). These values were linearly imaged in greyscale, from black to white. Shapes of projected objects, together with edges, are clearly visible in images. The object *C* interior seems to be homogeneous (Fig.7a) and spherical structures inside object *D* (Fig.7b) are hardly visible against the background of the ellipsoid projection (bright, round spots).

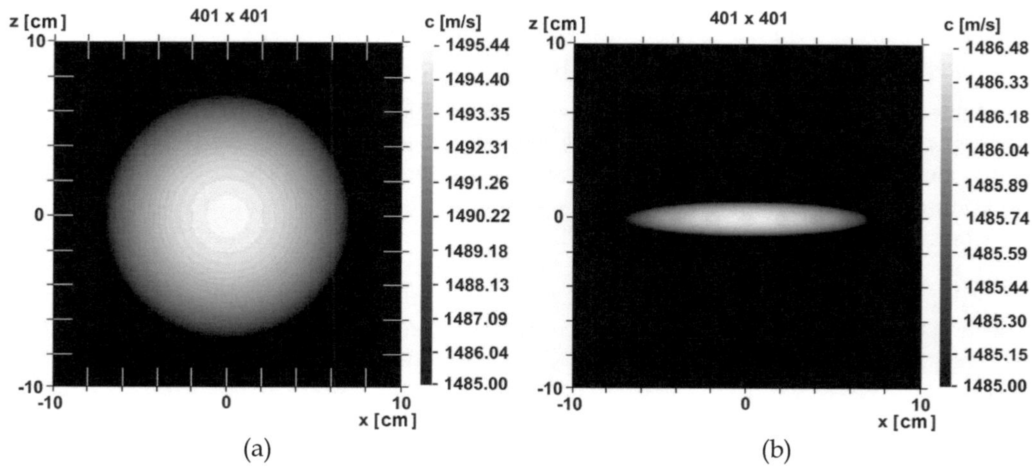

Fig. 7. Ultrasonic projection images of objects shown in Fig.6, obtained on the basis of computer-simulated measurements of mean values of ultrasonic wave propagation velocity: a) object *C*, b) object *D*

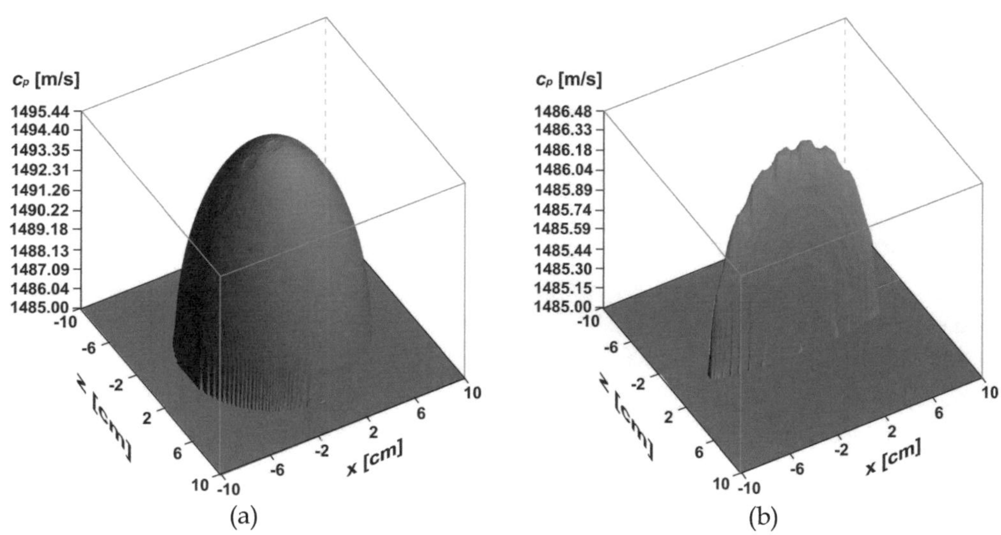

Fig. 8. Ultrasonic projection images of sphere – object *C* (a) and ellipsoid – object *D* (b), shown in pseudo-3-D

In order to improve the contrast of visualisation of heterogeneous structures, Figure 8 presents ultrasonic projection images of objects *C* and *D* in pseudo-three-dimensional way with an lighting, imaging mean values of propagation speed of an ultrasonic wave for the plane parallel to the axis, along which small balls are arranged in the Cartesian co-ordinate system. Here, projections of spherical heterogeneities are shown in the form of characteristic oval concavities in the object's structure. Difficulties in imaging inclusions in the object *D* structure result from the projective method of averaging values of ultrasonic wave speed on the propagation path. In the projection images (Fig.7), objects' shapes and their edges are clearly visible in the parallel projection. Small spherical heterogeneities inside the sphere are hardly evident on the projective pseudo-three-dimensional image (Fig.8) due to scant sizes in comparison to the diameter of surrounding sphere.

Conducted simulative calculations enable an initial estimation of the accuracy of the ultrasonic projective imaging in respect of detecting heterogeneities in the internal structure. In the contour image of object *C* (Fig.7a), structure heterogeneities are imperceptible and in the contour image of object *D* (Fig.7b) - hardly visible. The reason of such imaging in ultrasonic projection is mainly a small difference of propagation velocity of an ultrasonic wave in heterogeneities in comparison to propagation in their surrounding (1 m/s). Projection values of sound velocity of propagation of an ultrasonic wave, corresponding to central pixels of the image from Figures 7a and 7b, are equal to 1495.4185 m/s and 1486.4374 m/s, respectively. If structures of objects *C* and *D* were homogeneous, these values would be equal to 1495.4683 m/s and 1486.4865 m/s, what means that an inclusion in the form of a central ball changes the velocity by about 0.05 m/s in both cases. Thus, such small change is not visible in contour images, due to a limited human eye's ability of distinguishing shades; however, it is noticeable in pseudo-3-D images (Fig.8). It seems that in ultrasonic projection it is possible to image even smaller differences in velocity values, thanks to using additional operations of data processing in the form of algorithms of compression, expansion, gating, filtration and limitation of measurement data values (Opielinski & Gudra, 2000).

The change of mean speed caused by an inclusion is approximately the same for objects *C* and *D*, however, their widths are significantly various. It means that a deterioration of the dynamics of velocity values simulated for object *C* is caused by too big distance between sending and receiving transducers - sound speed in the measurement medium (water) has a negative influence on measurement results. Due to the above, such distance should be possibly slightly higher from the size of an object in the analysed projection.

6. Projection measurements

The developed, computer-assisted measurement setup for examining biological structures by the means of ultrasonic projection, in the set of single-element ultrasonic probes - sending and receiving - is presented in Fig.9 (Opielinski & Gudra, 2004c, 2005).

For the aims of scanning of objects immersed in water, there were used two ultrasonic probes of 5 mm diameter, which played roles of a source and a detector of ultrasonic waves, of work frequency of 5 MHz (Opielinski & Gudra, 2005). Probes, mounted on the axis opposite each other, are shifted with a meander move with a set step in a selected plane of an object. Movements of probes are controlled by the means of a computer software through RS232 bus, using mechanisms of shifting *XYZ* (Opielinski & Gudra, 2005). The sending probe is supplied by a burst-type sinusoidal signal and pulses received from the receiving probe by a digital oscilloscope are sent through RS232 bus to a computer and stored on a hard disk. Projective values of proper acoustic parameters, determined from particular receiving pulses, are imaged in the contour form in colour or grey scale and also in pseudo-3-D, using special software, which enables also advanced image processing (Opielinski & Gudra, 2005).

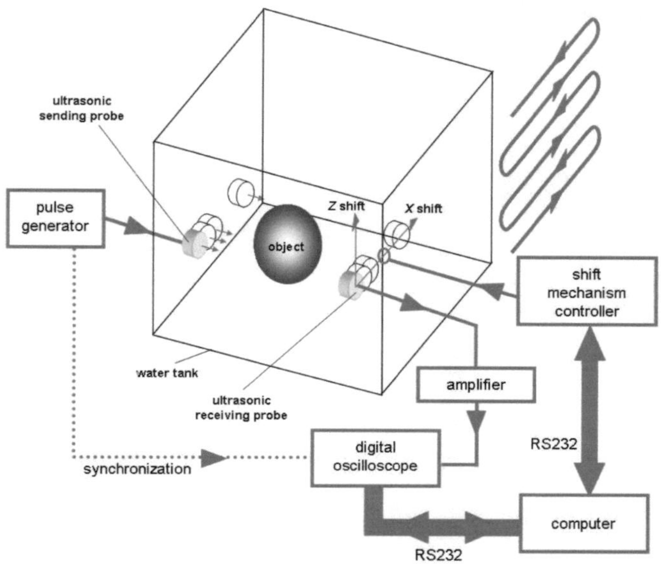

Fig. 9. Block scheme of the measurement setup for biological media structure imaging by means of the ultrasonic projection method using one-element ultrasonic probes

On the developed research setup, measurements of several 3-D biological objects were conducted in various scanning planes (Opielinski & Gudra, 2004b, 2004c, 2005, 2006). One of real biological media that was subjected to projective measurements was a hard-boiled chicken egg without a shell. Proteins composed of amino-acids are present in all living organisms, both animals and plants, and they are the most essential elements, being the basic structural material of tissues. A chicken egg is an easy accessible bio-molecular sample for ultrasonic examinations. Due to its oval shape (a possibility to transmit ultrasonic waves from numerous directions around it), structure and acoustic parameters (a relatively low attenuation and a slight refraction of beam rays of an ultrasonic waves at boundaries water/white/yolk), a boiled chicken egg without a shell is a great object, enabling testing of visualisation of biological structures by the means of ultrasonic projection and the method of ultrasonic transmission tomography (Opielinski, 2007). Figure 10 presents obtained projective images (the transmission method, f = 5 MHz) in comparison to an image obtained from an ultrasonography (the reflection method, f = 3.5 MHz) of a hard-boiled chicken egg without a shell. On the basis of the set of recorded receiving pulses in one scanning plane, with the step of 1.5 mm x 1.5 mm, images presenting the projection of a distribution of three various acoustic parameters in the object structure were obtained: propagation velocity, frequency derivative of ultrasonic waves attenuation and ultrasonic wave amplitude after transition (Opielinski & Gudra, 2004b, 2004c, 2005). In the images, distributions of particular parameters are marked by a solid line for pixels along marked broken lines. Negative values of a derivative of the ultrasonic wave attenuation on frequency in the image in Fig.10b result from errors of the measurement of a shift of mid-frequency of pulse after transition at edges of structures, where a signal is weakened or faded most often.

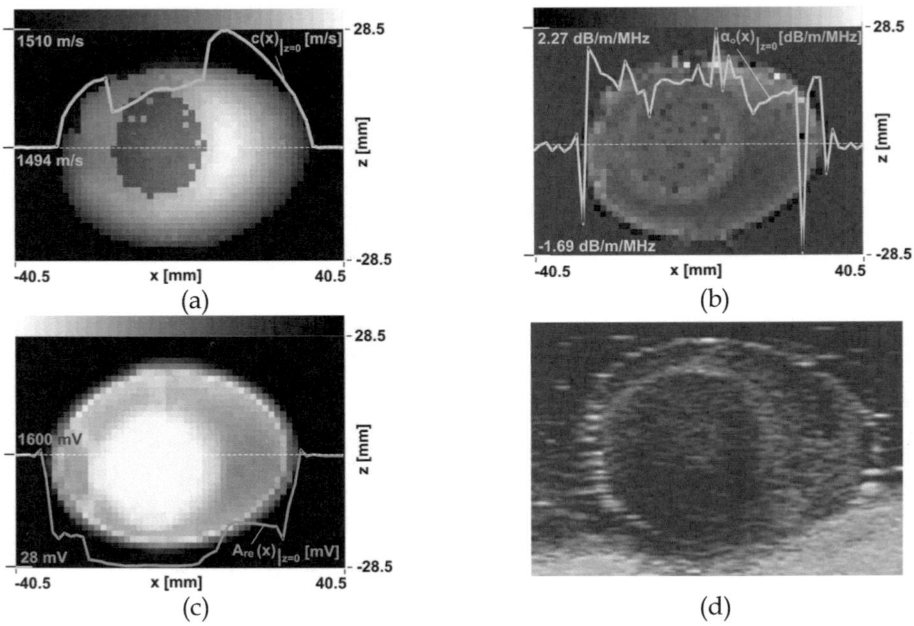

Fig. 10. Ultrasonic projection images of a hard-boiled hen's egg without shell, obtained from the following measurements of mean values: a) ultrasonic wave propagation velocity, b) derivative of the ultrasonic wave amplitude attenuation coefficient on frequency, c) ultrasonic wave pulse amplitude, in comparison to the ultrasonogram of the same egg (d)

In order to verify the correctness of imaging the internal structure of the analysed object, Fig.11 presents its optical image in the cross-section for the analysed plane, obtained after cutting an egg into half and scanning the examined section using an optical scanner. Comparing images presented in Figures 10 and 11, it can be unequivocally stated that a computer-assisted ultrasonic projection enables proper recognising of biological structures (a yolk is clearly visible in the egg structure). One advantage of the ultrasonic projection method is an availability to obtain several different images from a single measurement set, every of which characterises some other features of an object.

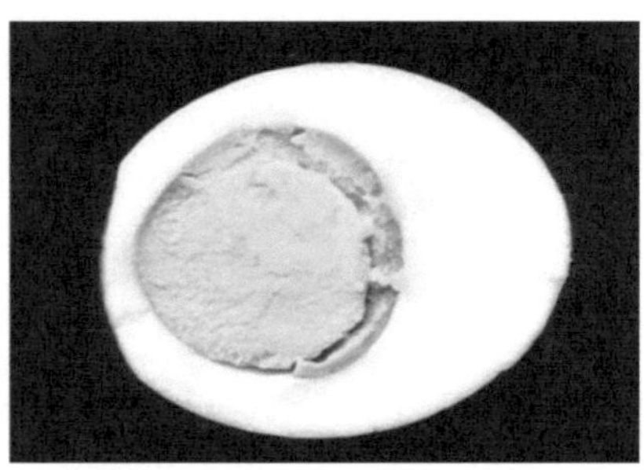

Fig. 11. The optical image (scan) of the measured egg cross-section structure

These images can be properly processed and correlated by the means of special software, what enables recognising structures, which are not visible in single images. The image of the distribution of the sound speed projection values clearly visualises constant changes of heterogeneity (Fig.10a), while the image of the distribution of sound attenuation frequency derivative projection values better visualises discrete changes (Fig.10b). In the image of the distribution of sound velocity projection values, it is also visible that sound speed in a yolk is higher than in water and lower than in white of an egg (Fig.10a). In the image of the distribution of the amplitude projection values, it is visible that attenuation in yolk is larger than in egg white and much higher than in water. The image of the distribution of amplitudes of receiving pulses is characterized by a large dynamics of value changes and in a similar rate visualises both continuous and step changes (Fig.10c). The ultrasonographic image of an egg (Fig.10d) visualises clearly only boundaries of yolk and white structures.

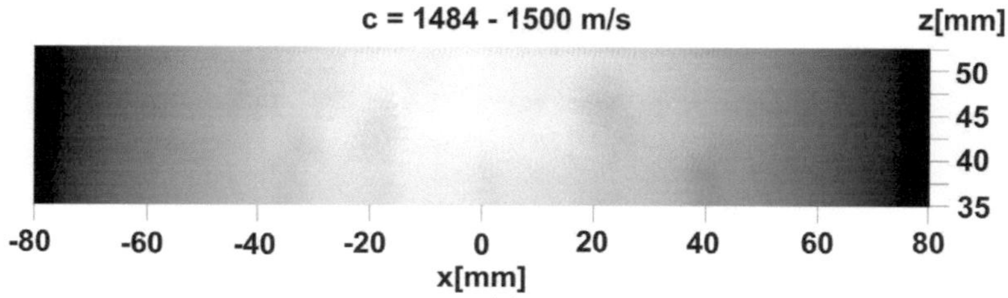

(a)

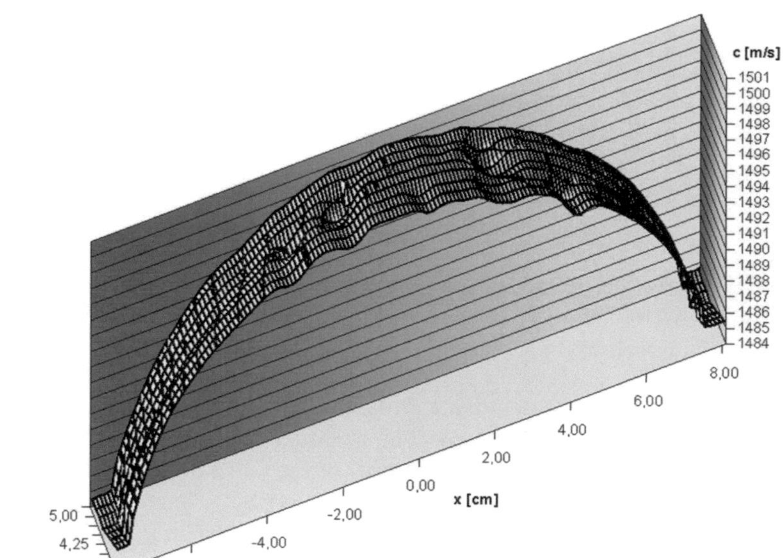

(b)

Fig. 12. Ultrasonic projection images of CIRS model 052 breast biopsy phantom in the range of measured altitudes $h = 3.50 \div 5.25$ mm along its longest dimension – length: a) image in gray scale, b) image in pseudo-3-D

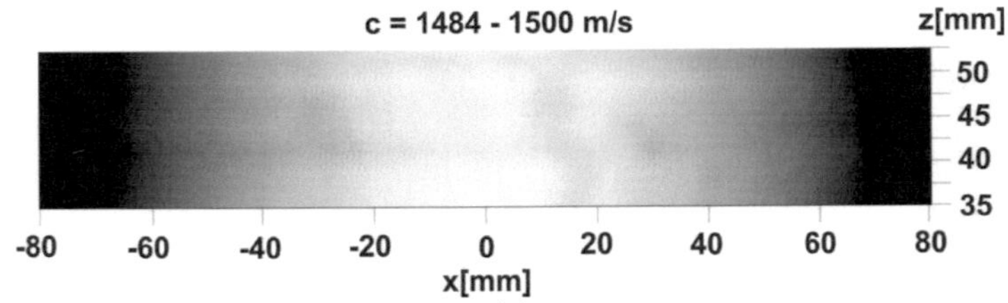

(a)

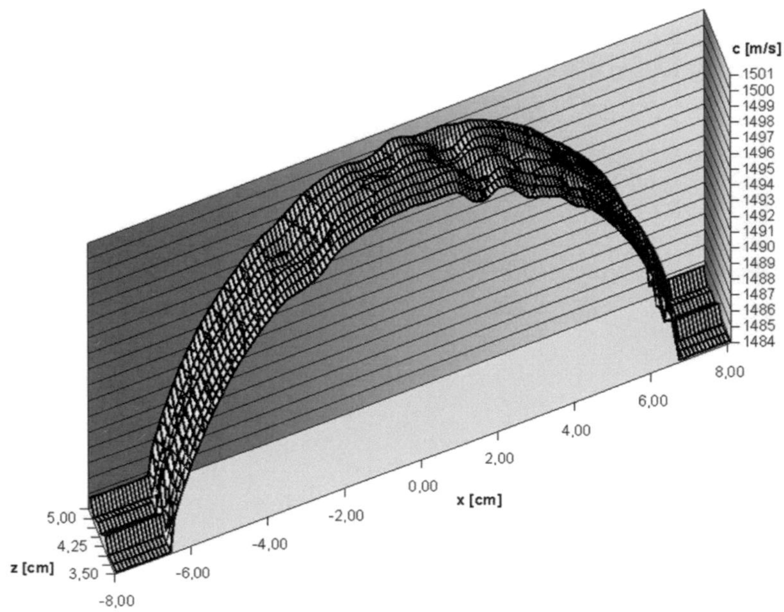

(b)

Fig. 13. Ultrasonic projection images of CIRS model 052 breast biopsy phantom in the range of measured altitudes $h = 3.50 \div 5.25$ mm, along its shortest dimension – width: a) image in gray scale, b) image in pseudo-3-D

A subject of projection studies was also a breast biopsy phantom of the American CIRS company, model 052, which simulates acoustic parameters of tissues that are present in a female breast. Ultrasonic projection measurements of the CIRS phantom were conducted on the research setup for ultrasonic transmission tomography (Opielinski & Gudra, 2010b), measuring mean values of transition times of an ultrasonic pulse in the geometry of parallel-ray projections, by the means of single-element ultrasonic probes of 5 mm diameters and work frequency of 5 MHz, located centrally opposite each other, in the distance of 160 mm (Opielinski & Gudra, 2004a, 2004c). There were used 161 steps of probes pair shift along the phantom, with the 1 mm step (161 measurement rays) for each of 100 turns around the phantom with the step of 1.8° (100 measurement projections) and for each of eight positions of probes pair in vertical direction, in distances from the phantom base of 35 mm, 37.5 mm, 40 mm, 42.5 mm, 45 mm, 47.5 mm 50 mm, 52.5 mm, respectively. Such data set enables obtaining 100 projection images in *XZ* planes (of resolutions 101 x 8 pixels each), from all sides of the object flank, in the angle range of 0° ÷ 178.2°. Figures 12 and 13 present images of sound speed projections for parallel ultrasonic rays, penetrating lateral surfaces of the CIRS phantom in the range of measured heights of h = 3.50 ÷ 5.25 mm, in greyscale and pseudo-3-D (Opielinski & Gudra, 2004c).

These images were obtained by temperature scaling and assembling all measurement projections (sets of values for full shifts of the probes pair along the object) for rotation angles of 0° (Fig.12) and 90° (Fig.13), extracted from projection measurement sets of cross-sections of the phantom for particular heights.

Figure 14 presents 2 geometrical projections of the CIRS breast phantom structure from the sides of its length and width, obtained by the means of 3-D reconstruction from tomographic images (Opielinski & Gudra, 2004a, 2004c). It can be observed that there is a good conformity of geometrical projections of the phantom with projection images obtained by the means of ultrasound transmission (compare Fig.12 and Fig.13 with Fig.14a,b). An interpretation of projection images requires a spatial intelligence and basic knowledge in spatial geometry. Particular images present projections of the distribution of local values of a measured acoustic parameter in the plane parallel to the scan surface and are a mirror reflection for projection planes from the back and front sides of the object. Projection imaging is not a full quantitative imaging; however, on the basis of values of particular pixels of an image it is possible to determine the diversification of parameters of an examined structure. At a proper resolution, step changes of values of an examined acoustic parameter in the projection plane are noticeable in an image, while continuous changes are, in general, not easy detectable. A step change of values of an examined parameter is visible in a contour image in the form of a discrete change of contrast. Pseudo-3-D images provide a better dynamics of projection visualisation. On the basis of several projection images of an examined structure from numerous directions, it is possible to undertake an attempt of a 3-D reconstruction of heterogeneity borders in its interior (Opielinski & Gudra, 2004a).

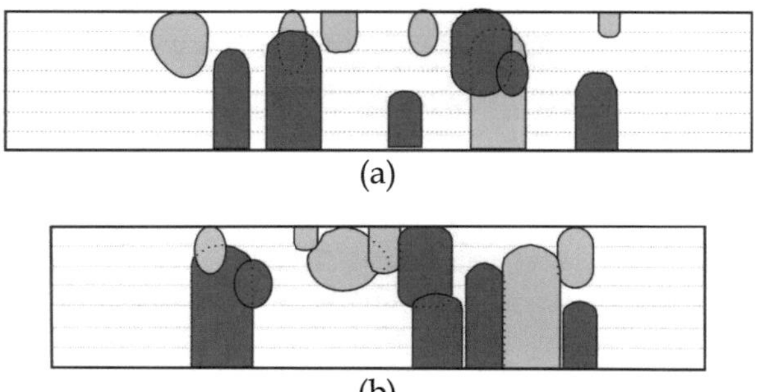

(a)

(b)

Fig. 14. Geometric projections of phantom CIRS structure along its length (a) and width (b), obtained on the basis of three-dimensional reconstruction from tomographic images

Ultrasonic projection imaging can be also used for non-invasive *in vivo* visualisation of injuries and lesions of human upper and lower limbs (Ermert et al., 2000) for the aims of e.g. diagnosing osteoporosis degree. However, it is necessary to consider specific distortions, which occur in an image due to refraction of ultrasonic wave rays in case of biological structures of a large refraction index (the ratio of velocity in water to velocity in an examined structure (Opielinski & Gudra, 2008)). In case of bones examinations, it is possible not only to record projection values of acoustic parameters described in Section 3, but also to determine images of the distribution o two main parameters of ultrasonic waves, used in medicine for diagnosing various stages of osteoporosis: SOS (speed of sound) and BUA (broadband ultrasound attenuation). On their basis, it is also to determine the distribution of the stiffness factor, which defines the state of osseous tissues in relation to a healthy female population, in the age of so called peak bone mass.

7. 2-D ultrasonic matrices

At designing multi-element ultrasonic matrices - sending and receiving one - for the aims of projective imaging of biological media, it is essential to achieve a compromise between the resolution (it depends on the type and size of elementary transducers, their work frequency, distance between them) and the efficiency and sensitivity, which grows along with the surface size of an elementary transducer. Thus, it is very important to specify the sizes of elementary transducers and distances between them already at the designing stage. The technique of mounting transducers on a matrix and the way of conducting electrodes has also a substantial influence on the way of transducers' work. The ongoing search for improved ultrasonic imaging performance will continue to introduce new challenges for beamforming design (Thomenius, 1996). The goal of beamformer is to create as narrow and uniform a beam with as low sidelobes over as long a depth as possible. Among already proposed imaging methods and techniques are elevation focusing (1.25-D, 1.5-D and 1.75-D arrays) (Wildes et al., 1997), beam steering, synthetic apertures, 2-D and sparse matrices, configurable matrices, parallel beamforming, micro-beamformers, rectilinear scanning, coded excitation, phased subarray processing, phase aberration correction, and others (Drinkwater & Wilcox, 2006; Johnson et al., 2005; Karaman et al., 2009; Kim & Song, 2006; Lockwood & Foster, 1996; Nowicki et al., 2009). The most common complication introduced by these is a significant increase in channel count. Generating narrow ultrasonic wave beams in biological media by multi-element probes, built as matrices of elementary piezoceramic transducers in a rectangular configuration, can be realised using transducers having a spherical surface, ultrasonic lenses, mechanical elements (e.g. complex system of properly rotated prisms), focusing devices or electronic devices, which control the system of activating and powering individual matrix elements in a proper manner (Drinkwater & Wilcox, 2006; Ermert et al., 2000; Granz & Oppelt, 1987; Green et al., 1974; Nowicki, 1995; Opielinski et al., 2009; Opielinski & Gudra, 2010a, 2010c; Opielinski et al. 2010a, 2010b; Ramm & Smith, 1983; Thomenius, 1996). Exciting individual transducers of the multi-element probe using pulses with various delay is a universal method of focusing and deflecting a beam (Johnson et al., 2005; Thomenius, 1996). Adequate delays between activations of each successive elementary transducer allow shaping of the wave front and the direction of its propagation. In case of multi-element ultrasonic matrices (Opielinski et al., 2009, 2010a, 2010b) used for projection imaging of internal structure of biological media (Ermert et al., 2000; Granz & Oppelt, 1987; Green et al., 1974; Opielinski & Gudra, 2005), introduction of delays of propagation of pulse ultrasonic wave for individual transducers can result in imaging errors associated with side lobes. However there are a number of mechanisms that suppress grating side lobes (a few wavelengths short excitation pulses, apodization weighting functions, a different transmit and receive geometries, random element spacing) and hence allow this criterion to be somewhat relaxed in practice (Drinkwater & Wilcox, 2006; Karaman et al., 2009; Kim & Song, 2006; Lockwood & Foster, 1996; Thomenius, 1996; Yen & Smith, 2004). This method of focusing makes it also necessary to develop synchronised delaying systems and sophisticated technologies of attaching a large number of electrodes to the surface of minute piezoceramic transducers by integration of some of the electronics with the transducer matrix enabling for miniaturization of the front-end and funnelling the electrical connections of a 2-D matrix consisted with hundreds of elements into reduced number of channels (Eames & Hossack, 2008; Opielinski et al., 2010a, 2010b; Wygant et al., 2006a; 2006b; Yen, Smith, 2004).

The experimental results have shown that using double ultrasonic pulse transmission of short coded sequences based on well-known Golay complementary codes allows considerably suppressing the noise level (Drinkwater & Wilcox, 2006). However this type of transmission is more time consuming.

Increase of directivity and intensity of the wave generated by the multi-element ultrasonic probe can be achieved by simultaneous in phase powering (no delays) of sequences of many elementary transducers (using one generator) in the sending system (which will however result in the probe's input impedance decrease) or by simultaneous receiving ultrasonic wave by means of sequences of many elementary transducers (Chiao & Thomas, 1996; Hoctor & Kassam, 1990; Karaman et al., 2009; Lockwood & Foster, 1996; Opielinski & Gudra, 2010c; Wygant et al., 2006a; Yen & Smith, 2004). Such sending probes are usually activated by low power generators of sinusoidal burst type pulses, the generated voltage values of which are low (a couple of tens of volts peak-to-peak) and output resistance of about 50 Ω. If the probe's impedance is close to output impedance of the generator, it results in a decrease of amplitude of activating pulse voltage which in turn causes decrease of intensity of the ultrasonic wave generated in the medium.

Recently in the medical imaging field a number of authors have suggested the use of matrices with large numbers of elements and investigated methods of selecting the optimal numbers and distribution of elements for the transmit and receive apertures (Drinkwater & Wilcox, 2006). The development of 2-D matrices for clinical ultrasound imaging could greatly improve the detection of small or low contrast structures. Using 2-D matrices, the ultrasound beam could be symmetrically focused and scanned throughout a volume (Lockwood & Foster, 1996; Yen & Smith, 2004).

This Section (7) presents an idea of minimising the number of connections between individual piezoelectric transducers in a row-column multi-element ultrasonic matrix system used for imaging of biological media structure by means of ultrasonic projection (Opielinski & Gudra, 2010c; Opielinski et al., 2010a, 2010b). It allows achieving significant directivity and increased wave intensity with acceptable matrix input impedance decrease without any complicated and expensive focusing or 2-D beamforming systems, what is a great advantage. This concept results in the necessity of creating several small models (e.g. square matrices of 16 transducers in 4 x 4 configuration) at the designing stage, in order to optimally select all essential parameters, together with developing a proper production technology (Opielinski & Gudra, 2010a). Such technology can be easily copied later with proper modifications and improvements, developing the matrix model by adding a greater number of elementary transducers. The following assumptions of constructing multi-element ultrasonic matrices were adopted (Opielinski & Gudra, 2010a):

a. transducer work frequency in a biological medium in the range of f_r = 1 MHz ÷ 2 MHz, in order to achieve a proper depth resolution (measurement accuracy) and relatively low attenuation of an ultrasonic wave,
b. dimensions of matrix piezoceramic transducers of $a \approx 1.5$ mm, $b \approx 1.5$ mm, in order to achieve a proper scanning resolution, identical all over matrix plane,
c. distance between transducers of $d \approx 1$ mm, in order to enable conducting electrodes and mounting matrices in laboratory conditions.

Desired dimensions of elementary piezoceramic transducers at possibly least material losses were achieved thanks to mechanical cutting with a diamond saw of 0.2 mm thickness. Cut plates were selected from a larger group due to repeatability of work frequency and electric conductance values.

7.1 Ultrasonic standard matrices

First of all, ultrasonic 512-element standard matrices (with separated electrodes connections) assigned for work in the sending and receiving character were constructed from SONOX P2 piezoceramic transducers, of the size of 1.5 x 1.5 mm (Opielinski & Gudra, 2010a). For a precise mounting of transducers, a mask made of engraving laminate of 0.8 mm thickness, with square-shaped holes for elementary transducers, cut using laser, at accuracy of 10 μm was used. At the back side (ground), transducers were stuck using conducting glue, to properly etched paths of a printed-circuit board, located at distances of 1 mm each other, in the layout of 16 rows and 32 columns. In order to connect a signal lead (electrode) to the active surface of each elementary transducer of the matrix, contact elastic connection was used. The lead, bent in the form of a hook, elastically adhering to the transducer surface, is conducted to the other side through the hole in the printed-circuit board, where it is stuck to its surface. The matching layer of matrices, which also serves the roles of isolation and stabilisation of contacts, was made by spraying a proper lacquer over their surfaces. A picture of the developed standard ultrasonic matrix is shown in Fig.15 (Opielinski & Gudra, 2010a). Detailed measurements of electro-mechanical parameters of elementary ultrasonic transducers of the developed standard matrix showed a presence of three resonance frequencies $f_{r1} \approx 1$ MHz, $f_{r2} \approx 1.8$ MHz and $f_{r3} \approx 2$ MHz. The reasons of occurrence of so many resonances is the assumed way of mounting transducers using conducting glue to proper patch of the board, without a back attenuation layer. The average efficiency of elementary transducers of the standard matrix, in the distance of 2.5 cm from the surface can be estimated on average at about 750 Pa/V for fr_1 = 1.1 MHz and at about 400 Pa/V for fr_2 = 2.1 MHz. Study results show that the developed standard matrices are suitable for projective imaging of biological media of parameters close to parameters of soft tissues, using the scanning method through switching proper pairs of sending and receiving elementary transducers (or proper synthetic apertures (Opielinski & Gudra, 2010c)).

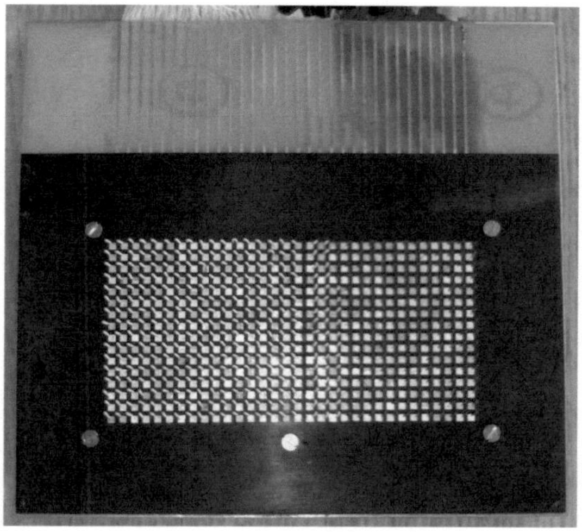

Fig. 15. Photo of the developed 512-element standard matrix for ultrasonic projection examinations

7.2 Ultrasonic passive matrices

An increase of directivity and intensity of the wave generated by the multi-element standard 2-D ultrasonic matrix (Section 7.1) can be obtained by simultaneous in phase supply (without delays) of a sequence of numerous elementary transducers, using one generator in the transmitting set-up (an aperture in transmitting and/or in receiving set-up - (Opielinski & Gudra, 2010c)), what causes a drop of the input impedance of the matrix. Such matrices are, usually, powered by low-power generators of burst-type sinusoidal pulses of low values of generated voltages (max. 20 V_{pp}) and output resistance of about 50 Ω. If the matrix sending transducer group impedance is close to the output impedance of the generator, a drop of the amplitude of an exciting pulse voltage occurs, what is followed by a drop of the intensity of the ultrasonic wave, which is generated to a medium. In the far field, in the plane perpendicular to axis Z, a distribution of the acoustic field generated by multi-element matrices with simultaneous supply to the group of elementary transducers has consecutive maxima and minima. Locations of maxima and minima and the acoustic pressure amplitude of the major and side lobes depend on the sizes of matrix's elementary transducers, distance between them and the length of the wave radiated into the medium. If the distance between adjacent transducers is small enough ($d < \lambda/2$), there are no side lobes in the area of 90°. This criterion is very difficult to be achieved as for the frequency of 2 MHz, the length of a wave in tissue is about 0.75 mm, what forces the use of elementary transducers of sizes below 0.375 mm. So, this Section (7.2) presents the concept of minimising the number of connections of particular piezoelectric transducers in the row-column arrangement of the multi-element ultrasonic passive matrix, assigned for imaging of biological media structures, by the means of the ultrasonic projection method (Opielinski et al., 2010a), what simultaneously enables achieving a large directivity and an increase of the wave intensity at acceptable drop of input impedance of the matrix.

So called passive matrix is the simplest solution for the matrix type controlling. This matrix includes two sets of keys (electronic switches), of which one K_y selects an active column and the second W_x – an active row. In this way, transducer P_{xy} is activated, which is located on the cross-cut of the selected row and column (Fig.16a). Piezoceramic transducers of the matrix are supplied with an alternating signal of high frequency (several MHz). Their structure (dielectric with sublimated electrodes) makes that, from the electronic circuit point of view, they have the capacitive character and an exciting signal can pass through inactive transducers to other rows and columns, despite disconnection of keys (Fig.16b). Directivity and efficiency/sensitivity of such matrix shall, therefore, depend to a high degree on a distribution of voltages of an exciting/receiving signal on all elementary transducers of the matrix, which in such way (with crosstalk) shall be excited to oscillations (Opielinski et al., 2010a).

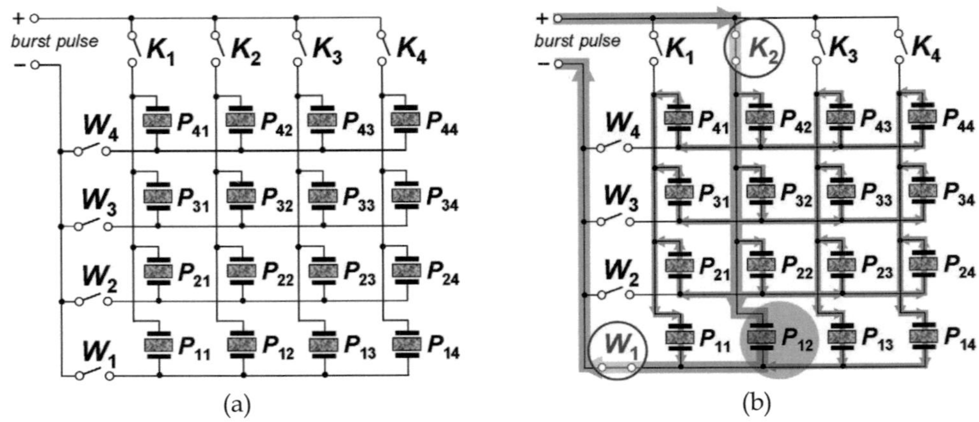

Fig. 16. Diagram of connections of transducers in a passive matrix (a) and a phenomenon of crosstalk formation (b)

Calculations and measurements of distributions of an acoustic field of the developed model of the 16-element passive matrix, in the 4 x 4 transducers layout, showed that after making a switch of one elementary transducer, all matrix transducers are excited by voltages, according to a specified pattern, thanks to which the matrix generated a directional beam. Exciting elementary transducers in a row-columns arrangement in a multi-element ultrasonic passive matrix enables a substantial minimisation of the number of connections of particular piezoelectric transducers. For example, for a 1024-element matrix, it is enough to use 64 paths, etched at a printed-circuit board, which conduct a signal exciting elementary transducers, instead of soldering 1024 separate electrodes in the form of thin, isolated leads to transducers surface and mounting multi-pin slots. The concept of a passive matrix enables also achieving a large directivity and an increase the wave intensity at an acceptable drop of the input impedance of the matrix. The developed concept of a passive matrix enabled designing of a full-sized matrix of 512 transducers, arranged in the structure of 16 rows and 32 columns. Due to a compact construction of the system, it was decided to put transducers and all electronic systems on a one printed circuit. Elements were arranged on the board with a division into two areas. The lower part is the proper matrix of transducers and the upper one includes switching circuits. Such strict division enables immersing the lower part in water, without the risk of damaging electronic circuits. Printed circuits include connections only for columns of the passive matrix. Connections of rows were realised using an additional upper board (microwave teflon-ceramic laminate) with a printed circuit, which is simultaneously a mask for positioning transducers. It is a single-sided circuit, composed of sixteen especially designed horizontal paths. Each path has square areas deprived of copper. These areas mark locations of ultrasonic transducers and were cut out by the means of a special punch. Such square openings enable a precise placing of transducers. Electrical connection of ultrasonic transducers with the back base board were realised by hot-air soldering. The upper board (mask) thickness of about 0.8 mm was selected in order to locate transducer surface evenly with the board surface. A small amount of conducting glue enabled connection of transducer metallization with the path around a hole. There were developed two 512-element passive matrices with Pz37 Ferroperm ceramic transducers of the size 1.6 x 1.6 mm, arranged at the distances of 0.9 mm. One matrix is assigned for sending work and the second for receiving. Figure 17 presents the view of the developed 512-element ultrasonic passive matrix. Large round holes at the sides of matrix surface are designed for mounting semi-conducting lasers in the sending matrix for positioning the sending matrix with the receiving one.

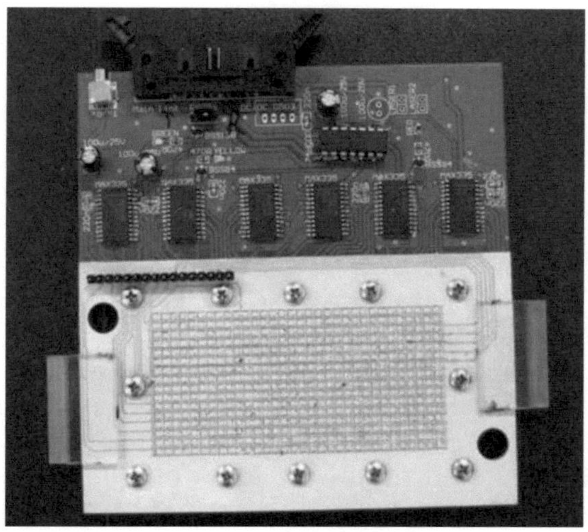

Fig. 17. The view of the developed 512-element ultrasonic passive matrix

One characteristic feature of the passive matrix is occurrence of crosstalk (Opielinski et al., 2010a), which causes a presence of some signal voltage, even on inactive transducers. This voltage is most often substantially lower than the voltage on a switched transducer but has an influence of the shape of obtained resultant characteristics of the matrix directivity. On the basis of conducted simulations and calculations, the following formula that describes the distribution of an excitement voltage on the passive 512-element matrix in the arrangement of N x M transducers was derived:

$$x = \frac{1}{(M-1)+(N-1)+1} \tag{7}$$

The switched on transducer of the passive matrix is supplied by voltage $+U$, all transducers in this (switched on) row are supplied by voltage $+U{\cdot}x{\cdot}(N{-}1)$, all transducers in this (switched on) column are supplied by $+U{\cdot}x{\cdot}(M{-}1)$, and all the other transducers (in switched off rows and columns) are supplied by voltage $-U{\cdot}x$. The simulated in this way distribution of voltages for the passive matrix of 16 x 32 sizes is presented in Fig.18. Elementary transducers of the passive matrix, in the frequency range of 1.4 ÷ 2.4 MHz, exhibit a possibility of working at three resonance frequencies $f_{r1} \approx 1.6$ MHz, $f_{r2} \approx 1.8$ MHz, $f_{r3} \approx 2$ MHz. The imaginary part of the amplitude-phase electrical admittance characteristics of transducers of the passive matrix reveals their capacitive character ($C_o \approx 160$ pF, including the capacity of connection leads of about 100 pF, $R_o \approx 3$ kΩ $|Z(f_r)| \approx 2$ kΩ).

-0.021	0.66	-0.021
0.32	1	0.32
-0.021	0.66	-0.021

Fig. 18. Simulated distribution of values of voltages exciting elementary transducers of the passive matrix of 16 x 32 transducers arrangement („–" marks a inversed phase)

Single transducers of the earlier developed multi-element matrices with separated electrode connections (standard matrices) exhibited electrical capacity $C_o \approx 10$ pF, resistance of electric losses $R_o \approx 30$ kΩ and electric impedance in resonance $|Z(f_r)| \approx 7$ kΩ (Opielinski & Gudra, 2010a) (for a comparison, at simultaneously excitement of 16 transducers of the standard matrix with the same voltage U, impedance of the circuit would be $|Z(f_r)| \approx 440$ Ω). These values suggest that in case of the passive matrix, the phenomenon of increasing electric capacity of transducers and reducing their losses resistance occurs due to a parallel connection of all matrix elements, what is confirmed by the presence of crosstalk in such arrangement. Measurements and calculations show that the developed passive matrix enable achieving much larger directivity and almost three time higher amplitude of an ultrasonic wave generated in a medium than in case of a single supply, electrically separated transducer of the standard matrix (Opielinski & Gudra, 2010a). The divergence angle of the beam generated by the developed 16-element model of a passive matrix in the 4 x 4 arrangement, with an activation of 1 element is about 6 ÷ 8°.

7.3 Ultrasonic active matrices

A large number of elementary transducer of a projection matrix forces a high number of electrical connections and substantially makes miniaturisation of the whole unit difficult. A solution to such inconvenience is using a row-column selection of active transducers (so called passive matrix), presented in Section 7.2. For the matrix of 512 elements, arranged in 16 rows and 32 columns, it is enough to connect 48 leads (16+32) this way, instead of 512. A solution that can enable elimination of crosstalk between transducers in the passive matrix (Fig.16) with the maintenance of the electrode minimisation is the use of active elements (keys) in matrix nodes (active matrix) (Opielinski et al., 2010b). In this solution, each transducer is switched by its own individual key T_{xy} (Fig.19).

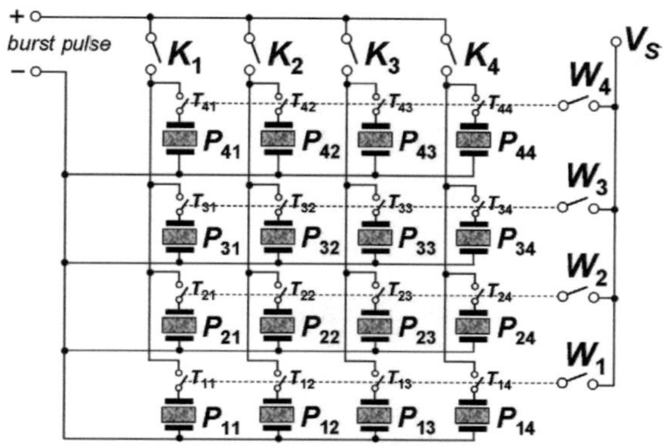

Fig. 19. Diagram of transducer connections in an active matrix

Key K_y in the signal path conducts voltage to a particular column, while a closing of keys W_x determines an activated row. Field-effect transistors play the role of individual keys T_{xy}. Their advantage is a small housing size, what enables a miniaturisation of connections.

On the basis of earlier conducted experiment, similarly as in case of the passive matrix, an active matrix, including 512 ultrasonic transducers was developed. Its construction assumes the use of two keys switching signal on for each of elementary transducer. Such solution complicates the matrix construction but enables elimination of crosstalk, present in the passive matrix. Matrix control is realised in a similar way as in the passive matrix (section 7.2), however, circuits of switching matrix columns were more developed. Only an active column is connected to the signal source and all other are shorted to ground. It eliminates a possibility of signal slips into inactive columns. Thanks to utilisation of individual transistors at each piezoceramic transducer, there is no need to key signal in matrix rows (Fig.20). This way, using additional keys K_{zy} (Fig.20), the active matrix is working as a standard matrix (see Section 7.1). Without using additional keys K_{zy} (Fig.20), after switching electrodes to a proper row (key W_x) and column (key K_y) of the matrix and after switching a transistor key for a certain transducer (key T_{xy}) (see Fig.19), it shall be excited by fed voltage U, and other transducers in this column shall be excited with a voltage below $0.3{\cdot}U$ (Opielinski et al., 2010b).

Fig. 20. The method of crosstalk elimination in a column of the active matrix using additional switches K_{zy}: (a) short-circuiting an inactive transducer, (b) open-circuiting when activating a transducer

Application of such sending and receiving matrix in a projection arrangement is possible in such way that one of them is switched by the angle of 90° in relation to the other one and sequences of switching sending and receiving transducers are so coupled that enables a proper control of scanning of a received beam (Opielinski & Gudra, 2010b).

Active matrices act similarly as earlier developed standard matrices (Section 7.1) but their construction is more simple and improved.

8. Models of ultrasonic transmission camera

Using especially developed multi-element ultrasonic projection matrices with quick electronic switching of elementary transducers (Section 7), it is possible to obtain images in pseudo-real time (e.g. with a slight constant delay, resulting from the need of data buffering and processing), thus, a device using this method can be called an ultrasonic transmission camera (UTC) (Ermert et al., 2000). There are several know works regarding such cameras, conducted at some centres in the world, as described in Introduction (Brettel et al., 1981; Ermert et al., 2000; Granz & Oppelt, 1987; Green et al., 1974; Lehmann et al., 1999).

In the device presented in the work (Green et al., 1974), an incoherent ultrasonic wave was used for sonification of an object and a one-dimensional linear array, mounted in a set position served the role of a receiver. Scanning of an object structure in the second dimension was realised by the means of a complex system of properly rotated prisms. A disadvantage of such system is the necessity to use mechanical elements.

The device presented in (Brettel et al., 1981) uses a similar concept as in work (Green et al., 1974). The difference consists in the fact that a projection image is projected on the water surface of a camera and next, it is detected by a linear array of transducers. What is more, scanning in the second dimension was realised using mechanical movements of an array, instead of rotating prisms. A disadvantage of such system is the necessity to use mechanical elements.

The device presented in work (Granz & Oppelt, 1987) functions in pseudo-real time and its construction is similar to the camera described in (Brettel et al., 1981). An incoherent ultrasonic wave passes through an object in a wide beam in order to reduce the effect of spots. An image is obtained using a spherical mirror, located behind an object and in front of a 2-D, electronically controlled receiving matrix of PVDF foil transducers, in the arrangement of 128 x 128 elements. The camera enables obtaining images in pseudo-real time (with data buffering) at the rate of 25 frames per second. A disadvantage of this system is a high production cost and difficulties related with a faultless construction of such matrix.

The device presented in the work (Ermert et al., 2000) uses a linear sending probe, composed of 128 piezoceramic transducers, of rectangular dimensions (width much larger than height), a linear receiving probe, composed of 128 PVDF transducers (height much larger than width) and a lens, which focuses an ultrasonic wave beam in one direction. Probes work in the frequency range of 2 ÷ 4 MHz. Sweeping of the structure of an object located in water, between probes, is realised by the means of phase focusing of an ultrasonic wave beam in the object plane, in the form of horizontal lines, 1.25 mm wide, which, after a transition through the object and the lens, are received by proper linear transducers of the receiving probe. One disadvantage in this case is beam focusing in a particular area of the object, what causes image spreading and occurrence of artefacts outside the focus.

The device presented in work (Lehmann et al., 1999) is a model of an acoustic holograph used the through-transmission signal. This approach uses coherent sound and coherent light to produce real-time, large field-of-view images with pronounced edge definition in soft tissues of the body.

In most of known studies, the projection parameter is the signal amplitude, not its transition time, which seems to more attractive due to the simplicity and precision of measurements.

This Section (8) includes a description of the results of author's studies, which aim is a construction of a special measuring setup for visualisation of biological structures by the means of ultrasonic projection, which enables simultaneous measurements of several acoustic parameters in pseudo-real time (multi-parameter ultrasonic transmission camera), using a pair of electronically controlled multi-element matrices of piezoceramic transducers, described in Section 7.

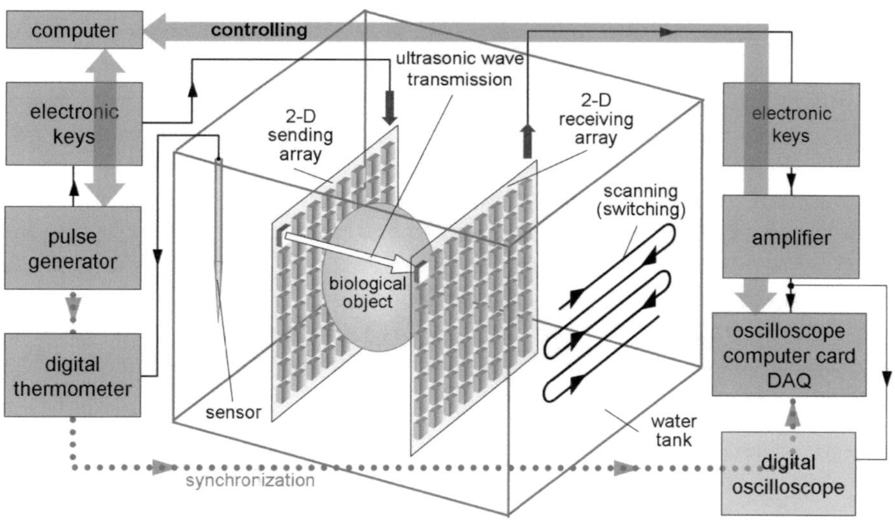

Fig. 21. Block scheme of the laboratory measurement setup for visualization of the distribution of acoustic parameters of biological media internal structure by means of ultrasonic projection method

Due to diversified ways of conducting electrodes to transducers of matrices and different controlling methods, two laboratory setups were developed, for imaging internal structures of biological media, by the means of ultrasound projection: using a pair of ultrasonic 512—element standard matrix - sending and receiving one - and using a pair of ultrasonic 512-element passive or active matrices (Opielinski & Gudra, 2010a; Opielinski et al., 2010a, 2010b). A general block scheme of the laboratory measurement setup is presented in Fig.21. Elementary piezoceramic transducers of the sending matrix (Opielinski & Gudra, 2010a) are excite by a burst type pulses generator, controlled by a computer through the GPIB connection. A computer is also used for switching transducers of sending and receiving matrices, using developed and software-assisted systems of electronic keys. Signals received by elementary piezoceramic transducers of the receiving matrix (Opielinski & Gudra, 2010a) after an amplification and acquisition by the means of a computer oscilloscope data acquisition (DAQ) card, are recorded on the hard drive. Additionally, a digital oscilloscope is used for visualisation and control of measurement signals. Projection values of acoustic parameters assigned for visualisation (e.g. transition time of an ultrasonic wave or a projective value of sound velocity) are determined from recorded signals, by the means of a specially developed software. Water temperature is controlled by a digital thermometer of the resolution at 0.1 °C. The acceleration of measurements, e.g. by decreasing sampling frequency of receiving signal, enables to obtain images in pseudo-real time (several to ten-odd frames per second). It can be used simple square burst pulse generator as integrated electronic circuit instead sinusoidal one, what enables to achieve the amplitude of ultrasonic matrix transducer exciting signal more than 60 V_{pp}, as well.

One of objects that were subjected to projection examinations on the developed measurement setups, was phantom *AC_Blue* in the form of a cylinder made of agar gel, of the diameter of 55 mm and the length of 50 mm, made of 3 % (by weight) agar solution, with an additive of propylene glycol (by volume), in order to achieve a greater diversification of sound velocity values in the structure in relation to the values in water (Fig.22a). Two cylinders, made of about 4 % agar solution with an additive of propylene glycol, were decentrally put into the cylinder. Diameters of internal cylinders were 15 mm and 8 mm. Additionally, in the *AC_Blue* cylinder, a through-and-through round hole, of 6 mm diameter, was made. This hole was filled with water during measurements. Sound velocities in the structure of *AC_Blue* cylinder are different from the speed in water, by the following values: +10 m/s - surroundings, +13 m/s larger internal cylinder, +12 m/s smaller internal cylinder. Figure 22b presents, in grey scale, an image of the distribution of projection values of sound velocity in a lateral projection of agar phantom *AC_Blue*, obtained from measurements, made using a pair of 512-element standard matrices (Section 7.1) in the resolution of 32 x 16 pixels.

Figure 23a presents an image (in grey scale) of the distribution of the projection values of sound velocity in longitudinal projection of a hard-boiled chicken egg, without shell, immersed in water, obtained from measurements, made using a pair of 512-element passive matrices, in the resolution of 32 x 16 pixels. A similar image of the distribution of frequency of ultrasonic wave pulses in a longitudinal projection of the same egg is presented in Figure 23b.

Heterogeneities, differing in sound velocities values even by about several m/s are clearly visible in projection images (Fig.22). On the basis of these images, it is possible to estimate sound speeds in the structure of examined objects. The projection image of a hard-boiled egg (Fig.23) distinctly visualises borders of the yolk area. Images obtained from projection measurements of a frequency down shift of receiving pulses (Fig.23b) are of a worse quality than images visualising projections of sound velocity (Fig.23a), however, it is not difficult to clearly recognise heterogeneity boundaries in both images. The higher is attenuation in a medium on the way of an ultrasonic wave beam, the larger is drop of frequency of a receiving pulse (formula (4); see Fig.23b).

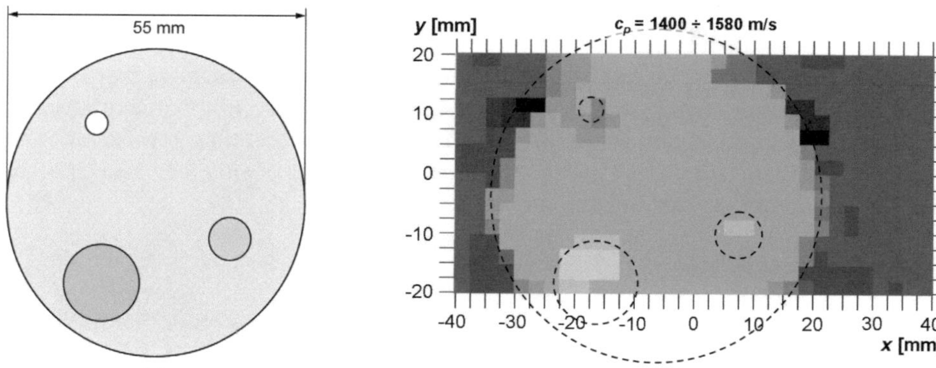

(a) (b)

Fig. 22. Lateral projection of the agar phantom *AC_Blue*: structure (a), image of the distribution of projection values of sound velocity in grey scale (b)

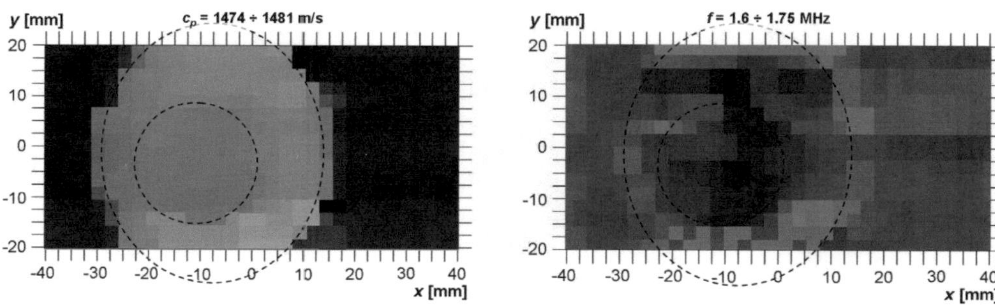

(a) (b)

Fig. 23. An image of the distribution of the projection values of sound velocity in a longitudinal projection of a hard-boiled chicken egg (a) and an image of the distribution of the projection values of frequency of receiving pulses in a longitudinal projection of the same egg (b)

Good image quality without artefacts caused by the beam focusing is a great advantage in that solution in comparison with UTC described in literature (Brettel et al., 1981; Ermert et al., 2000; Granz & Oppelt, 1987; Green et al., 1974). The disadvantage is not so good scanning resolution, at present, what have to be improved.

9. Summary

The results obtained confirm that the designed multi-element ultrasonic 2-D matrices are suitable for projection imaging of biological media (especially soft tissue) with the use of scanning method through switching of the right pairs (or groups) of sending and receiving elementary transducers. Using appropriate combinations of apertures of the sending and receiving matrix (with accordance to the scanning method) allows to increasing the directivity and the acoustic pressure level of the ultrasonic wave beam and ensures its apodization in the transmission system. Additionally, rotation of a pair of probes around the studied biological object submerged in water will allow tomographic, three dimensional reconstruction of its internal structure.

In order to improve the quality of ultrasound projection images, there are currently conducted research works on developing constructions of ultrasonic matrices and the ultrasonic transmission camera set-up. In order to increase scanning resolution, it is planned to develop ultrasonic matrices with elementary transducers of smaller sizes, circular-shaped, in a higher amount and arranged more densely.

10. References

Brettel, H.; Roeder, U. & Scherg, C. (1981). Ultrasonic Transmission Camera for Medical Diagnosis, *Biomedizinische Technik*, Vol. 26, No. s1, pp. 135-136, ISSN 0013-5585

Brettel, H.; Denk, R.; Burgetsmaier, M. & Waidelich, W. (1987). Transmissionssonographie, In: *Ultraschalldiagnostik des Bewegungsapparates*, T.Struhler & A.Feige (Ed.), pp. 246-251, Springer Verlag, ISBN 3540166920, Berlin, Heidelberg, New York, London, Paris, Tokyo

Chiao, R.Y. & Thomas, L.J. (1996). Aperture formation on reduced-channel arrays using the transmit-receive apodization matrix, *1996 IEEE Ultrasonics Symposium Proceedings*, ISBN 0-7803-3615-1, San Antonio, USA, November 1996

Drinkwater, B.W. & Wilcox, P.D. (2006). Ultrasonic arrays for non-destructive evaluation: A review, *NDT&E International*, Vol. 39, No. 7, pp. 525-541, ISSN 0963-8695

Eames, M.D.C. & Hossack, J.A. (2008). Fabrication and evaluation of fully-sampled, two-dimensional transducer array for "Sonic Window" imaging system, *Ultrasonics*, Vol. 48, pp. 376-383, ISSN 0041-624X

Ermert, H.; Keitmann, O.; Oppelt, R.; Granz, B.; Pesavento, A.; Vester, M.; Tillig, B. & Sander, V. (2000). A New Concept For a Real-Time Ultrasound Transmission Camera, *IEEE Ultrasonics Symposium Proceedings*, ISBN 0-7803-6365-5, San Juan, Puerto Rico, October 2000

Granz, B. & Oppelt, R. (1987). A Two Dimensional PVDF Transducer Matrix as a Receiver in an Ultrasonic Transmission Camera, *Acoustical Imaging*, Vol. 15, pp. 213-225, ISSN 0270-5117

Green, P.S.; Schaefer, L.F.; Jones, E.D. & Suarez, J.R. (1974). A New High Performance Ultrasonic Camera, *Acoustical Holography*, Vol. 5, pp. 493-503, ISSN 0065-0870

Green, P.S.; Schaefer, L.F.; Frohbach, H.F. & Suarez, J.R. (1976). *Ultrasonic Camera System and Method*, US Patent 3937066.

Greenleaf, J.F. & Sehgal, Ch.M. (1992). *Biologic System Evaluation with Ultrasound*, Springer-Verlag, ISBN 0387978518, 3540978518, New York, USA

Hoctor, R.T. & Kassam, S.A. (1990). The Unifying Role of the Coarray in Aperture Synthesis for Coherent and Incoherent Imaging, *Proceedings of the IEEE*, Vol. 78, No. 4, pp. 735-752, ISSN 00189219

Johnson, J.A.; Karaman, M. & Khuri-Yakub B.T. (2005). Coherent-Array Imaging Using Phased Subarrays. Part I: Basic Principles, *IEEE Transactions on Ultrasonics, Ferroelectrics, and Frequency Control*, Vol. 52, No. 1, pp. 37-50, ISSN 0885-3010

Kak, A.C & Slaney, M. (1988). *Principles of Computerized Tomographic Imaging*, IEEE Press, ISBN/ASIN 0879421983, New York, USA

Karaman, M.; Wygant, I.O.; Oralkan, O. & Khuri-Yakub, B.T. (2009). Minimally Redundant 2-D Array Design for 3-D Medical Ultrasound Imaging, *IEEE Transactions on Medical Imaging*, Vol. 28, No. 7, pp. 1051-1061, ISSN 0278-0062

Keitmann, O.; Benner, L.; Tillig, B.; Sander, V. & Ermert, H. (2002). New Development of an Ulrasound Transmission Camera, *Acoustical Imaging*, Vol. 26, pp. 397-404, ISBN 0306473402 / 0-306-47340-2

Kim, J-J. & Song, T-K. (2006). Real-Time High-Resolution 3D Imaging Method Using 2D Phased Arrays Based on Sparse Synthetic Focusing Technique, *2006 IEEE Ultrasonics Symposium Proceedings*, ISBN 9781424402014, Vancouver, Canada, October 2006

Lehmann, C.D.; Andre, M.P.; Fecht, B.A.; Johannsen, J.M.; Shelby, R.L. & Shelby, J.O. (1999). Evaluation of Real-Time Acoustical Holography for Breast Imaging and Biopsy Guidance, *SPIE Proceedings*, Vol. 3659, pp. 236-243, ISBN 0-8194-3131-1, San Diego, USA, May 1999

Lockwood, G.R. & Foster, F.S. (1996). Optimizing the Radiation Pattern of Sparse Periodic Two-Dimensional Arrays, *IEEE Transactions on Ultrasonics, Ferroelectrics, and Frequency Control*, Vol. 43, No. 1, pp. 15-19, ISSN 0885-3010

Narayana, P.A. & Ophir, J. (1983). A Closed Form Method for the Measurement of Attenuation in Nonlinearly Dispersive Media, *Ultrasonic Imaging*, Vol. 5, No. 1, pp. 17-21, ISSN 0161-7346

Nowicki, A. (1995). *Basis of Doppler Ultrasonography*, PWN, ISBN 83-01-11793-1, Warsaw, Poland (in Polish)

Nowicki, A.; Wójcik, J. & Kujawska, T. (2009). Nonlinearly Coded Signals for Harmonic Imaging, *Archives of Acoustics*, Vol. 34, No. 1, pp. 63-74, ISSN 0137-5075

Opielinski, K. & Gudra, T. (2000). Ultrasound Transmission Tomography Image Distortions Caused by the Refraction Effect, *Ultrasonics*, Vol. 38, No. 1-8, pp. 424-429, ISSN 0041-624

Opielinski K.J. & Gudra T. (2004). Three-Dimensional Reconstruction of Biological Objects' Internal Structure Heterogeneity from the Set of Ultrasonic Tomograms, *Ultrasonics*, Vol. 42, No. 1-9, pp. 705-711, ISSN 0041-624

Opielinski, K.J. & Gudra, T. (2004). Ultrasonic Transmisson Camera: *Proceedings of LI Open Seminar on Acoustics*, ISBN 83-87280-35-6, Gdansk-Sobieszewo, Poland, September 2004 (in Polish)

Opielinski, K.J. & Gudra, T. (2004). Biological Structure Imaging by Means of Ultrasonic Projection, In: *Acoustical Engineering: Structures – Waves – Human Health*, Vol. 13, No. 2, R.Panuszka (Ed.), pp. 97-106, Polish Acoustical Society, Krakow, Poland

Opielinski, K.J. & Gudra, T. (2005). Computer Recognition of Biological Objects' Internal Structure Using Ultrasonic Projection, *Computer Recognition Systems*, pp. 645-652, ISSN 1615-3871

Opielinski, K.J. & Gudra, T. (2006). Multi-Parameter Ultrasound Transmission Tomography of Biological Media, *Ultrasonics*, Vol. 44, No. 1-4, pp. e295-e302, ISSN 0041-624X

Opielinski, K.J. (2007). Ultrasonic Parameters of Hen's Egg, *Molecular and Quantum Acoustics*, Vol. 28, pp. 203-216, ISSN 1731-8505

Opielinski, K. & Gudra, T. (2008). Nondestructive Tests of Cylindrical Steel Samples using the Ultrasonic Projection Method and the Ultrasound Transmission Tomography

Method, *Acoustics '08*, ISBN 9782952110549 2952110549, Paris, France, June/July 2008

Opielinski, K.J.; Gudra, T. & Pruchnicki, P. (2009). *The Method of a Medium Internal Structure Imaging and the Device for a Medium Internal Structure Imaging*, Patent Application No. P389014, Wroclaw University of Technology, Poland (in Polish).

Opielinski, K.J. & Gudra, T. (2010). Multielement Ultrasonic Probes for Projection Imaging of Biological Media, *Physics Procedia*, Vol. 3, No. 1, pp. 635-642, ISSN 18753892

Opielinski, K.J. & Gudra, T. (2010). Ultrasonic Transmission Tomography, In: *Industrial and Biological Tomography – Theoretical Basis and Applications*, J.Sikora & S.Wojtowicz (Ed.), Electrotechnical Institute, ISBN 978-83-61956-04-4, Warsaw, Poland

Opieliński, K.J. & Gudra, T. (2010). Aperture Synthesis on 2-D Ultrasonic Transducer Arrays for Projection Imaging of Biological Media, *Proceedings of 20th International Congress on Acoustics ICA 2010*, Sydney, Australia, August 2010

Opielinski, K.J.; Gudra, T. & Pruchnicki, P. (2010). Narrow Beam Ultrasonic Transducer Matrix Model for Projection Imaging of Biological Media, *Archives of Acoustics*, Vol. 35, No. 1, pp. 91-109, ISSN 0137-5075

Opielinski, K.J.; Gudra T. & Pruchnicki, P. (2010). A Digitally Controlled Model of an Active Ultrasonic Transducer Matrix for Projection Imaging of Biological Media, *Archives of Acoustics*, Vol. 35, No. 1, pp. 75-90, ISSN 0137-5075

Ramm von, O.T. & Smith, S.W. (1983). Beam Steering with Linear Arrays, *IEEE Transaction Biomedical Engineering*, Vol. BME-30, No. 8, pp. 438-452, ISSN 0018-9294

Thomenius, K.E. (1996). Evolution of Ultrasound Beamformers, *1996 IEEE Ultrasonics Symposium Proceedings*, Vol. 2, ISBN 0-7803-3615-1, San Antonio, USA, November 1996

Wildes, D.G.; Chiao, R.Y.; Daft, Ch.M.W.; Rigby K.W.; Smith L.S. & Thomenius, K.E. (1997). Elevation Performance of 1.25D and 1.5D Transducer Arrays, *IEEE Transactions on Ultrasonics, Ferroelectrics, and Frequency Control*, Vol. 44, No. 5, pp. 1027-1037, ISSN 0885-3010

Wygant, I.O.; Karaman, M.; Oralkan, O. & Khuri-Yakub, B.T. (2006). Beamforming and Hardware Design for a Multichannel Front-End Integrated Circuit for Real-Time 3D Catheter-Based Ultrasonic Imaging, *SPIE Medical Imaging*, Vol. 6147, pp. 61470A-1-8, ISBN 9780819471048

Wygant, I.O.; Lee, H.; Nikoozadeh, A; Yeh, D.T.; Oralkan, O.; Karaman, M. & Khuri-Yakub, B.T. (2006). An Integrated Circuit with Transmit Beamforming and Parallel Receive Channels for Real-Time Three-Dimensional Ultrasound Imaging, *2006 IEEE Ultrasonics Symposium Proceedings*, ISBN 9781424402014, Vancouver, Canada, October 2006

Yen, J.T. & Smith, S.W. (2004). Real-Time Rectilinear 3-D Ultrasound Using Receive Mode Multi-plexing, *IEEE Transactions on Ultrasonics, Ferroelectrics, and Frequency Control*, Vol. 51, No. 2, pp. 216-226, ISSN 0885-3010

Zhang, D.; Gong, X.; Rui, B.; Xue, Q. & Li, X. (1997). Further Study on the Nonlinearity Parameter Tomography for Pathological Porcine Tissues, *Proceedings of Ultrasonics World Congress WCU'97*, ISBN 4-9900616-0-8, Yokohama, Japan, August 1997

3

3-D Modelings of an Ultrasonic Phased Array Transducer and Its Radiation Properties in Solid

Kazuyuki Nakahata[1] and Naoyuki Kono[2]
[1]*Ehime University*
[2]*Hitachi, Ltd*
Japan

1. Introduction

Ultrasonic phased array technology has become popular in the field of industrial nondestructive testing (NDT). The technology is notable for its capability to provide images of the inside of a target in real time. Since the image quality results from rapid steering and focusing of the radiation beam from an array transducer, it is essential to have well understood characteristics of the radiated beam. Even though fundamental concepts of the array transducer in medical fields were introduced over 30 years ago (Macovski, 1979), the beam modeling and optimization of the array transducer for the NDT have only been investigated intensively in the past decade (Azar et al., 2000; Song & Kim, 2002). The characteristics of the radiated beam from the ultrasonic phased array transducer vary according to the transducer parameters such as frequency, aperture size, number of elements, pitch(inter-element spacing), element width, layout dimensions and so forth. If these parameters are not chosen properly, spurious grating lobes or side lobes with high amplitude will exist in the radiation beam field. Consequently, the image quality will be deteriorated significantly.

In this chapter, we first show principles of electronic scanning with phased array transducers for the NDT and effective settings of the delay time for the beam steering and focusing. And then mathematical models of linear and matrix array transducers and simulation tools to predict ultrasonic beams from the transducers are outlined. We explain three-dimensional (3-D) numerical simulation tools in both frequency and time domains. This chapter is arranged as follows:

- In Section **2**, we show various ultrasonic array transducers for the NDT and electronic control of the array transducer. First the appropriate delay time setting for beam steering and focusing is described, and then causes and prevention of the grating lobe are explained.
- In Section **3**, modelings of the radiated beam field in the frequency domain are introduced. Basically, the Rayleigh Sommerfeld model is used for the expression of the radiated wave field from an array transducer. Here the multi-Gaussian beam (MGB) model (Schmerr, 2000) accelerates the 3-D simulation of the radiated wave field. The MGB model has an ability to calculate the radiated beam by superposing a small number of Gaussian beams. Based on the simulation results, some characteristics about beam steering and focusing of the linear and matrix array transducers are described.

- In Section **4** the finite integration technique (FIT) (Fellinger et al., 1995) is introduced as a time domain calculation tool. The FIT represents very stable and straightforward schemes to investigate the wave propagation in complex fields such as inhomogeneous and anisotropic media. For example, a dissimilar welding part includes metal grains of different size and orientation, therein the FIT is of assistance to predict the ultrasonic propagation direction as well as scattering and attenuation in the welding part. Here a transient wave simulation of the array ultrasonic testing (UT) for a welded T-joint is demonstrated.
- In Section **5**, we summarize our research and discuss prospects for array UT.

2. Array transducer for NDT

Ultrasonic phased array techniques have been applied to the NDT. In the phased array techniques, angled beams with various focal lengths are generated by the array transducers composed of small ultrasonic transducer elements. The type of beams is electrically controlled through delay time control for excitation of transmitted waves and signal processing of received waves.

The array transducers have two advantages for NDT in comparison with the conventional single or dual element transducers. Firstly, the various angled beams can cover wider range of an inspection target. Secondly, NDT results can be visualized immediately as a cross section images (B-scan) or 3-D images.

2.1 Principle of electric scanning

General configurations of the array transducers are classified into one-dimensional (1-D) arrays and two-dimensional (2-D) arrays shown in Fig.1. Linear arrays (Fig.1(a)) and annular arrays (Fig.1(c)) are typical configurations of the 1-D array transducers. The linear arrays provide beam-steering and focusing in a plane. The Annular arrays adjust focal length along an axis. In recent years, the 2-D arrays have been applied to the NDT field for maximization of the advantages of the ultrasonic phased array technology (Drinkwater & Wilcox, 2006). Matrix arrays (Fig.1(b)) and segmented annular arrays (Fig.1(d)) can steer beams in 3-D volume. Ultrasonic beams are electrically scanned by delay-time control for each transducer element composed of array transducers as shown in Fig.2. Patterns of the delay time for transmitting and receiving (hereinafter referred to as "delay law") are stored on the delay-time controller which shifts excitation pulses in transmitting and signals of reflection waves in receiving.

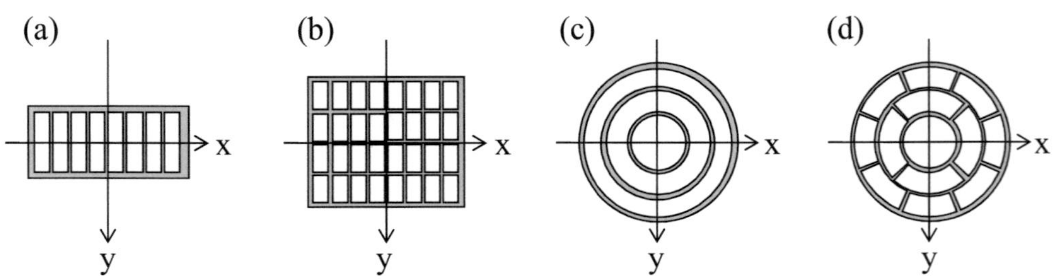

Fig. 1. General element configurations of array transducers: (a) 1-D linear array, (b) 2-D matrix array, (c)1-D annular array, and (d) 2-D segmented annular array.

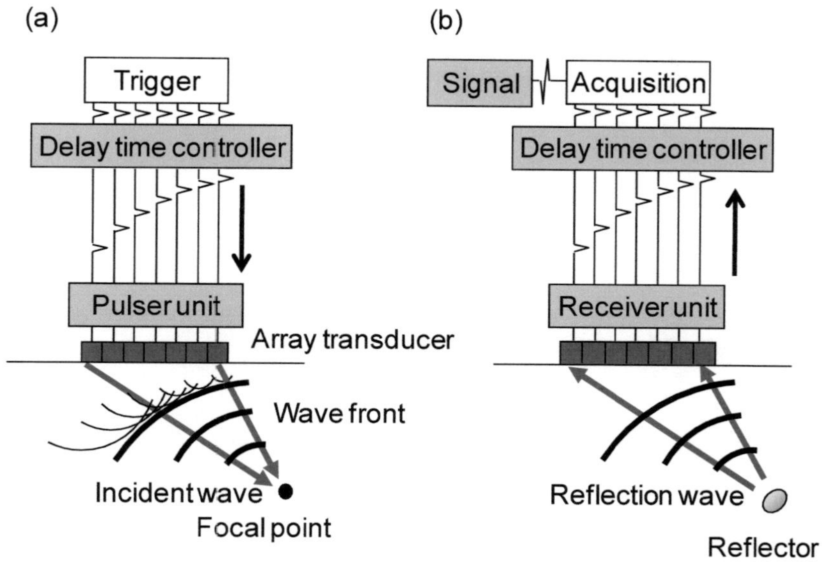

Fig. 2. Schematic of the electrical scanning. (a) Transmitting: ultrasonic waves are transmitted at each transducer element with delayed excitation; each transmitting waves are phased near the focal point. (b) Receiving: received waves at each transducer element are synthesized with time-delay signal processing to a single receiving wave.

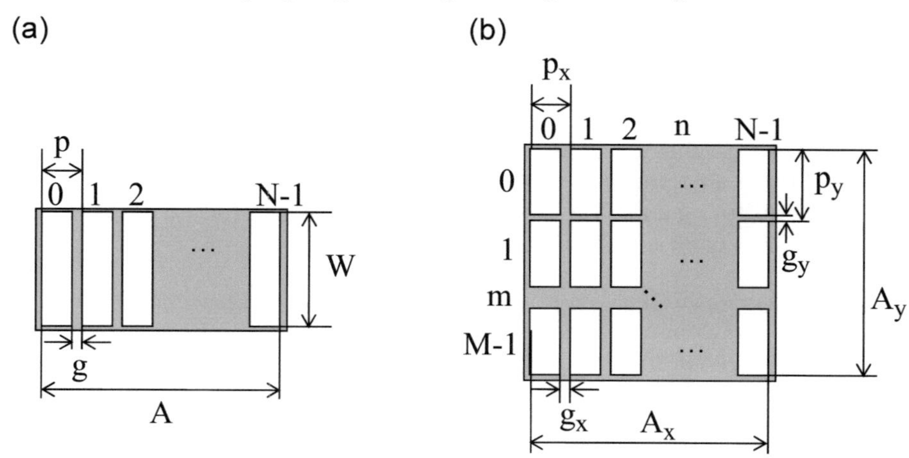

Fig. 3. Main parameters of array transducers: element pitches p, element numbers N, transducer apertures A, element gaps g and element widths W; (a) linear array and (b) matrix array

The main parameters of the linear and matrix arrays are the aperture, the element pitch, and the element number, shown in Fig.3. The apertures A are expressed as a function of the element pitch p, the element number N and the element gap g in Equation (1). For the matrix array transducers, subscripts of parameters correspond to the x and y axes.

$$A = Np - g. \quad (1)$$

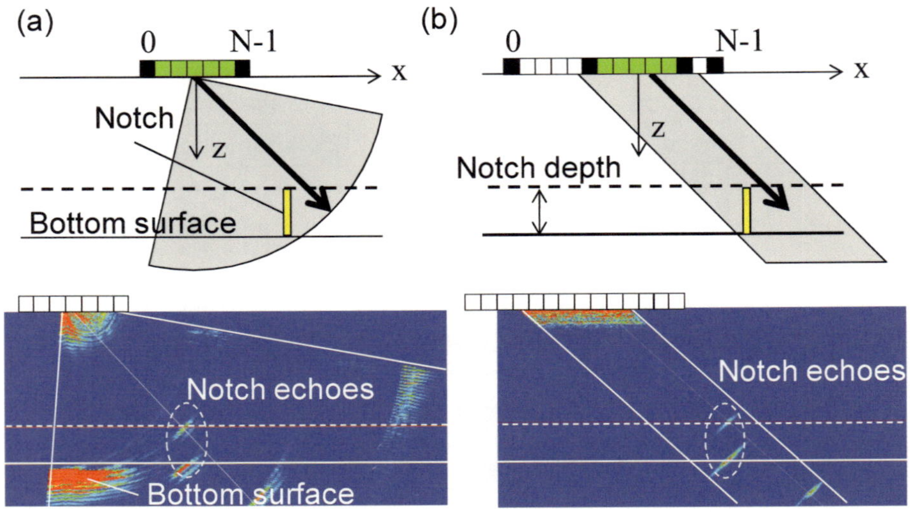

Fig. 4. Schematics of the electric scanning and B-scan results for electrically discharged notches: (a) sectorial scanning and (b) linear scanning.

Electrical scanning patterns widely-used in the NDT field are sectorial and linear scanning as schematized in Fig.4. B-scan results for electrically discharged notch of the sector and the linear scanning are displayed also in Fig.4. The through-wall depth of the notch is measured as the difference of the z-axis between echoes from the root and the tip of the notch. NDT data in the sectorial scanning is displayed as a sector-form B-scan image (cross section) by changing propagation directions. In the sectorial scanning, imaging area can be wider than an aperture of an array transducer. Element number is generally from 15 to 63. The linear scanning provides a parallelogram B-scan image by switching positions of active elements. In the linear scanning, imaging area can be wider by a large number of transducer elements, e.g. 63 to 255. Substituting a motion axis of mechanical scanning for the linear scanning can shorten scanning time of an automated UT.

2.2 Delay setting for electrical scanning

The coordinate systems for the linear array transducer and the matrix transducer are defined as Fig.5. The delay laws imposed on each transducer element positioned at $\boldsymbol{x} = (x_n, y_m)$ for an angled beam with focal point $\boldsymbol{F}$ in a material characterized by velocity c is expressed by the difference of the propagation time:

$$\Delta\tau_{nm}(\boldsymbol{x}; \boldsymbol{F}) = (|\boldsymbol{F} - \boldsymbol{x}_0| - |\boldsymbol{F} - \boldsymbol{x}|)/c, \tag{2}$$

where $\boldsymbol{x}_0$ is the center position of the array transducer. In this section, the center position is set to the origin. The focal point $\boldsymbol{F}$ is expressed by the incident angle (zenith angle) θ and rotational angle (azimuth angles) ϕ : $\boldsymbol{F} = (F_1, F_2, F_3) = (R\sin\theta\cos\phi, R\sin\theta\sin\phi, R\cos\theta)$. Therefore the delay law $\Delta\tau_{nm}$ is obtained as

$$\Delta\tau_{nm}(\boldsymbol{x}; \boldsymbol{F}) = R\left[1 - \sqrt{(\sin\theta\cos\phi - x_n/R)^2 + (\sin\theta\sin\phi - y_m/R)^2 + \cos^2\theta}\right]/c. \tag{3}$$

In the case of the linear array transducers, Equation (3) is simplified to

$$\Delta\tau_n(\boldsymbol{x};\boldsymbol{F}) = R\left[1-\sqrt{(\sin\theta - x_n/R)^2+\cos^2\theta}\right]. \tag{4}$$

In the finite limitation of the focal length R, Equations (3) and (4) are simplified to the following expressions, respectively:

$$\begin{aligned}\Delta\tau_{nm}(\boldsymbol{x};\theta,\phi) &= \Delta\tau_n(\boldsymbol{x};\theta,\phi)+\Delta\tau_m(\boldsymbol{x};\theta,\phi)\\ &= x_n\sin\theta\cos\phi/c + y_m\sin\theta\sin\phi/c, \qquad (5)\\ \Delta\tau_n(\boldsymbol{x};\theta) &= x_n\sin\theta/c. \qquad (6)\end{aligned}$$

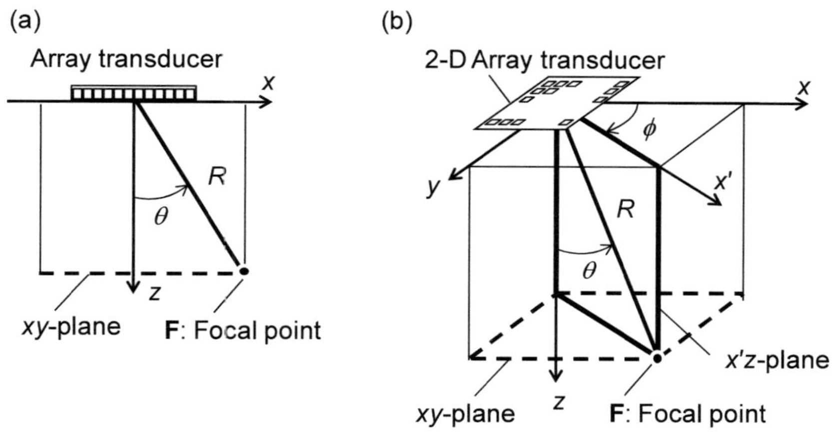

Fig. 5. Definition of the coordinate system and focal points $\boldsymbol{F}$: (a) the 1-D array transducer and (b) 2-D array transducer.

2.3 Effects of beam focusing and steering

Characteristics of the focusing beams , such as a beam width, peak amplitude and focus length, depend on both parameters of the array transducers (shown in Fig.3) and delay laws. As examples, delay laws of normal and angled focused beams for the linear array transducers are plotted in Fig.6. The horizontal axis is position of the transducer element and the vertical axis is delay time. Negative values of the delay time correspond to transmitting ultrasonic beam before the relative origin of the time. Therefore the delay time in an actual ultrasonic phased array equipment is shifted to nonnegative value by adding constant values to the theoretical delay time in Equation (3).

Beam profiles for the delay laws in Fig.6 are calculated based on a Rayleigh-Sommerfeld model (Kono et al., 2010), as shown in Fig.7. The calculation conditions are as follows: the array configuration is the linear array transducer with 16 elements; element pitches, a half of a wavelength λ of 2.95mm; an element width, 8λ; center frequency, 2.0MHz; focal lengths, 20-80mm and incident angles, 0–45°. Peak levels of the amplitude have a propensity to decrease with increased focal length from the results of the normal beams in Fig.7(a)-(c). For the angled beam, peak levels decrease with higher incident angles from the results in Fig.7(b),(e) and (f).

2.4 Generation of grating lobes

Grating lobes can be generated under the condition that difference of the delay time between the adjacent transducer elements is a multiple of a period T. For the matrix array transducers, Equation (5) is separated into x-dependent and y-dependent terms:

$$\Delta\tau_{nm}(\boldsymbol{x};\theta,\phi) - \Delta\tau_{nm}(\boldsymbol{x};\theta',\phi') = (i+j)\,T, \tag{7}$$

$$\Delta\tau_{n}(\boldsymbol{x};\theta,\phi) - \Delta\tau_{n}(\boldsymbol{x};\theta',\phi') = p_x\left(\sin\theta\cos\phi - \sin\theta'\cos\phi'\right)/c = iT, \tag{8}$$

$$\Delta\tau_{m}(\boldsymbol{x};\theta,\phi) - \Delta\tau_{m}(\boldsymbol{x};\theta',\phi') = p_y\left(\sin\theta\sin\phi - \sin\theta'\sin\phi'\right)/c = jT. \tag{9}$$

where the position of the transducer element is set to element pitches: $\boldsymbol{x} = (p_x, p_y)$. The prime mark indicates the direction of grating lobes, m and n are integer indexes for the grating lobe with phase difference $2(i+j)\pi$.

Therefore, the propagation direction of grating lobes for the matrix array transducers are calculated from the delay time differences in Equations (8) and (9) (Kono & Mizota, 2011).

$$\tan\phi' = \frac{p_x}{p_y}\left(\frac{p_y\sin\theta'\sin\phi'/c}{p_x\sin\theta'\cos\phi'/c}\right) = \frac{p_x}{p_y}\left(\frac{p_y\sin\theta\sin\phi/c - jT}{p_x\sin\theta\cos\phi/c - iT}\right), \tag{10}$$

$$\sin\theta' = \frac{p_x\sin\theta\cos\phi/c - iT}{p_x\cos\phi'/c} = \frac{p_y\sin\theta\sin\phi/c - jT}{p_y\sin\phi'/c}. \tag{11}$$

Similarly, the generation conditions of grating lobes for the linear array transducers are derived from Equation (11) substituted p for x_n:

$$\Delta\tau_n(p;\theta) - \Delta\tau_n(p;\theta') = p\left(\sin\theta - \sin\theta'\right)/c = iT, \tag{12}$$

where p is an element pitch and i is an integer index for the grating lobe with phase difference $2i\pi$. The incident angle of grating lobes for the linear array transducer is transformed from Equation (12) and the relation between cycle T and wavelength $\lambda : T = \lambda/c$:

$$\sin\theta' = \sin\theta - i\lambda/p. \tag{13}$$

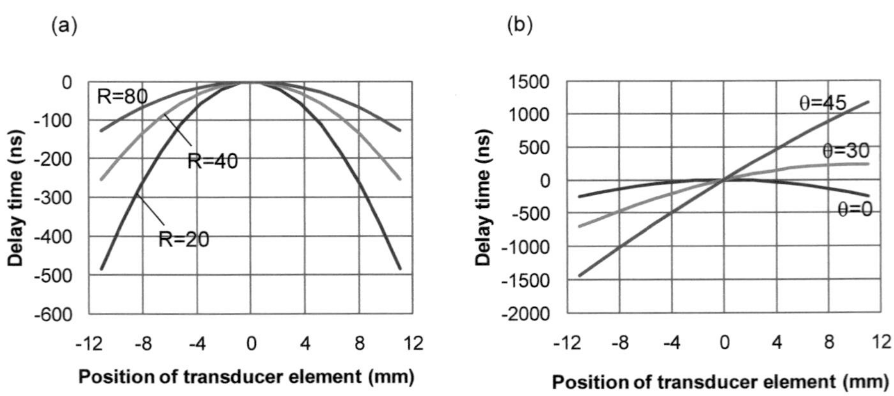

Fig. 6. Delay laws for the linear array transducer; number of element N=16; element pitch $p = 1.475$mm; velocity c=5900m/s (carbon steel): (a) normal beams and (b) angled beams with the focal length R=40 mm.

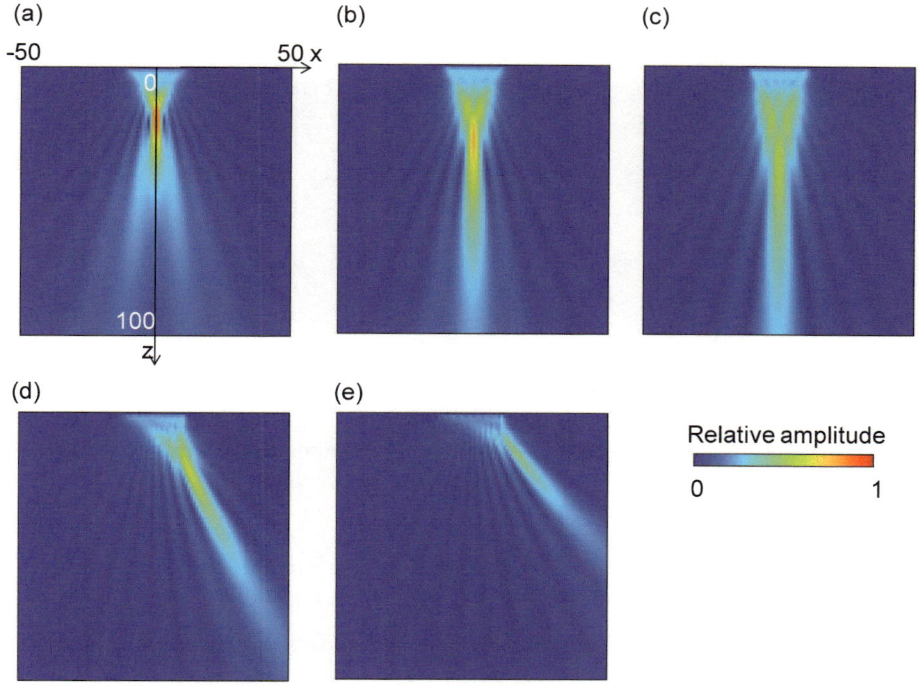

Fig. 7. Calculation results of the beam profiles with the delay laws in Fig.6 and calculation region: 100mm square in the xy-plane: (a) normal beam with the focal length R=20mm, (b) normal beams with R=40mm, (c) normal beams with R=80mm, (d) 30° angled beam with R=40mm, and (e) 45° angled beam with R=40mm.

Equation (13) is consistent with Equation (11) substituted zero for the rotation angles of the main lobe and grating lobes. Grating lobes are generated on the condition that the absolute value of the left side in Equation (13) is less than or equal to one; therefore the grating lobes can be suppressed when an element pitch is less than one-half wave length.

Beam profiles for angled beams generated by two kinds of linear array transducers with different element pitch are compared in Fig.8. The position of the focal point $\boldsymbol{F}$ is the same each other. No grating lobe is observed for the linear array transducer with element pitch equal to one-half wavelength. A grating lobe is generated in the case of the array transducer with one-wavelength element pitch. The incident angle of the grating lobe is obtained as $-17.0°$ from Equation (13) for i=1 and θ=45°.

3. Modeling of phased array transducer in frequency domain

Fundamentally, ultrasonic beam can be modeled by superposing spherical waves according to the Huygens' principle. The Rayleigh–Sommerfeld integral (RSI) model is also based on the Huygens' principle and well known as an expression of wave field in liquid from an immersion piston transducer. On the other hand, the wave field in solid from a contact piston transducer can be expressed with the Vezzetti's model (Vezzetti, 1985). This is a little bit complicated because the Vezzetti's model remains the integral form of the fundamental

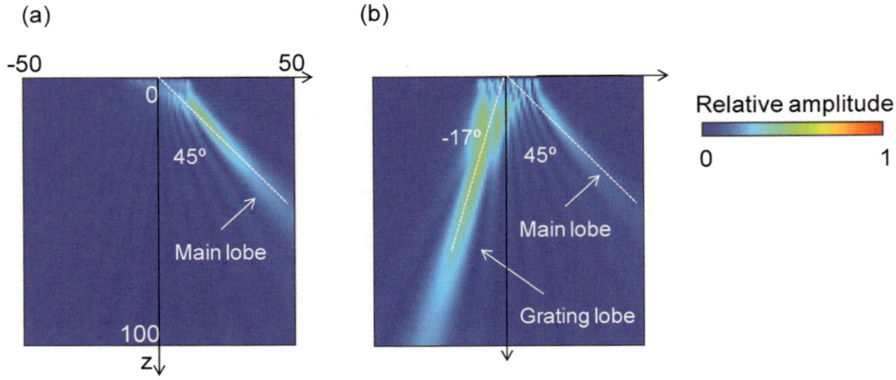

Fig. 8. Calculation results of 45° angled beam with focal length R=40mm: (a) the element pitch p=1.475mm (one-half wavelength); the element number N=16, and (b) the element pitch p=2.95mm (one wavelength) ; the element number N=8.

solution in elastic half space. Schmerr (Schmerr, 1998) has introduced an explicit expression of the Vezzetti's model using the stationary phase method. Here we show numerical modelings of the ultrasonic phased array based on the Schmerr's expression.

Here we assume the harmonic wave of $\exp(-i\omega t)$ time dependency. First of all, consider a single planar piston transducer which generates a P-wave in solid. To model the radiation field from the transducer, we use the geometry shown in Fig.9, where the solid region is the half-space $x_3 \geq 0$. On the x_1–x_2 plane, we assume that velocity in the x_3 direction is zero everywhere except for a finite region S. Here the displacement in solid due to the P-wave is written as (Schmerr, 1998)

$$\boldsymbol{u}(\boldsymbol{x},\omega) = \frac{P_0}{2\pi\rho c_p^2} \int_S K_p(\theta) \boldsymbol{d}_p \frac{\exp(ik_p r)}{r} dS(\boldsymbol{y}) \tag{14}$$

where, $r = |\boldsymbol{x} - \boldsymbol{y}|$, c_p and $k_p(= \omega/c_p)$ are the velocity and the wave number of the P-wave, respectively. Also ρ is the density and P_0 is a constant known pressure in S. In Equation (14),

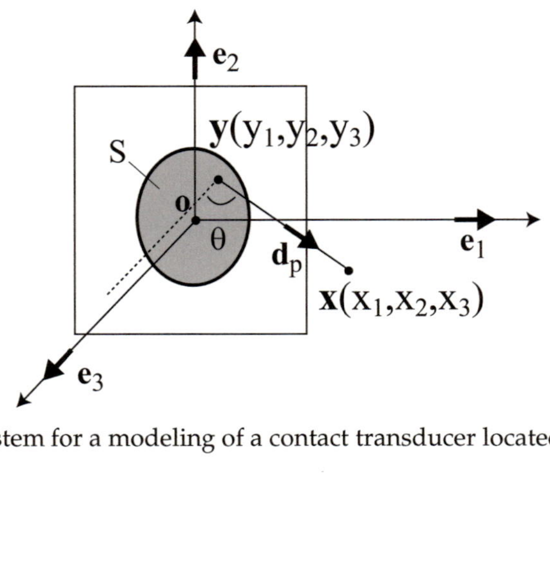

Fig. 9. Coordinate system for a modeling of a contact transducer located on solid.

$\boldsymbol{d}_p$ is the polarization vector:

$$\boldsymbol{d}_p = \frac{x_1 - y_1}{r}\boldsymbol{e}_1 + \frac{x_2 - y_2}{r}\boldsymbol{e}_2 + \frac{x_3 - y_3}{r}\boldsymbol{e}_3 \tag{15}$$

and $K_p(\theta)$ is the directivity function:

$$K_p(\theta) = \frac{\cos\theta\kappa^2(\kappa^2/2 - \sin^2\theta)}{2G(\sin\theta)}$$
$$G(x) = (x^2 - \kappa^2/2)^2 + x^2\sqrt{1 - x^2}\sqrt{\kappa^2 - x^2} \tag{16}$$

where c_s is S-wave velocity, $\kappa = c_p/c_s$ and θ is the angle between the vector $(\boldsymbol{x} - \boldsymbol{y})$ and the normal of the transducer surface.

In the next section, we show two methods to obtain the solution $\boldsymbol{u}$ in Equation (14). One is based on a numerical integral method and the other is an approximation method using superposition of the Gaussian beam. The former method needs some numerical calculation but is effective even if the transducer element has a complicated shape. The latter case is a fast method, but only applicable to circular and rectangular element of the transducer.

3.1 Rayleigh–Sommerfeld numerical integral (RSNI)

It is not easy to solve the integral in Equation (14) analytically except for simple element shape such as a rectangle or a circle. Therefore let us consider a numerical integral over S in Equation (14). For convenience we use the local coordinate system (ξ, η) on the transducer surface. Using the four-node two-dimensional element, we can interpolate the coordinate $\boldsymbol{y}$ as

$$\boldsymbol{y}(\xi,\eta) = \sum_{\alpha=1}^{4} N_\alpha(\xi,\eta)\boldsymbol{y}^\alpha \tag{17}$$

where

$$N_1 = \frac{1}{2}(1-\xi)\cdot\frac{1}{2}(1-\eta),\quad N_2 = \frac{1}{2}(1+\xi)\cdot\frac{1}{2}(1-\eta),$$
$$N_3 = \frac{1}{2}(1+\xi)\cdot\frac{1}{2}(1+\eta),\quad N_4 = \frac{1}{2}(1-\xi)\cdot\frac{1}{2}(1+\eta). \tag{18}$$

In Equation (17), $\boldsymbol{y}^\alpha$ is the coordinate of the four element nodes. Substituting Equation (17) into (14), we can rewrite the integral form as follows.

$$\int_S K_p(\theta)\boldsymbol{d}_p\frac{\exp(ik_pr)}{r}dS(\boldsymbol{y}) = \int_{-1}^{1}\int_{-1}^{1} K_p(\theta)\boldsymbol{d}_p\frac{\exp(ik_p|\boldsymbol{x} - \boldsymbol{y}(\xi,\eta)|)}{|\boldsymbol{x} - \boldsymbol{y}(\xi,\eta)|}|Y|d\xi d\eta \tag{19}$$

where $|Y|$ is the Jacobian operator:

$$|Y| = \left[\left(\frac{\partial y_2}{\partial\xi}\frac{\partial y_3}{\partial\eta} - \frac{\partial y_3}{\partial\xi}\frac{\partial y_2}{\partial\eta}\right)^2 + \left(\frac{\partial y_3}{\partial\xi}\frac{\partial y_1}{\partial\eta} - \frac{\partial y_1}{\partial\xi}\frac{\partial y_3}{\partial\eta}\right)^2 + \left(\frac{\partial y_1}{\partial\xi}\frac{\partial y_2}{\partial\eta} - \frac{\partial y_2}{\partial\xi}\frac{\partial y_1}{\partial\eta}\right)^2\right]^{1/2}. \tag{20}$$

Equation (19) can be calculated with numerical integration of the Gauss quadrature.

$$\int_{-1}^{1}\int_{-1}^{1} f(\xi,\eta)d\xi d\eta \cong \sum_{i=1}^{I}\sum_{j=1}^{I} f(\xi_i,\eta_j)w_iw_j \tag{21}$$

where w is the integration weight and I is the number of the integration points.

In the above procedure, the wave field from a piston transducer, namely the wave field from an element of the array transducer, can be calculated. When the array transducer has elements N in the x_1 direction and M in the x_2 direction, we can obtain total wave field by appropriately adding up $N \times M$ wave components. Using the time delay $\Delta\tau_{nm}$ for the element positioned at n-th in the x_1 direction and m-th in the x_2 direction, the total wave field can be expressed as

$$\boldsymbol{u}(\boldsymbol{x},\omega) = \sum_{n=0}^{N-1}\sum_{m=0}^{M-1} \frac{1}{2\pi\rho c_p^2} \int_{S_{nm}} P_0 \exp\left(\mathrm{i}\omega\Delta\tau_{nm}\right) \boldsymbol{d}_p K_p(\theta) \frac{\exp(\mathrm{i}k_p r)}{r} dS(\boldsymbol{y}) \tag{22}$$

Equation (22) can be applied to not only a flat array element but also a curved one.

3.2 Multi-Gaussian beam (MGB) model

The multi-Gaussian beam (MGB) model (Schmerr, 2000) is adopted for simulation of beam fields radiated from circular and rectangular transducers. However, the MGB model is based on the assumption of the paraxial approximation, and will lose its accuracy at the high-refracted angle, which is a condition that paraxial approximation is not satisfied (Park et al., 2006). Such problem often happens in simulating the steering beam of phased array transducers with the conventional MGB model. To address such a problem, the MGB without the nonparaxial approximation has been developed by Zhao and Gang (Zhao & Gang, 2009). The paper had some print errors, so here we show the correct formulation of the nonparaxial MGB. As below, we abbreviate "the nonparaxial MGB model" to "the MGB model".

A uniform normal pressure on the transducer surface can be expanded as the superposition of Gaussian beams (Wen & Breazeale, 1988). For a rectangular element with length $2a_1$ and width $2a_2$ in the x_1- and x_2-directions, respectively, the pressure over the transducer surface can be expressed as

$$\begin{aligned} P_0 &= \begin{cases} P_0 & |y_1| \le a_1, |y_2| \le a_2, y_3 = 0 \\ 0 & \text{otherwise} \end{cases} \\ &= P_0 \sum_{k}^{10}\sum_{l}^{10} A_k A_l \exp(-B_k y_1^2/a_1^2 - B_l y_2^2/a_2^2) \end{aligned} \tag{23}$$

where A and B are ten complex coefficients obtained with an optimization method. Now the coordinate of wave field to be calculated is defined as $\boldsymbol{x}' = (x_1', x_2', x_3')$. In order to obtain more accurate solution beyond the paraxial approximation region, the distance factor r is approximated as

$$r = \sqrt{(x_1' - y_1)^2 + (x_2' - y_2)^2 + (x_3')^2} \approx R + \frac{y_1^2 + y_2^2 - 2x_1'y_1 - 2x_2'y_2}{2R} \tag{24}$$

where $R = \sqrt{(x_1')^2 + (x_2')^2 + (x_3')^2}$. The directivity function $K_p(\theta)$ is not sensitive to a small element, so it can be substituted by $K_p(\theta_o)$ and we move it to outside of the integral. Here θ_o is the angle between the vector $(\boldsymbol{x} - \mathbf{o})$ and the surface normal of transducer. Introducing

Equations (23) and (24) into (14), we can obtain:

$$\begin{aligned}\boldsymbol{u}(\boldsymbol{x}',\omega) &= \frac{P_0 \boldsymbol{d}_p K_p(\theta_o)}{2\pi\rho c_p^2} \frac{\exp(\mathrm{i}k_p R)}{R} \sum_k^{10} \sum_l^{10} A_k A_l \\ &\times \int_{-\infty}^{\infty} \exp\left[-\left(\frac{B_k}{a_1^2} - \frac{\mathrm{i}k_p}{2R} \right) y_1^2 - \frac{\mathrm{i}k_p x_1' y_1}{R} \right] dy_1 \\ &\times \int_{-\infty}^{\infty} \exp\left[-\left(\frac{B_l}{a_2^2} - \frac{\mathrm{i}k_p}{2R} \right) y_2^2 - \frac{\mathrm{i}k_p x_2' y_2}{R} \right] dy_2. \end{aligned} \quad (25)$$

Using the known integral formula

$$\int_{-\infty}^{\infty} \exp\left(-ax^2 + bx\right) dx = \sqrt{\frac{\pi}{a}} \exp\left(\frac{b^2}{4a}\right), \quad \Re\{a\} > 0 \quad (26)$$

the MGB model for a rectangular element can be obtained as

$$\begin{aligned}\boldsymbol{u}(\boldsymbol{x}',\omega) &= \frac{\mathrm{i}P_0 \boldsymbol{d}_p K_p}{\rho k_p c_p^2} \sum_k^{10} \sum_l^{10} \frac{A_k A_l \exp(\mathrm{i}k_p R)}{\sqrt{1 + \mathrm{i}B_k R/D_1}\sqrt{1 + \mathrm{i}B_l R/D_2}} \\ &\times \exp\left(-\frac{\mathrm{i}k_p}{2} \frac{(x_1')^2/R}{1 + \mathrm{i}B_k R/D_1} \right) \exp\left(-\frac{\mathrm{i}k_p}{2} \frac{(x_2')^2/R}{1 + \mathrm{i}B_l R/D_2} \right) \end{aligned} \quad (27)$$

where $D_1 = k_p a_1^2/2$ and$D_2 = k_p a_2^2/2$. To predict the wave field in solid from a rectangular element, we only perform the summation of 10×10 in Equation (27). Therefore fast calculation is possible but it is noted that the MGB is not applicable to complicated element shapes.

The total beam field of a phased array transducer can be calculated by simple superposition of individual wave fields with the corresponding time delay $\Delta\tau_{nm}$. The P-wave displacement field can be calculated by

$$\boldsymbol{u}(\boldsymbol{x},\omega) = \sum_{n=0}^{N-1} \sum_{m=0}^{M-1} \boldsymbol{u}(\boldsymbol{x}',\omega) \exp(\mathrm{i}\omega\Delta\tau_{nm}) \quad (28)$$

where $\boldsymbol{x}$ is the position vector of the global coordinate. In Equation (27), the origin of $\boldsymbol{x}'$ is always set on the center of each element. The total wave field at the global coordinate can be obtained by adding up the results of Equation (27) in consideration of the delay and location of each element.

3.3 Numerical comparison of the RSNI and MGB model

Here the wave fields from an array transducer in stainless steel (c_p=5800m/s, c_s=3200m/s, $\rho = 7800\mathrm{kg}/m^3$) are calculated with both the RSNI and the MGB model. The array transducer used in numerical comparisons is a linear phased array with 24 elements and center frequency f=3.0MHz. The element pitch p is 1.0mm and the gap g is 0.1mm. We show the magnitude of displacement $|\boldsymbol{u}|$ which is normalized by multiplying $\rho c_p^2/P_0$.

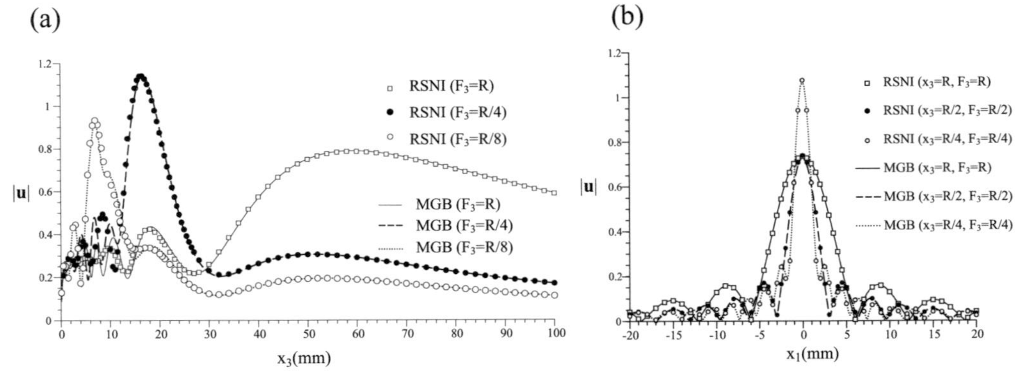

Fig. 10. Displacements $|\boldsymbol{u}|$ due to radiated ultrasonic wave: (a) on-axis in the case of different desired beam focal lengths (F_3=R, $R/4$ and $R/8$), (b) off-axis in the case of different beam focal lengths (F_3=R, $R/2$ and $R/4$).

In order to investigate the applicability of the MGB model to the UT modeling, the magnitudes of displacement fields are compared in both methods. Here we introduce the Rayleigh distance*:

$$R = \frac{A^2 f}{4c_p} (= 73.66\text{mm}) \tag{29}$$

Fig.10(a) shows the on-axis (x_1=0, x_2=0) displacement when we vary F_3=R, $R/4$ and $R/8$ with keeping the steering angle 0°. From Fig.10(a), the result calculated by the MGB model can conform to the RSNI results. Similar behaviors are given in the off-axis results as shown in Fig.10(b). Fig.10(b) shows displacements on the x_1-direction on x_3=R, $R/2$ and $R/4$ in the case of F_3=R, $R/2$ and $R/4$, respectively. It is understood that the MGB model is an effective tool in simulating the beam field radiated from phased arrays over a wide range of steering angle.

3.4 Visualization of 3-D wave field from a matrix array transducer

The following numerical examples are carried out with the MGB model because the MGB model can keep good accuracies in both the on-axis and off-axis regions. Here we show a visualization of 3-D wave field from a matrix array transducer. It is assumed that the array transducer has parameters as p_x=p_y=1.0mm and $N \times M$=16 × 16=256. The 3-D wave fields in the cases of f=2.5 and 5.0MHz when we set the focal point to (F_1,F_2,F_3)=(10mm,10mm,20mm) are illustrated in Fig.11. The main lobe only appears in the case of f=2.5MHz, however not only a main lobe but also grating lobes are generated in the case of f=5.0MHz. According to Equations (10) and (11), the directions of the grating lobe in f=5.0MHz are estimated to (θ', ϕ')=$(58.8°, -61.5°)$ and $(58.8°, 151.5°)$, on the other hand the grating lobe in f=2.5MHz is nonexistent. These estimation results show good agreements with the visualization results in Fig.11.

* Note that it is well known that the actual focal length is shorter than the desired (designed) one and the focal point is never beyond the Rayleigh distance (Schmerr, 1998).

3.5 Properties of linear and matrix array transducers

Here we investigate properties of both linear and matrix array transducers for the beam focusing and the beam steering.

3.5.1 Beam focusing

First, the property of beam focusing in the case of a linear array transducer is investigated. As a linear array transducer, the specification of p=1.0mm, g=0.1mm, N=24 and W/A=0.3 is assumed. The Rayleigh distance R of this transducer is 98.48mm in the case of f=4.0MHz. In Fig.12(a), the actual focal lengths F_A are plotted in varying the desired focal length F_3. The vertical axis in Fig.12(a) shows the parameter $\alpha(= F_A/F_3)$ which represents the efficiency of the beam focus. When the parameter of α comes close to 1, it shows that we can transmit ultrasonic beam to the intended position. The parameter $\gamma(= R/F_3)$ represents the closeness of focal length for the Rayleigh distance. From Fig.12(a), the focal length F_A in the range of small γ is much shorter than the desired length F_3. The peak of α appears at $\gamma = 4$, and the distance between F_A and F_3 increases when γ becomes large more than $\gamma = 4$.

Next let us look at the property of beam focusing for a matrix array transducer. Here a matrix array transducer with $p_x = p_y$=1.0mm, $g_x = g_y$=0.1mm, N=M=24 and A_y/A_x=1.0 is modeled. From Fig.12(b), the focal length F_A comes closer and closer to F_3 when γ becomes larger and larger. Through the all range of γ, the difference between F_A and F_3 is smaller than the case of the linear array transducer. Also ultrasonic beam can be focused efficiently even at the region near the surface of array transducer.

3.5.2 Beam steering

Figure 13(a) shows the deviation between the desired and actual focal points in a linear array transducer. Here the Rayleigh distance is R=98.48mm and we keep constant focal length $|\boldsymbol{F}|$=$R/4$=24.62mm through all steering angles. The actual focus locations are less than the desired focal length and the deviation of the actual focus from desired focal position also increases as the steering angle increases. And Fig.13(b) shows maximal amplitude of normalized displacement and actual focal lengths F_A versus the steering angle. These simulation results point out that the peak-to-peak amplitude will decrease as the steering angle increases. These changes will reduce the angular sensitivity of the phased array transducer.

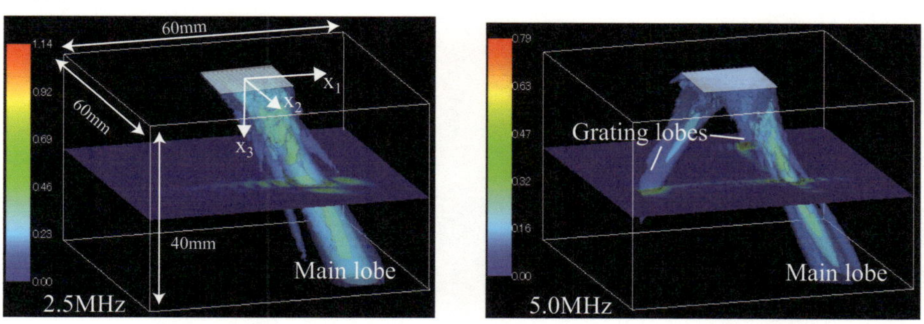

Fig. 11. 3-D wave fields radiated from a matrix array transducer with p_x=p_y=1.0mm and $N \times M$=16 $\times$ 16=256 in the case of 2.5MHz(left side figure) and 5.0MHz(right side figure).

In contrast, the behavior of the matrix array transducer is different from the linear array transducer. Figure 14(a) shows that the deviation distance between the desired and actual focal points is almost constant as the steering angle increases. However the peak-to-peak amplitude becomes lower as the steering angle increases.

From these results, the effective steering angle for the linear array transducer is up to approx. 40°. Although the beam steering for the matrix array transducer is possible in higher angle than the linear array transducer, we suggest that the effective steering angle is up to approx. 55° from the standpoint of the 6dB down of the peak-to-peak amplitude.

4. Modeling of phased array transducer in time domain

We here introduce a time domain simulation tool to predict the wave propagation from the array transducer. The method is a combined technique of the finite integration technique (FIT) (Marklein, 1998) and an image-based modeling (Terada et al., 1997). In the FIT, the finite difference equations of the stress and the particle velocity are derived from integral forms of the governing equations. Computational grids of these quantities are arranged in a staggered configuration in space, and the finite difference equations can be solved by marching time steps in the leap-frog manner. The FIT has an advantage that it can treat different boundary conditions without difficulty. This is essential to model the ultrasonic wave propagation in

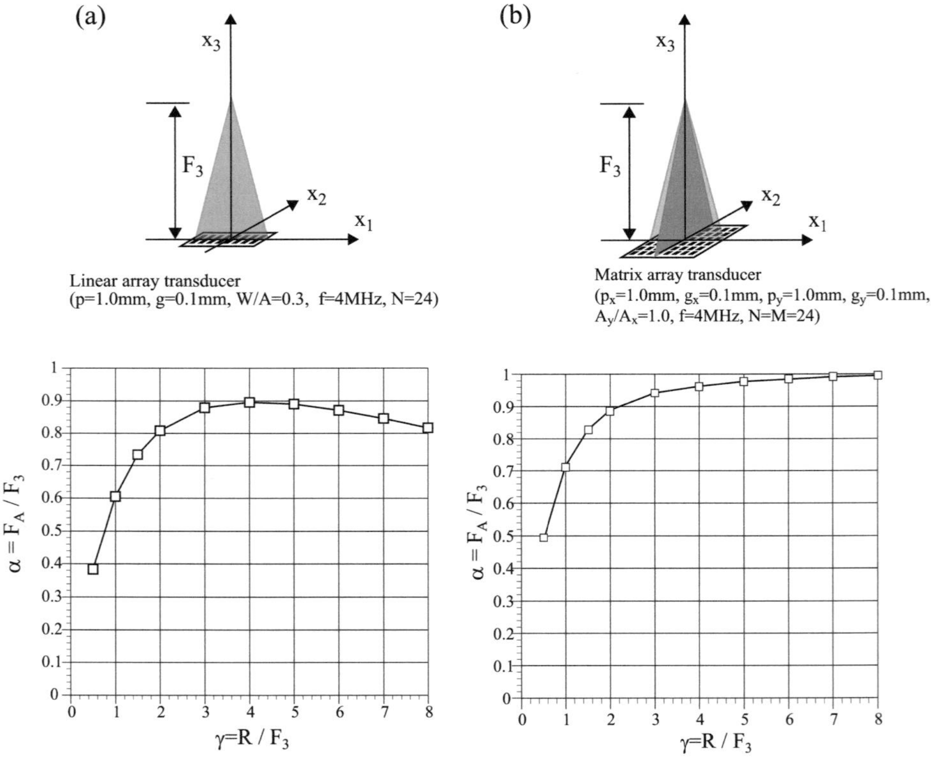

Fig. 12. Ratios of actual focal length F_A and desired one F_3 in the case of the linear array transducer (a) and matrix array transducer (b).

heterogeneous material (Schubert & B. Köhler, 2001). In order to perform simulations of the UT accurately, a realistic shape data of the target is required. Here the image-based modeling is adopted as a pre-processing of the FIT. Using the image-base modeling, we can make geometries of targets directly from a digital image such as X-ray photograph, captured curve data of surface, CAD data, etc. Here we describe the 3-D image-based FIT (Nakahata et al., 2011) and show a numerical example of the phased array UT.

4.1 3-D finite integration technique (FIT)

We show the formulation of the FIT and the calculation flow as below. We consider the Cartesian coordinates (x_1, x_2, x_3). The governing equations of elastic waves are the Cauchy equation of motion and the equation of deformation rate. These equations are given in integral forms for a finite volume V with the surface S by

$$\frac{\partial}{\partial t}\int_V \tau_{kl}(\boldsymbol{x},t)dV = \int_S C_{klij}v_i(\boldsymbol{x},t)n_j dS \tag{30}$$

$$\frac{\partial}{\partial t}\int_V \rho v_i(\boldsymbol{x},t)dV = \int_S \tau_{ij}(\boldsymbol{x},t)n_j dS + \int_V f_i(\boldsymbol{x},t)dV \quad (i = 1,2,3) \tag{31}$$

where $\boldsymbol{v}$ is the particle velocity vector, $\boldsymbol{\tau}$ is the stress second rank tensor, ρ is the density, $\boldsymbol{n}$ is the outward normal vector on the surface S, and $\boldsymbol{f}$ is the body force vector. In Equation (30), $\boldsymbol{C}$ is the stiffness tensor of rank four. In the case of isotropic materials, $\boldsymbol{C}$ can be written as:

$$C_{ijkl} = \lambda\delta_{ij}\delta_{kl} + \mu(\delta_{ik}\delta_{jl} + \delta_{il}\delta_{jk}) \tag{32}$$

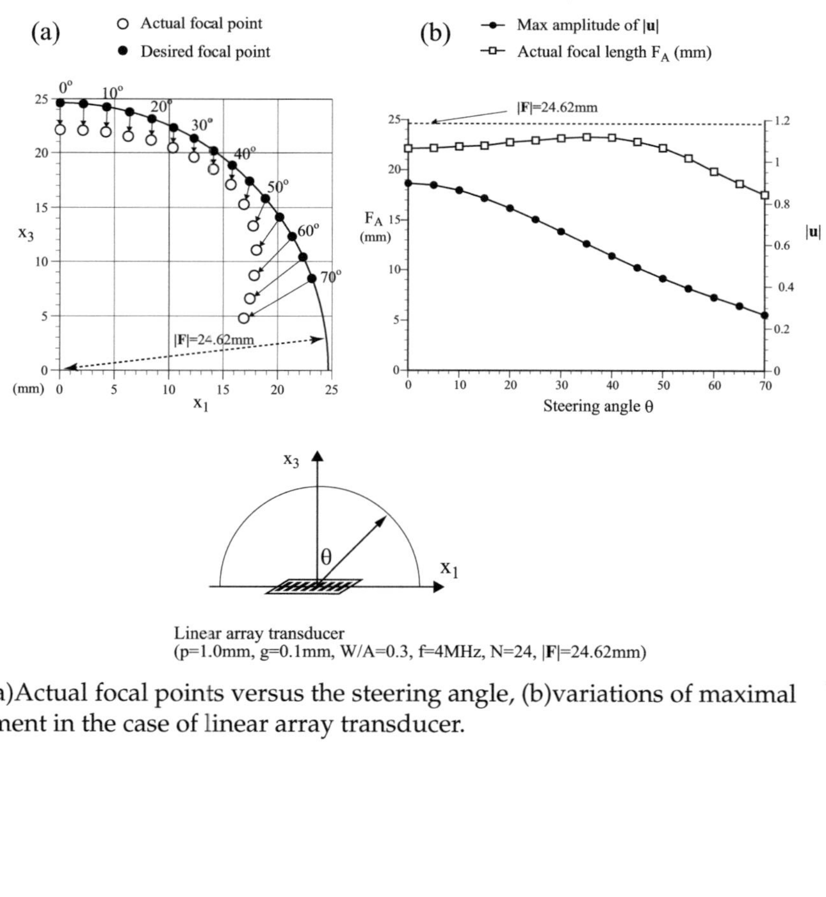

Fig. 13. (a)Actual focal points versus the steering angle, (b)variations of maximal displacement in the case of linear array transducer.

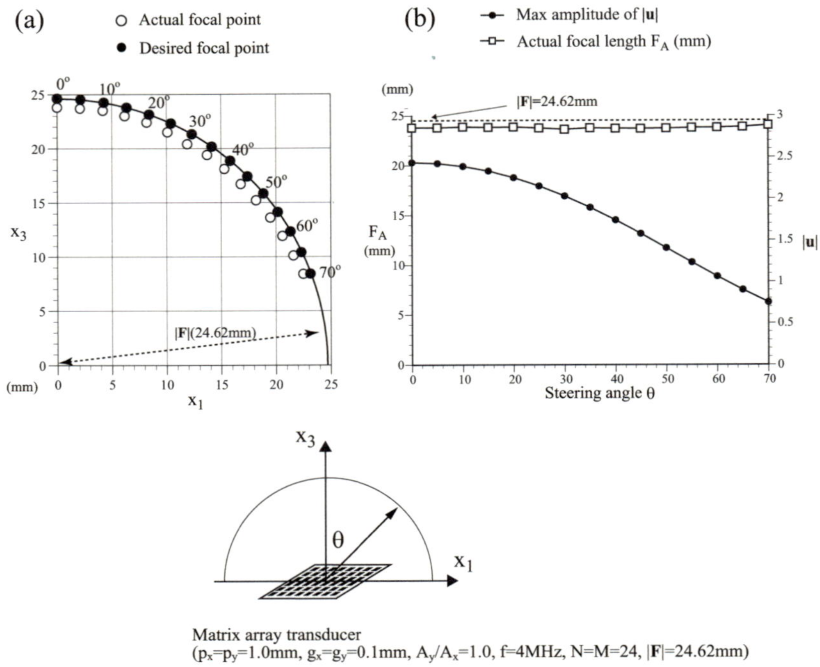

Fig. 14. (a)Actual focal points versus the steering angle, (b)variations of maximal displacement in the case of matrix array transducer.

in terms of the two Lamé constants, λ and μ. In Equation (32), δ_{ij} is the Kronecker delta tensor. P and S wave velocities are given as:

$$c_P = \sqrt{\frac{(\lambda + 2\mu)}{\rho}}\ ,\ c_S = \sqrt{\frac{\mu}{\rho}}. \tag{33}$$

In the case of an acoustic problem, we can use above equations but have to set $\tau_{ij} = 0\ (i \neq j)$ in Equation (31) and $\mu = 0$ in Equation (32).

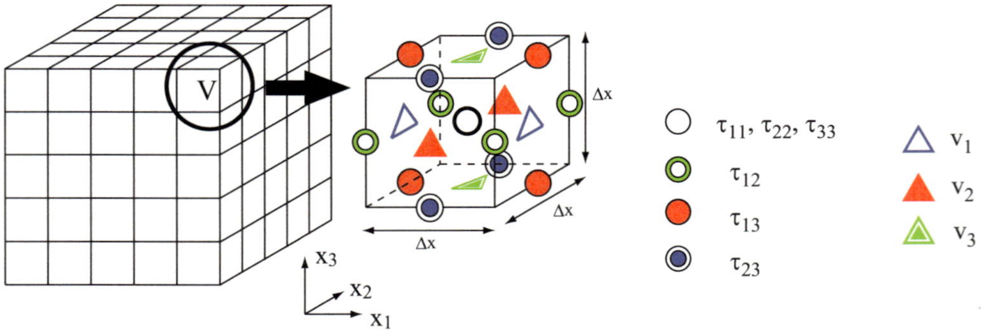

Fig. 15. Finite volume V and grid arrangement in the 3-D FIT.

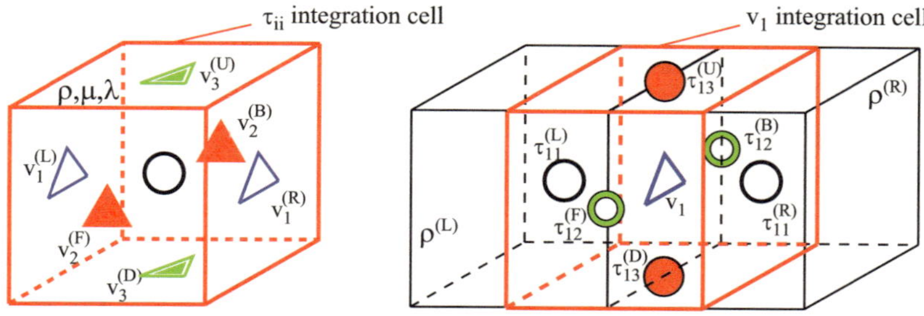

Fig. 16. τ_{ii}–integration cell (left figure) and v_1–integration cell (right figure). All material parameters are defined in the τ_{ii}–integration cell.

We consider a discretization form of Equation (30) using a cube as an integral volume V in Fig.15. Assuming that τ_{11} is constant in V, we have

$$\begin{aligned}\dot{\tau}_{11}\Delta x^3 &= (\lambda + 2\mu)\left[v_1^{(R)} - v_1^{(L)}\right]\Delta x^2 + \lambda\left[v_2^{(B)} - v_2^{(F)} + v_3^{(U)} - v_3^{(D)}\right]\Delta x^2 \\ \dot{\tau}_{11} &= \frac{\lambda + 2\mu}{\Delta x}\left[v_1^{(R)} - v_1^{(L)}\right] + \frac{\lambda}{\Delta x}\left[v_2^{(B)} - v_2^{(F)} + v_3^{(U)} - v_3^{(D)}\right]\end{aligned} \tag{34}$$

where $\dot{\tau}_{11} = \partial\tau_{11}/\partial t$ and Δx is the length on a side of V. In Equation (34), superscript () expresses positions of physical quantities in the integral volume as shown in Fig.16. Similarly a discretization of Equation (31) becomes

$$\begin{aligned}\overline{\rho}\dot{v}_1\Delta x^3 &= \left[\tau_{11}^{(R)} - \tau_{11}^{(L)} + \tau_{12}^{(B)} - \tau_{12}^{(F)} + \tau_{13}^{(U)} - \tau_{13}^{(D)}\right]\Delta x^2 \\ \dot{v}_1 &= \frac{1}{\overline{\rho}\Delta x}\left[\tau_{11}^{(R)} - \tau_{11}^{(L)} + \tau_{12}^{(B)} - \tau_{12}^{(F)} + \tau_{13}^{(U)} - \tau_{13}^{(D)}\right]\end{aligned} \tag{35}$$

where we let $\boldsymbol{f} = \boldsymbol{0}$. Since the density ρ is given in the τ_{ii}-integral volume as shown in Fig.16, the average value $\overline{\rho} = (\rho^{(L)} + \rho^{(R)})/2$ is used in Equation (35).

In the time domain, stress components $\boldsymbol{\tau}$ are allocated at half-time steps, while the velocities $\boldsymbol{v}$ are at full-time steps. The following time discretization yields an explicit leap-frog scheme:

$$\{v_i\}^z = \{v_i\}^{z-1} + \Delta t\{\dot{v}_i\}^{z-\frac{1}{2}} \tag{36}$$

$$\{\tau_{ij}\}^{z+\frac{1}{2}} = \{\tau_{ij}\}^{z-\frac{1}{2}} + \Delta t\{\dot{\tau}_{ij}\}^z \tag{37}$$

where Δt is the time interval and the superscript z denotes integer number of the time step. Therefore the FIT repeats the operations of Equations (36) and (37) by means of adequate initial and boundary conditions. A specific stability condition and adequate spatial resolution must be fulfilled to calculate the FIT accurately (Nakahata et al., 2011; Schubert & B. Köhler, 2001).

4.2 Image based modeling

We show a procedure of the 3-D image-based modeling with a 3-D curve measurement system based on the coded pattern projection method (Nakahata et al., 2012). The system consists of a projector that flashes narrow bands of black and white light onto the model's face and a camera that captures the pattern of light. The distortion pattern of light leads to the

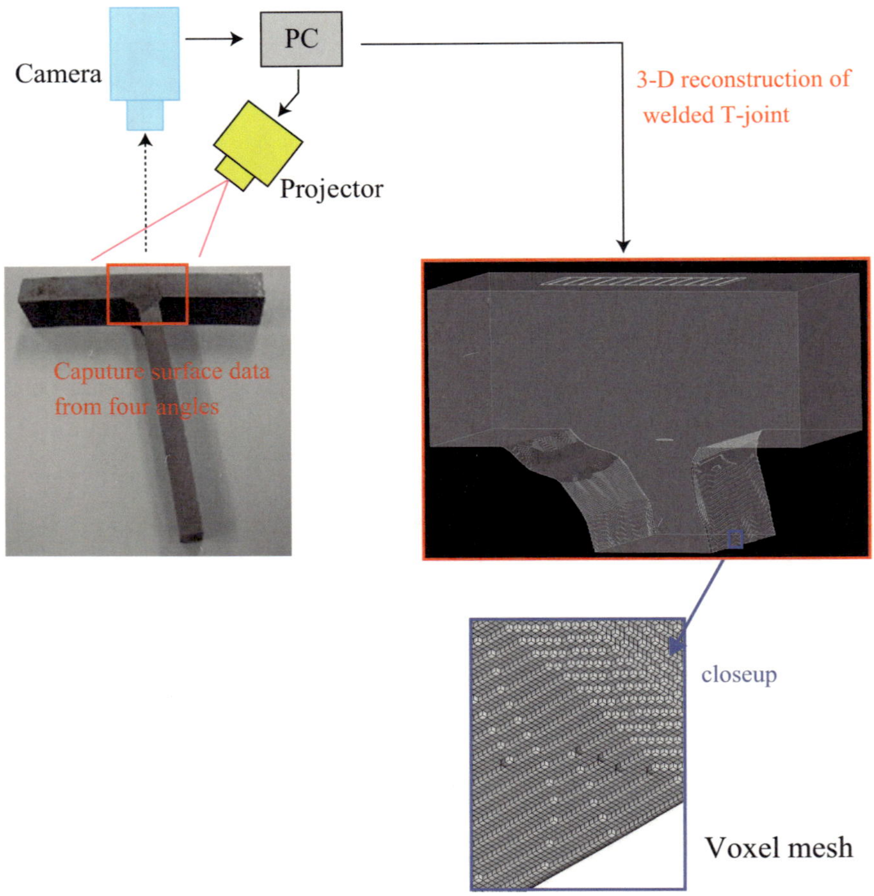

Fig. 17. Image-based modeling of a welded T-joint using a 3-D curve measurement system.

reconstruction of the surface of model. As shown in Fig.17, a 3-D shape of a welded T-joint is reconstructed from surface profiles which are measured at multiple angles. After unnecessary parts for calculation are trimmed away, the rest part is discretized into a voxel data. Here voxel is "volumetric pixel", representing a value on a regular grid in a 3-D space. The voxel data is directly used as the computational cell in the FIT, therefore the voxel size is set to be equivalent to the cell size of the FIT.

4.3 3-D numerical simulation

A 3-D simulation of ultrasonic wave propagation is demonstrated using the model in Fig.17. Here an artificial defect which is assumed to be a lack of weld penetration is introduced at the inside of the welding part(see in Fig.18). The diameter of the defect is approx. 3mm and thickness is 0.1mm. Material parameters of the welded T-joint are set as c_P = 5800m/s, c_S = 3100m/s and ρ = 7800kg/m^3. The linear array transducer (total element number N=24, the pitch p=0.6m, the element width W=7mm and the gap g=0.1mm) is located on the top of the model. In the simulation, we choose Δx = 0.02mm, Δt=1.0 nano-second (ns) and the total time step is 6500. Total number of the voxel is about 1 billion (1000 millions). The P-wave with the

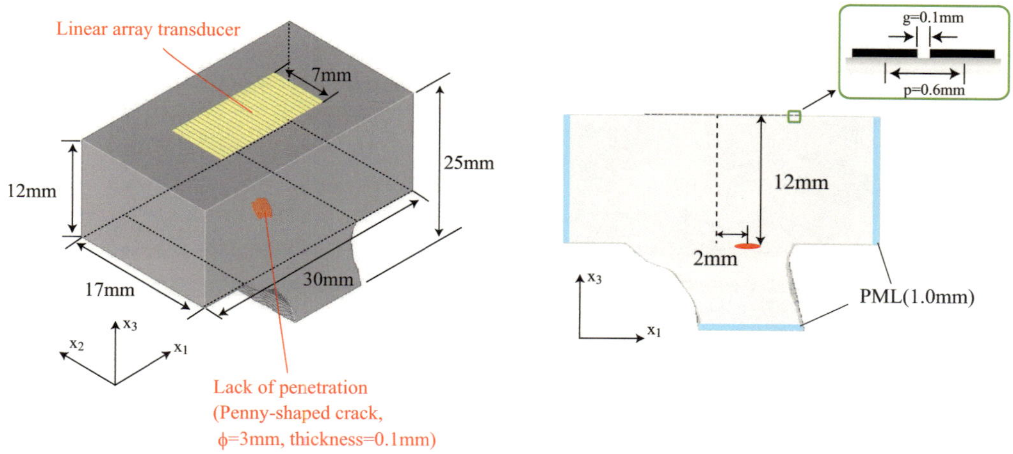

Fig. 18. Numerical model of the welded T-joint including an artificial defect (penny-shaped crack).

center frequency 4.0MHz is transmitted from the array transducer toward the defect. Here the ultrasonic wave is generated into solid by giving a time-dependent normal stress at the surface of the transducer.

Figure 19 shows the isosurface of absolute value of the displacement vector $|\boldsymbol{u}(=\int \boldsymbol{v}dt)|$ and Fig.20 shows $|\boldsymbol{u}|$ at a cross section of the model. These results show that ultrasonic wave propagates toward the artificial defect and scattered waves are generated. After some time steps, the echo from the defect arrives at the array transducer. In the general array UT, the echoes at the receiving process are also synthesized with the appropriate delay time in the same way to the transmitting process. Figure 21 shows the synthesized echo at receiving process, and waveform around 5 μs is the wave component from the defect. The calculation

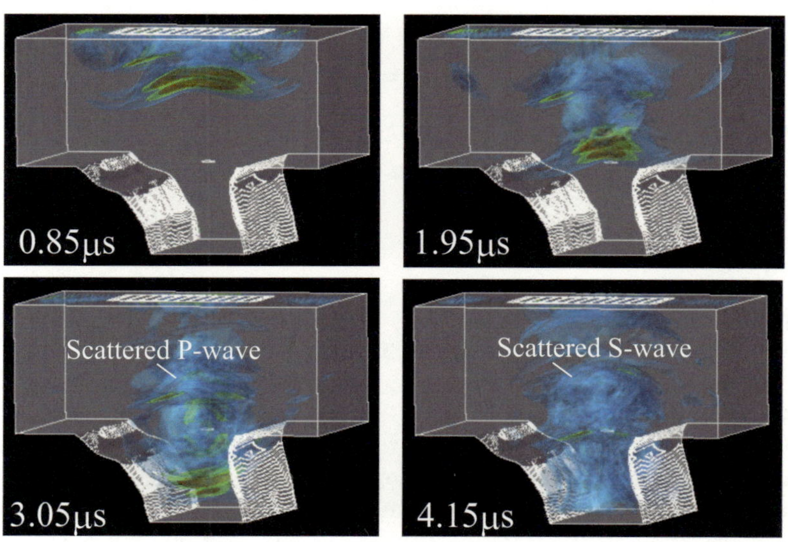

Fig. 19. Snapshots of ultrasonic wave propagation in welded T-joint (isosurface rendering).

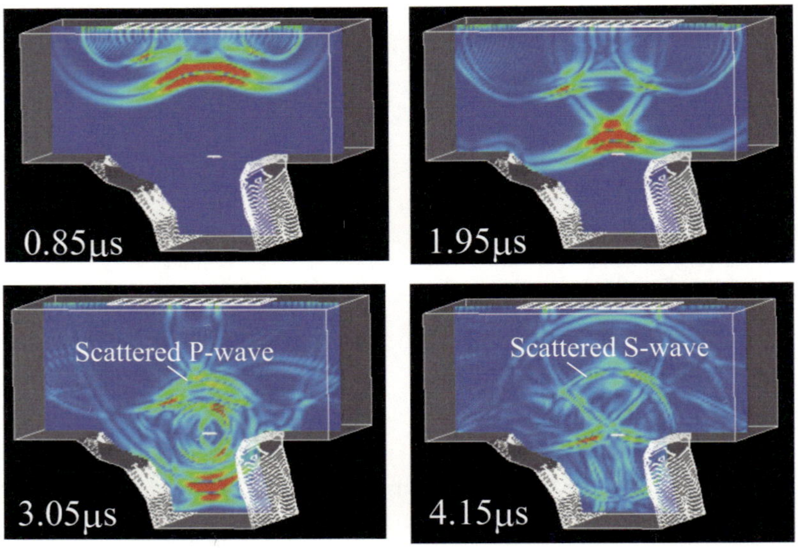

Fig. 20. Snapshots of ultrasonic wave propagation in welded T-joint (cross-sectional view)

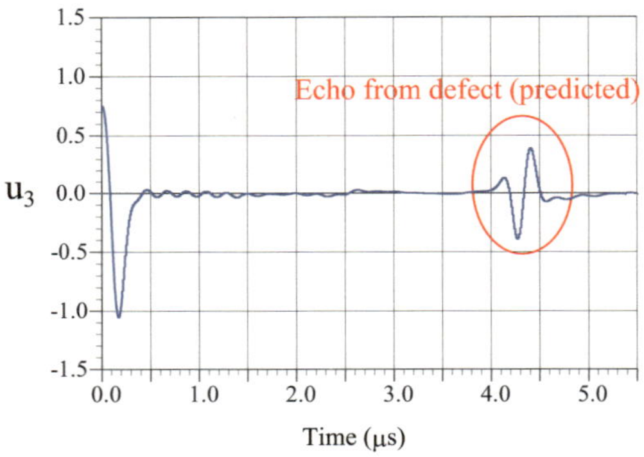

Fig. 21. Predicted echo which is observed at the array transducer.

time of this 3-D model was about 12 hours with a parallel computer system with 64 CPUs (Flat MPI).

5. Summary

For the reliable application of phased array techniques to NDT, it is essential to have thorough understanding on the properties of the phased array transducer and characteristics of radiation beam in solid. Here we show principle of electronic scanning with phased array transducers and an effective setting of the delay time to avoid grating and side lobes. In order to predict the wave field due to the array transducer, 3-D calculation tools in both frequency and time domains are introduced. From the simulation results with the MGB model in the frequency domain, characteristics about beam steering and focusing are compared between the linear and matrix array transducers. As the time domain calculation tool, the 3-D image-based FIT modeling of the UT for a welded T-joint is demonstrated.

Phased array transducer has a great potential to detect defects in real time. In the future, it is expected that the detectability of defects with the phased array UT can be improved by using above presented useful numerical tools.

6. References

Azar, L.; Shi, Y. & Wooh, S.C. (2000). Beam focusing behavior of linear phased arrays, *NDT & E International*, Vol.33, No.3, 189–198.

Drinkwater, B.W. & Wilcox, P.D. (2006). Ultrasonic arrays for non-destructive evaluation: A review, *NDT & E International*, Vol.39, No.7, 525–541.

Fellinger, P.; Marklein, R.; Langenberg, K.J. & Klaholz, S. (1995). Numerical modeling of elastic wave propagation and scattering with EFIT - elastodynamic finite integration technique, *Wave motion*, Vol.21, No.1, 47–66.

Kono, N. & Mizota H. (2011). Analysis of characteristics of grating lobes generated with Gaussian pulse excitation by ultrasonic 2-D array transducer, *NDT & E International*, Vol.44, No.6, 477–483.

Kono, N.; Baba, A. & Ehara, K. (2010). Ultrasonic multiple beam technique using single phased array probe for detection and sizing of stress corrosion cracking in austenitic welds, *Materials Evaluation*, Vol.68, No.10, 1163–1170.

Macovski, A. (1979). Ultrasonic imaging using arrays, *Proceedings of the IEEE*, Vol.67, No.4, 484–495.

Marklein, R. (1998). *Numerical Methods for the Modeling of Acoustic, Electromagnetic, Elastic, and Piezoelectric Wave Propagation Problems in the Time Domain Based on the Finite Integration Technique*, Shaker Verlag, Aachen, ISBN 3826531728. (in German)

Nakahata, K.; Schubert, F. & Köhler, B. (2011). 3-D image-based simulation for ultrasonic wave propagation in heterogeneous and anisotropic materials, *Review of Quantitative Nondestructive Evaluation*, Vol.30, pp.51-58.

Nakahata, K.; Ichikawa, S.; Saitoh, T. & Hirose, S. (2012). Acceleration of the 3D image-based FIT with an explicit parallelization approach, *Review of Quantitative Nondestructive Evaluation*, Vol.31, accepted for publication.

Park, J.S.; Song, S.J. & Kim, H.J. (2006). Calculation of radiation beam field from phased array transducers using expanded multi-Gaussian beam model, *Solid State Phenomena*, Vol.110, 163–168.

Schmerr, L.W. (1998). *Fundamentals of Ultrasonic Nondestructive Evaluation*, Plenum Press, New York, ISBN 0306457520.

Schmerr, L.W. (2000). A multigaussian ultrasonic beam model for high performance simulation on a personal computer, *Materials Evaluation*, 882–888.

Schubert, F. & Köhler, B. (2001). Three-dimensional time domain modeling of ultrasonic wave propagation in concrete in explicit consideration of aggregates and porosity, *Journal of Computational Acoustics*, Vol.9, No.4, 1543–1560.

Song, S.J. & Kim, C.H. (2002). Simulation of 3-D radiation beam patterns propagated through a planar interface from ultrasonic phased array transducers, *Ultrasonics*, Vol.40, 519–524.

Terada, K.; Miura, T. & Kikuchi, N. (1997). Digital image-based modeling applied to the homogenization analysis of composite materials, *Computational Mechanics*, Vol.20, No.4, 331–346.

Vezzetti, D.J. (1985). Propagation of bounded ultrasonic beams in anisotropic media, *Journal of Acoustical Society of America*, Vol.78, No.3, 1103–1108.

Wen, J.J & Breazeale, M.A. (1988). A diffraction beam field expressed as the superposition of Gaussian beams, *Journal of Acoustical Society of America*, Vol.83, 1752–1756.

Zhao, X. & Gang, T.(2009). Nonparaxial multi-Gaussian beam models and measurement models for phased array transducer, *Ultrasonics*, Vol.49, 126–130.

4

Goldberg's Number Influence on the Validity Domain of the Quasi-Linear Approximation of Finite Amplitude Acoustic Waves

Hassina Khelladi
University of Sciences and Technology Houari Boumediene
Algeria

1. Introduction

Nonlinear propagation occurs widely in many acoustic systems, especially in the field of medical ultrasound. Despite the widespread use of ultrasound in diagnosis and therapy, the propagation of ultrasound through biological media was modeled as a linear process for many years. The invalidity of infinitesimal acoustic assumption, at biomedical frequencies and intensities, was demonstrated by Muir and Carstensen (Muir & Carstensen, 1980). It was realized that nonlinear effects are not negligible and must therefore be taken into account in theoretical developments of ultrasound in biomedical research. Indeed, increasing the acoustic frequency or intensity in order to enhance resolution or penetration depth may alter the beam shape in a way not predicted by linear theory.

Nonlinear effects occur more strongly when ultrasound propagates through slightly dissipative liquids such as water or amniotic fluid. As in medical sonography, the full bladder or the pregnant uterus, which may be filled with amniotic fluid, is used as an acoustic window in many types of diagnoses; a special attention is given to slightly dissipative liquids where the possibility of signal distortions has several implications. However, within soft tissues, the tendency for wave distortion to occur is limited by dissipation.

In absorbing medium, nonlinear effects cannot be examined without considering dissipation. The absorption limits the generation of harmonics by decreasing their amplitudes gradually. In addition, as the absorption coefficient increases with frequency, the energy transformation towards frequencies higher than the fundamental frequency (generation of harmonics) can also lead to significant acoustic losses. Nonlinear effects create all higher harmonics from the energy at the insonation frequency, but, due to the absorption of high frequency components, only the lower harmonic orders and the fundamental remain. So, the tendency for wave distortion to occur is limited by dissipation.

Dissipation can have various origins (Sehgal & Greenleaf, 1982): viscosity (resulting from shear motions between fluid particles), thermal conduction (due to the energy loss resulting from thermal conduction between particles) or molecular relaxation (where the molecular equilibrium state is affected by the pressure variations of the acoustic wave propagation).

Nonlinear effects and dissipation are antagonistic phenomena. The nonlinearity mechanism shocks the wave by generating harmonics while dissipation increases with frequency and attenuates the harmonics resulting from nonlinear effects. The shock length l_s (Enflo & Hedberg, 2002; Naugolnykh & Ostrovosky, 1998) quantifies the influence of the nonlinear phenomena, and it is necessary to define another parameter, denoted Goldberg's number Γ (Goldberg, 1957), when dissipation is added. Γ represents the ratio of the absorption length l_a (the inverse of the absorption coefficient α and corresponds to the beginning of the old age region) to the shock length l_s at which the waveform would shock if absorption phenomena were absent:

$$\Gamma = \frac{l_a}{l_s} = \frac{k\beta M}{\alpha} \tag{1}$$

where k, M, β and α are, respectively, the wave number, the acoustic Mach number, the acoustic nonlinearity parameter and the absorption coefficient.

It should be noted that higher harmonics may turn the wave into shock state. On the other hand, dissipation attenuates higher harmonics much more than lower harmonics, thus making it more difficult for the waves to go into shock.

The dimensionless parameter Γ measures the relative importance of the nonlinear and dissipative phenomena, which are in perpetual competition. Thus, the Goldberg's number is a reliable indicator for any analysis including these two phenomena. An analysis based on the Goldberg's number is important since it is an essential step for solving general problems involving ultrasound waves of finite amplitude.

Nowadays, Tissue Harmonic Imaging (THI) or second harmonic imaging offers several advantages over conventional pulse-echo imaging. Both harmonic contrast and lateral resolution are improved in harmonic mode. Tissue Harmonic Imaging also provides a better signal to noise ratio which leads to better image quality in many applications. The major benefit of Tissue Harmonic Imaging is artifact reduction resulting in less noisy images, making cysts appear clearer and improving visualization of pathologic conditions and normal structures. Indeed, Tissue Harmonic Imaging is widely used for detecting subtle lesions (e.g., thyroid and breast) and visualizing technically- challenging patients with high body mass index.

In order to create images exclusively from the second harmonic, a theoretical review with some mathematical approximations is elaborated, in this chapter, to derive an analytical expression of the second harmonic. The performance of the simplified model of the second harmonic is interesting, as it can provide a simple, useful model for understanding phenomena in diagnostic imaging.

Despite the significant advantages offered by Tissue Harmonic Imaging, theory has been partially explained. A number of works were elaborated over recent decades. Among these, are Trivett and Van Buren (Trivett & Van Buren, 1981) work which have presented an analysis of the generated harmonics based on the generalized Burgers' equation. Significant differences in the calculated harmonic content were found by Trivett and Van Buren when compared with those obtained by Woodsum (Woodsum, 1981). No explanation was given by Trivett and Van Buren to justify their results. In an author's reply, Woodsum seemed to attribute these differences to the high number of terms retained by Trivett and Van Buren in the Fourier series.

Similarly, Haran and Cook (Haran & Cook, 1983) have used the Burgers' equation to elaborate an algorithm for calculating harmonics generation by a finite amplitude plane wave of ultrasound propagating in a lossy and nondispersive medium. Their algorithm accounts for an absorption coefficient of any desired frequency dependence. The variation effect of the absorption coefficient on the second harmonic was demonstrated in a medium similar to carbon tetrachloride. Calculations for several types of tissue and biological fluids were presented. It was shown that for some biological media having a low absorption coefficient, a significant distortion of the plane wave can be observed for large propagation ranges.

Recently, D'hooge et al. (D'hooge et al., 1999) have analyzed the nonlinear propagation effects of pulses on broadband attenuation measurements and their implications in ultrasonic tissue characterization by using a simple mathematical model based on the numerical solution, in the time domain, of the Burgers' equation. The developed model has been validated by measuring the absorption coefficient of both a tissue-mimicking phantom in vitro and a liver in vivo at several pressure amplitudes using transmission and reflection measurements, respectively.

In the present chapter, the intensity effects on the behavior of the fundamental and the generated second harmonic, by using both the numerical solution of the Burgers' equation and the analytical expressions established with the quasi-linear approximation are examined. An analysis on the validity domain of the fundamental and the second harmonic analytical expressions established with the quasi-linear approximation is elaborated. The deviations resulting from the analytical expressions established with the quasi-linear approximation and the numerical solution of the Burgers' equation are estimated. This investigation is based on Krassilnikov et al. (Krassilnikov et al., 1957) experimental results. These experimental data concern water and glycerol that correspond, respectively, to a weakly dissipative liquid approaching the characteristics of urine or amniotic fluid (Bouakaz et al., 2004) and a strongly dissipative liquid with some similarities to soft tissues.

It should be noted that in this study all derivations are developed entirely in the frequency domain, thus avoiding both the steep waveform problems and the use of FFT, which alternates between time and frequency domains. The utility of the method resides in the ease with which it can be implemented on a digital computer.

2. Theoretical formulation

The description of acoustic waves in a liquid is founded on the theory of motion of a liquid, which is considered to be continuous. In the present investigation, the viscosity and the heat conduction coefficients, although in general are functions of the state variables, are assumed to be constant. The theoretical formulation of the propagation of finite amplitude plane progressive waves in a homogeneous and dissipative liquid is elaborated in section 2.1, and the theoretical model is based on the derivation of a nonlinear partial differential equation in which the longitudinal particle velocity is a function of time and space. In section 2.2, the dimensionless Burgers' equation is presented, which is considered to be among the most exhaustively studied equations in the theory of nonlinear waves.

2.1 Basic equations

The propagation of finite amplitude plane progressive waves in a homogeneous and dissipative liquid is governed by the Burgers' equation. Here, it is assumed that the ultrasonic wave propagates in the positive z direction, and the differential change of the longitudinal particle velocity with respect to z is given by (Enflo & Hedberg, 2002; Naugolnykh & Ostrovosky, 1998):

$$\frac{\partial u(z,\tau)}{\partial z}=\frac{\beta}{c_0^2}u(z,\tau)\frac{\partial u(z,\tau)}{\partial \tau}+\frac{D}{2c_0^3}\frac{\partial^2 u(z,\tau)}{\partial \tau^2} \tag{2}$$

$D=\frac{1}{\rho_0}\left[\left(\frac{4}{3}\mu+\xi\right)+\kappa\left(\frac{1}{c_v}-\frac{1}{c_p}\right)\right]$ is the diffusivity of the sound for a thermoviscous fluid. This parameter is a function of the fluid shear viscosity μ, the fluid bulk viscosity ξ, the thermal conductivity κ, the specific heat at constant volume c_v and the specific heat at constant pressure c_p. The acoustic nonlinearity parameter $\beta=1+B/2A$ is function of the nonlinearity parameter of the medium B/A, which represents the ratio of quadratic to linear terms in the isentropic pressure-density relation (Hamilton & Blackstock, 1988; Khelladi et al., 2007, 2009). $\tau=t-z/c_0$ is the retarded time, c_0 is the infinitesimal sound speed and ρ_0 is the undisturbed density of the liquid.

The term on the left hand side of equation (2) is the linear wave propagation. The first term on the right hand side of equation (2) is the nonlinear term that accounts for quadratic nonlinearity producing cumulative effects in progressive plane wave propagation, while the second term represents the loss due to viscosity and heat conduction or any other agencies of dissipation.

Nonlinear propagation in a dissipative liquid is considered using Fourier series expansion. By assuming that the solution of equation (2) is periodic in time with period $2\pi/\omega_0$, the solution can be written as the sum of the fundamental and the generated harmonics. Thus $u(z,\tau)$ can be developed in Fourier series, with amplitudes that are functions of the spatial coordinate z:

$$u(z,\tau)=\sum_{n=1}^{+\infty}[v_n(z)\cos(n\omega_0\tau)+u_n(z)\sin(n\omega_0\tau)] \tag{3}$$

ω_0 is the characteristic angular frequency and v_n, u_n are the Fourier coefficients of the nth harmonic.

When complex notation is used, equation (3) changes to (Haran & Cook, 1983; Ngoc & Mayer, 1987):

$$u(z,\ \tau)=\sum_{n=-\infty}^{+\infty}W_n(z)\,e^{in\omega_0\tau} \tag{4}$$

The complex amplitude can be expressed as $W_n = w_n e^{in\varphi_n}$ where w_n, φ_n correspond respectively to the amplitude and the phase of the nth harmonic, and $i^2 = -1$. Note that $W_{-n}^* = W_n$, * symbolizes the complex conjugate.

For the easiest derivation, equation (4) is substituted into equation (2) (Haran & Cook, 1983; Ngoc & Mayer, 1987):

$$\frac{\partial W_n}{\partial z} = i\frac{\beta\omega_0}{c_0^2}\sum_{m=-\infty}^{+\infty}(n-m)W_m W_{n-m} - \alpha n^2 W_n \tag{5}$$

where $\alpha = \dfrac{D\omega_0^2}{2c_0^3}$

Equation (5) describes the amplitude variation of the nth harmonic in the propagation direction z. The summation over m expresses nonlinear interactions among various spectral components caused by the energy transfer process, while the other term accounts for loss due to dissipation relative to the nth harmonic.

Equation (5) is rewritten in another form (Haran & Cook, 1983; Ngoc & Mayer, 1987):

$$\frac{\partial W_n}{\partial z} = i\frac{\beta\omega_0}{c_0^2}\left[\sum_{m=1}^{n-1} m W_m W_{n-m} + \sum_{m=n}^{+\infty} n W_m W_{m-n}^*\right] - \alpha n^2 W_n \tag{6}$$

By using the real notation, knowing that $W_n = \dfrac{v_n - iu_n}{2}$ and $W_{-n} = \dfrac{v_n + iu_n}{2}$, equation (6) yields two coupled partial differential equations governing the behavior of the components v_n and u_n as a function of the spatial coordinate z (Aanonsen et al., 1984; Hamilton et al., 1985):

$$\frac{\partial v_n}{\partial z} = \frac{\beta\omega_0}{2c_0^2}\left[\sum_{m=1}^{n-1} m(u_m v_{n-m} + v_m u_{n-m}) - \sum_{m=n}^{+\infty} n(v_m u_{m-n} - u_m v_{m-n})\right] - \alpha n^2 v_n \tag{7}$$

$$\frac{\partial u_n}{\partial z} = \frac{\beta\omega_0}{2c_0^2}\left[\sum_{m=1}^{n-1} m(u_m u_{n-m} - v_m v_{n-m}) - \sum_{m=n}^{+\infty} n(u_m u_{m-n} + v_m v_{m-n})\right] - \alpha n^2 u_n \tag{8}$$

For a sinusoidal source condition, $u(0,\tau) = u_0 \sin(\omega_0 \tau)$ (Aanonsen et al., 1984; Hamilton et al., 1985; Hedberg, 1999; Menounou & Blackstock, 2004), equation (3) becomes:

$$u(z,\tau) = \sum_{n=1}^{+\infty} u_n(z)\sin(n\omega_0\tau) \tag{9}$$

Equation (8) is then written more simply as:

$$\frac{\partial u_n}{\partial z} = \frac{\beta\omega_0}{2c_0^2}\left[\sum_{m=1}^{n-1} m u_m u_{n-m} - \sum_{m=n}^{+\infty} n u_m u_{m-n}\right] - \alpha n^2 u_n \tag{10}$$

The incremental change of the particle velocity can be approximated by the first order truncated power series (Haran & Cook, 1983; Ngoc et al., 1987):

$$u(z+\Delta z,t)=u(z,t)+\frac{\partial u(z,t)}{\partial z}\Delta z \tag{11}$$

By combining equations (10) and (11), an iterative description of finite amplitude plane wave propagation in a homogeneous and dissipative liquid, is obtained:

$$u_n(z+\Delta z)=u_n(z)+\frac{\beta\omega_0}{2c_0^2}\left[\sum_{m=1}^{n-1} m u_m(z)u_{n-m}(z)-\sum_{m=n}^{+\infty} n u_m(z)u_{m-n}(z)\right]\Delta z-\alpha n^2 u_n(z)\Delta z \tag{12}$$

The first summation term on the right hand side of equation (12) represents the contribution of lower order harmonics to the nth harmonic, while the second one is associated with the contribution of higher order harmonics. According to the sign of each contribution the nth harmonic energy can be enhanced or decreased. The last term in this equation represents losses undergone by the nth harmonic.

Generally, the absorption coefficient α depends on the propagation medium characteristics and the insonation frequency. For the considered viscous fluids, this frequency dependence is quadratic with frequency and can be represented by (Smith & Beyer, 1948; Willard, 1941):

$$\alpha=\alpha_0 f^2 \tag{13}$$

where α_0 depends upon the nature of the liquid, and $f=\omega_0/2\pi$ is the insonation frequency.

Therefore the Goldberg's number Γ, increases with the amplitude of excitation and decreases with frequency.

Equation (12) becomes:

$$u_n(z+\Delta z)=u_n(z)+\frac{\beta\omega_0}{2c_0^2}\left[\sum_{m=1}^{n-1} m u_m(z)u_{n-m}(z)-\sum_{m=n}^{+\infty} n u_m(z)u_{m-n}(z)\right]\Delta z-\alpha_n u_n(z)\Delta z \tag{14}$$

where $\alpha_n=\alpha_0 n^2 f^2$

Equation (14) allows the determination of the nth harmonic amplitude at the location $z+\Delta z$ in terms of all harmonics at the preceding spatial coordinate z. This derivation requires an appropriate truncation of the finite series on the right hand side of equation (14) to ensure a negligibly small error in the highest harmonic of interest and to maintain some acceptable accuracy.

In the hypothesis of the quasi-linear approximation, all the harmonics of higher order than two can be neglected in the numerical solution of the Burgers' equation, so equation (14) changes to:

$$\begin{cases} \dfrac{\partial u_1(z)}{\partial z} = -\dfrac{\beta\omega_0}{2c_0^2} u_1(z) u_2(z) - \alpha_1 u_1(z) \\ \dfrac{\partial u_2(z)}{\partial z} = \dfrac{\beta\omega_0}{2c_0^2} u_1^2(z) - \alpha_2 u_2(z) \end{cases} \tag{15}$$

where $\alpha_1 = \alpha_0 f^2$ and $\alpha_2 = 4\alpha_0 f^2 = 4\alpha_1$ denote the absorption coefficients of the fundamental and the second harmonic, respectively.

In many situations, the experimental studies are based on pressure measurements. Knowing that the ratio of the nth harmonic pressure to the associated particle velocity is given by $p_n(z,t) = \rho_0 c_0 u_n(z,t)$ (Germain et al., 1989); equation (15) is rewritten as:

$$\begin{cases} \dfrac{\partial p_1(z)}{\partial z} = -\dfrac{\beta\omega_0}{2\rho_0 c_0^3} p_1(z) p_2(z) - \alpha_1 p_1(z) \\ \dfrac{\partial p_2(z)}{\partial z} = \dfrac{\beta\omega_0}{2\rho_0 c_0^3} p_1^2(z) - \alpha_2 p_2(z) \end{cases} \tag{16}$$

If $p_2(z) << \dfrac{2P_0}{\Gamma}$, then $\dfrac{\beta\omega_0}{2\rho_0 c_0^3} p_1(z) p_2(z)$ can be neglected comparatively to $\alpha_1 p_1(z)$. The acoustic pressure of the fundamental can be written as (Gong et al., 1989; Thuras et al., 1935):

$$p_1(z) = P_0 e^{-\alpha_1 z} \tag{17}$$

where P_0 is the characteristic pressure amplitude (the value of the fundamental pressure at $z = 0$).

Equation (16) becomes:

$$\frac{\partial p_2(z)}{\partial z} = h P_0^2 e^{-2\alpha_1 z} - \alpha_2 p_2(z) \tag{18}$$

with $h = \dfrac{\beta\omega_0}{2\rho_0 c_0^3}$

The solution of equation (18) is easily obtained. Knowing that for $z = 0$ $p_2(0) = 0$, the acoustic pressure of the second harmonic component can be expressed as (Cobb, 1983; Thuras et al., 1935):

$$p_2(z) = h P_0^2 \left(\frac{e^{-\alpha_2 z} - e^{-2\alpha_1 z}}{2\alpha_1 - \alpha_2} \right) \tag{19}$$

Moreover, if the term $(\alpha_2 - 2\alpha_1) z << 1$, an approximation of equation (19) can be made (Bjørnø, 2002; Cobb, 1983; Zhang et al.,1991):

$$p_2(z) = h P_0^2 z e^{-(\alpha_1 + \alpha_2/2) z} \tag{20}$$

2.2 Dimensionless equations

For theoretical analysis as well as for numerical implementation, it is more convenient to define dimensionless variables, by using the characteristic particle velocity U_0, the characteristic time $1/\omega_0$ and the lossless plane wave shock formation length l_s:

$$U = \frac{u}{U_0}, \ \theta = \omega_0 \tau \ \text{and} \ \sigma = \frac{z}{l_s} \tag{21}$$

where U, θ and σ are, respectively, the dimensionless longitudinal particle velocity, the dimensionless time and the dimensionless propagation path.

Insertion of equation (21) into the Burgers' equation (equation (2)), gives the dimensionless equation (Bjørnø, 2002; Fenlon, 1971; Hedberg, 1994):

$$\frac{\partial U(\sigma,\theta)}{\partial \sigma} = U(\sigma,\theta)\frac{\partial U(\sigma,\theta)}{\partial \theta} + \Gamma^{-1}\frac{\partial^2 U(\sigma,\theta)}{\partial \theta^2} \tag{22}$$

The dimensionless amplitude of the nth harmonic at the dimensionless location $\sigma + \Delta\sigma$ in terms of all harmonics at the preceding dimensionless location σ can be written as:

$$U_n(\sigma + \Delta\sigma) = U_n(\sigma) + \frac{1}{2}\left[\sum_{m=1}^{n-1} m U_m(\sigma) U_{n-m}(\sigma) - \sum_{m=n}^{+\infty} n U_m(\sigma) U_{m-n}(\sigma)\right]\Delta\sigma - n^2\Gamma^{-1}U_n(\sigma)\Delta\sigma \tag{23}$$

With this dimensionless notation, the acoustic pressure of the fundamental and the second harmonic can be expressed as:

$$p_1(\sigma) = P_0 e^{-\alpha_1 l_s \sigma} \tag{24}$$

$$p_2(\sigma) = \frac{1}{2}P_0\left(\frac{e^{-\alpha_2 l_s \sigma} - e^{-2\alpha_1 l_s \sigma}}{(2\alpha_1 - \alpha_2)\ l_s}\right) \tag{25}$$

In the case of $(\alpha_2 - 2\alpha_1)l_s\sigma << 1$, equation (25) becomes:

$$p_2(\sigma) = \frac{1}{2}P_0 \sigma e^{-(\alpha_1 + \alpha_2/2)l_s\sigma} \tag{26}$$

3. Numerical experiments and discussions

Krassilnikov et al. (Krassilnikov et al., 1957) experimental data for water and for glycerol are used in order to simulate the amplitude of the first two harmonics, by using both the numerical solution of the Burgers' equation (equation (23)) and the analytical expressions established with the quasi-linear approximation (equations (24), (25) and (26)). Table 1 lists material properties.

According to Krassilnikov et al. (Krassilnikov et al., 1957) experimental work, the absorption coefficient is a quadratic function of frequency. The absorption coefficient is that obtained from an infinitesimal acoustic excitation, even though the acoustic intensity increases. In the case of water $\alpha_0 = 0.23\ 10^{-13}\ Np.\,m^{-1}.\,Hz^{-2}$ and for glycerol $\alpha_0 = 26\ 10^{-13}\ Np.\,m^{-1}.\,Hz^{-2}$ (Krassilnikov et al., 1957).

Nonlinear effects occur more strongly when ultrasound propagates through slightly dissipative liquids, so a special attention is given to a propagation medium characterized by a Goldberg number greater than unity. In this case, when the waveform approaches the shock length, nonlinear effects dominate dissipation phenomena. The amplitude of the generated harmonics increases at the expense of the fundamental component. After the shock length, absorption limits the generation of harmonics by decreasing theirs amplitudes gradually with the propagation path. For this reason, all the simulations of the first two harmonics are plotted as a function of the dimensionless location σ up to unity and for several values of the acoustic intensity. Moreover, all the shock lengths for several intensities are greater than $19.8\ \text{cm}$ (Table 2). As in biomedical diagnosis the region of interest (ROI) is about $20\ \text{cm}$, it is absolutely useless to explore beyond $\sigma = 1$ and the selected range $0 \le \sigma \le 1$ is amply appropriate for this kind of investigation.

It should be pointed out that the shock length l_s depends on the medium characteristics ρ_0, c_0, β and on the external parameters such as the insonation frequency and the amplitude of excitation. In this study, the insonation frequency is fixed at $2\ \text{MHz}$, thus the shock length for a given medium will depend only upon the amplitude of excitation.

Among all the configurations presented in this study, including various acoustic intensities and two analyzed mediums, only one case is sensitive in biomedical diagnostic and must be analyzed with extreme caution. Indeed, a more favorable situation where nonlinear effects have sufficient time to be entirely established corresponds to the case of water, for which the acoustic intensity is equal to $4.7\ \text{W/cm}^2$ and as a consequence a shock length equal to $19.8\ \text{cm}$. As the generation of harmonics occurs while moving away from the source and approaching the shock length, the greatest signal distortion may occur in the range of interest. Moreover, the irradiation of living tissue with shock waves in diagnostic processes appears risky since the damage and exposure criteria for these radiations have not been delineated.

It should be noted that all the simulations are made with intensities of $0.2 - 4.7\ \text{W/cm}^2$ (Table 2), which correspond to breast lesion diagnosis (Nightingale et al., 1999).

It will be stated by the derivation of the Goldberg number that water surpasses any tissue in its ability to produce extremely distorted waveforms even at relatively low intensity. So, a special attention is given to this liquid where the possibility of distortion occurring has several implications. Indeed, water can generate extreme waveform distortion compared to glycerol, as indicated by the Goldberg's number for water, which is 200 times larger than that of glycerol for an acoustic intensity of about $0.2\ \text{W/cm}^2$ (Table 2).

Parameters	Temperature (°C)	Density $\rho_0(kg/m^3)$	Sound velocity $C_0(m/s)$	Acoustic nonlinearity parameter β
Water	20	998	1481	3.48
Glycerol	20	1260	1980	5.4

Table 1. Material properties.

Water			
Temperature (°C)	Intensity (W/cm²)	Shock length (m)	Goldberg Number
20	0.34	0.739	14.7
20	2	0.304	35.7
20	4.7	0.198	54.9
Glycerol			
Temperature (°C)	Intensity (W/cm²)	Shock length (m)	Goldberg Number
22	0.2	1.443	0.07
21	2.5	0.408	0.24
19	4.5	0.304	0.32

Table 2. Goldberg's number for water and glycerol with intensities of $0.2 - 4.7\,\mathrm{W/cm^2}$ and an insonation frequency of $2\,\mathrm{MHz}$.

Initially, the ultrasonic wave is taken to be purely sinusoidal with a frequency of $2\,\mathrm{MHz}$ in the two considered media. Only the fundamental wave exists at the starting location $\sigma = 0$, and the other harmonic modes are generated as the wave propagates from the source. Through an iterative method, the value of the Goldberg number is inserted into the Burgers' equation in order to determine its numerical solution (Table 2). 40 harmonics are retained to simulate the numerical solution of the Burgers' equation which is considered, in the deviation calculus, as an exact solution.

For a better readability and interpretation of the obtained numerical data, a symbol with a defined shape and type is inserted on the graphic layout of the analyzed functions. All the following simulations exploit equations (23), (24), (25) and (26) corresponding respectively to the numerical solution of the Burgers' equation, the quasi-linear approximation of the acoustic pressure of the fundamental, the quasi-linear approximation of the second harmonic and the quasi-linear approximation of the approximated second harmonic.

The simulations relating to water and glycerol are represented in all figures (a) and (b), respectively.

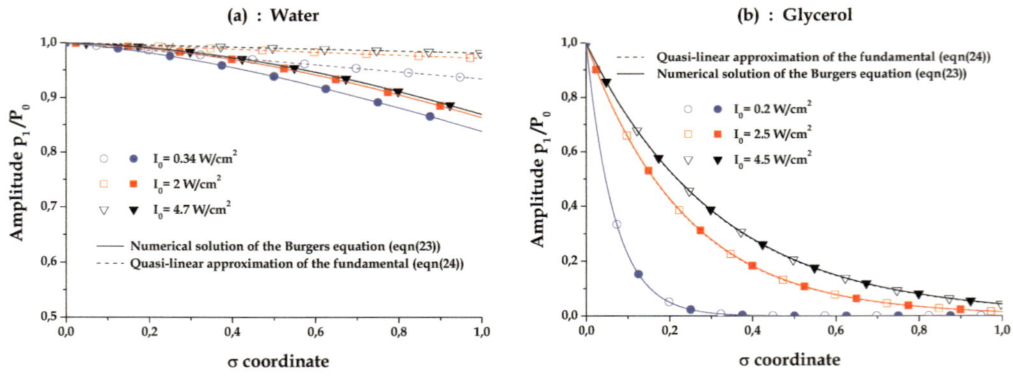

Fig. 1. Pressure amplitude p_1/P_0 versus the σ coordinate.

The amplitude p_1/P_0 (figure 1a, figure 1b) and the amplitude p_2/P_0 (figure 2a, figure 2b) increase with Γ (Table 2). So, the effect of the increased acoustic intensity is to enhance the amplitude of the fundamental and also that of the second harmonic.

In the hypothesis of linear acoustics, increasing the absorption coefficient leads systematically to a decrease of the wave amplitude. The finite amplitude waves do not obey to the same principle because nonlinear effects and dissipation are two phenomena in perpetual contest. The interplay between these two phenomena developed along the propagation path is not simply an additive effect as normally assumed in linear acoustics. Therefore, a measure of whether nonlinear effects or absorption will prevail is the Goldberg's number Γ. The larger Γ is, the more nonlinear effects dominate. Whereas for values of $\Gamma < 1$, absorption is so strong that no significant nonlinear effects occur. Thus the calculation of the Goldberg's number is required to quantify the amplitude of the generated harmonics.

By taking water as an example, the most significant amplitude of the generated harmonic, for various values of intensity, corresponds to the highest Goldberg's number (figure 2a). This is in perfect agreement with physical phenomena that take place in the analyzed medium. Indeed, a high Goldberg number corresponds to a predominance of the nonlinearity phenomenon as compared to dissipation, which represents the main factor of amplitude decrease. This situation is also apparent for glycerol (figure 2b).

For a slightly dissipative liquid, it can be seen that the second harmonic component grows cumulatively with increasing the normalized length σ at the expense of the fundamental (figure 1a, figure 2a). Its growth begins to taper off at the location of the initial shock formation, beyond this location the curves decay as expected. So, the nonlinearity mechanism is a bridge that facilitates the energy exchange among different harmonic modes. An increase of the Goldberg's number enhances the transfer of energy from the fundamental to higher harmonics and between harmonics themselves. Thus, the generated harmonics can only follow the evolution of the fundamental which gives them birth.

However, for a strongly dissipative medium, the absorption is so strong that significant nonlinear effects do not occur. Indeed, the old age region begins at a range smaller than the shock length and once nonlinear effects take place, absorption dominates the behavior of the fundamental and the generated harmonic (figure 1b, figure 2b). In absorbing media, the exchange of energy is more complicated, because absorption diminishes amplitude with increasing the propagation path and acts as a low pass filter that reduces the energy of higher harmonics (figure 2b).

The evaluation of the relative deviation, for each analytical expression in relation to the numerical solution of the Burgers' equation, is carried out in the following way:

$$Deviation(\%) = \frac{|\text{analytical expression- numerical solution(Burgers)}|}{\text{numerical solution(Burgers)}} \times 100 \qquad (27)$$

The relative deviation, on the selected range, of the analytical expression of the fundamental component (equation (24)) in relation to the numerical solution of the Burgers' equation is less than 4% for glycerol (figure 3b).

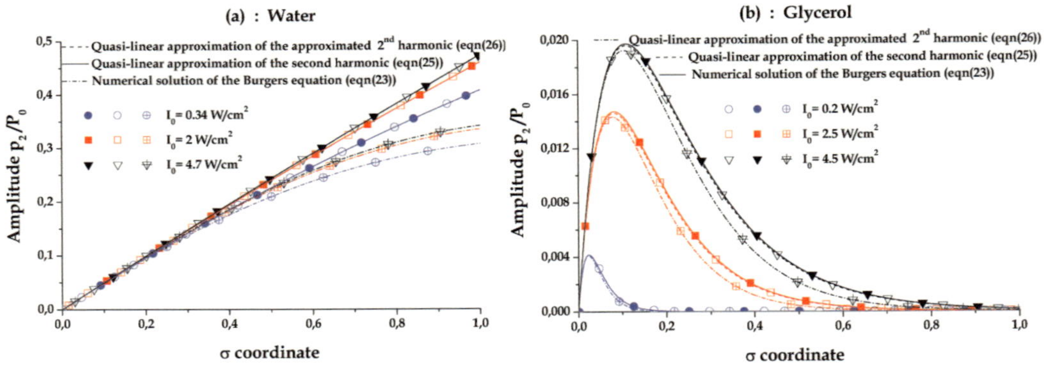

Fig. 2. Pressure amplitude p_2/P_0 versus the σ coordinate.

Thus, for a strongly dissipative liquid, equation (24) can be considered as a good approximation of equation (23). In fact, in this case the Goldberg's number is lower than unity (Table 2); then dissipation becomes important and dominates nonlinear effects.

As for water, the relative deviation of the analytical expression of the fundamental component (equation (24)) in relation to the numerical solution of the Burgers' equation is about 12% at $\sigma = 1$ (figure 3a). It should be noted that for water, the deviations increase with Γ (figure 3a). Indeed, in this case nonlinear effects become important ($p_2(z)$ much greater than $2P_0/\Gamma$) and the analytical expression of the fundamental established with the quasi-linear approximation is not valid.

For glycerol, the relative deviation of the analytical expression of the second harmonic (equation (25)) in relation to the numerical solution of the Burgers' equation is much weaker than that resulting from equation (26) (figure 4b). As an example, for $\sigma \approx 0.1$ the deviation obtained from equation (25) is lower than 1% , and that produced by equation (26) can reach 40% .

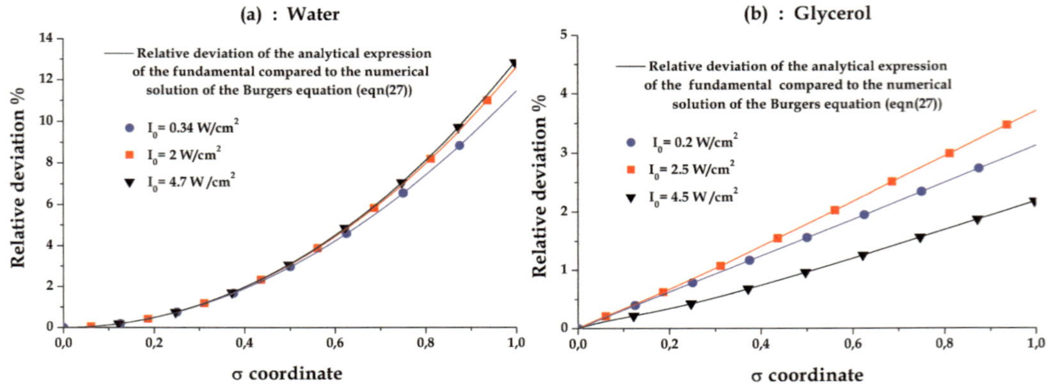

Fig. 3. Relative deviation of the analytical expression of the fundamental compared to the numerical solution of the Burgers' equation versus the σ coordinate.

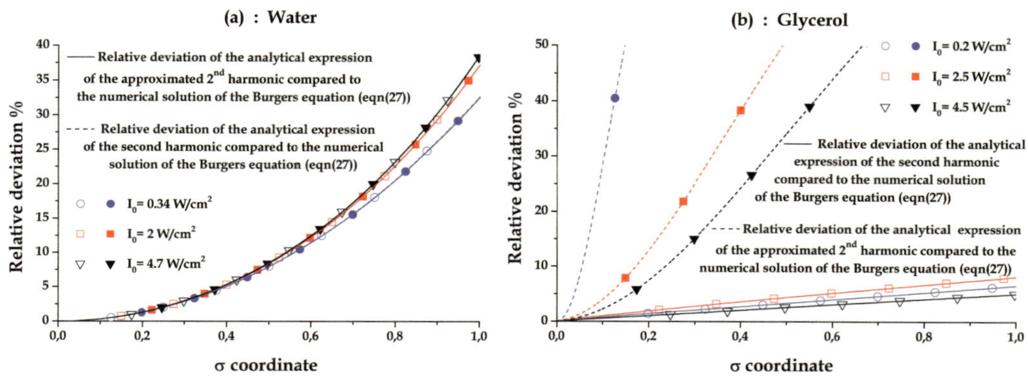

Fig. 4. Relative deviation of the respectively analytical expression of the second harmonic and the approximated second harmonic compared to the numerical solution of the Burgers' equation versus the σ coordinate.

So, for a strongly dissipative liquid, equation (25) is a good approximation of the numerical solution of the Burgers' equation (figure 4b). But, the equivalence of equations (25) and (26) is not checked (figure 4b). Indeed, equation (26) is a good approximation of equation (25) only if $(\alpha_2 - 2\alpha_1)l_s\sigma$ is weak comparatively to unity.

In the case of water, the relative deviation of the analytical expression of the second harmonic (equation (25)) in relation to the numerical solution of the Burgers' equation is about 40% at $\sigma = 1$ (figure 4a). In fact, the determination of the analytical expression of the second harmonic is based on the analytical expression of the fundamental. As in the case of a slightly dissipative medium a noticeable deviation between $p_1(\sigma)$ and the numerical solution of the Burgers' equation is observed, the deviation of the analytical expression of the second harmonic in relation to the numerical solution of the Burgers' equation becomes more significant. These deviations increase with Γ (figure 4a). Moreover in this case, $(\alpha_2 - 2\alpha_1)l_s\sigma$ is weak comparatively to unity and equations (25) and (26) are equivalent. Consequently, the preceding comments are also applicable for the analytical expression of the approximated second harmonic (equation (26)) (figure 4a).

According to this study, all these obtained solutions are valid, since the measurement is made near the source; otherwise some assumptions must be taken into account in the analysis of the propagation of finite amplitude acoustic waves in liquids. In addition, the analytical expressions precision depends essentially on the Goldberg's number value.

Moreover, for a strongly dissipative medium, the analytical expressions of the fundamental and second harmonic (equations (24) and (25)) can constitute a good approximation of the numerical solution of the Burgers' equation.

For a slightly dissipative medium, the analytical expressions established show discrepancies when compared to the numerical solution of the Burgers' equation. Indeed, equation (24) assumes that the differential variation of the fundamental component with respect to the spatial coordinate is only proportional to the product of the absorption coefficient and the acoustic pressure of the fundamental ($\frac{\partial p_1(z)}{\partial z} = -\alpha_1 p_1(z)$). This hypothesis is not always checked (equation (16)).

As mentioned at the beginning of this chapter, the performance of the simplified model (equation (26)) is interesting, as it can provide a simple, useful model for understanding phenomena in diagnostic imaging. In fact, tissue harmonic imaging offers several unique advantages over conventional imaging. The greater clarity, contrast and details of the harmonic images are evident and have been quantitatively verified, like the ability to identify suspected cysts... Despite the significant advantages offered by harmonic imaging, theory has been only partially explained. According to the theoretical development established in this chapter, equation (26) is valid only if $p_2(\sigma) << 2P_0/\Gamma$ and $(\alpha_2 - 2\alpha_1)l_s\sigma << 1$. Not taking into account these assumptions can generate erroneous numerical results.

On the other hand, as the finite amplitude method is based on pressure measurements of the finite amplitude wave distortion during its propagation, the analytical expressions of the fundamental (equation (24)), the second harmonic (equation (25)) and the approximated second harmonic (equation (26)) lead also to the measurement of the acoustic nonlinearity parameter β. However, this method necessitates an accurate model taking into account diffraction effects (Labat et al., 2000; Gong et al., 1989; Zhang et al., 1991). The omission of this phenomenon can explain the discrepancies observed of the nonlinearity parameter values measured by the finite amplitude method compared to those achieved by the thermodynamic method (Law et al. 1983; Plantier et al., 2002; Sehgal et al., 1984; Zhang & Dunn, 1991). The latter is potentially very accurate. The major advantage of the thermodynamic method is that it does not depend on the characteristics of the acoustic field (Khelladi et al., 2007, 2009).

4. Conclusion

The validity domain of the fundamental and the second harmonic analytical expressions established with the quasi-linear approximation can be preset only on the derivation of the Goldberg's number, which can be considered as a reliable indicator for any analysis incorporating nonlinear effects and dissipation.

The obtained numerical results illustrate that the analytical expressions of the fundamental and the second harmonic established with the quasi-linear approximation provide a good approximation of the numerical solution of the Burgers' equation for a propagation medium characterized by a Goldberg number that is small compared to unity.

In the other hand, for a propagation medium characterized by a Goldberg number greater than unity, the analytical expressions of the fundamental and the second harmonic already established with the quasi-linear approximation are not checked and must be redefined.

For that purpose, future studies will concentrate on a new mathematical formulation of the fundamental and second harmonic for a propagation medium characterized by a Goldberg number that is large compared to unity.

5. References

Aanonsen, S. I., Barkve, T., Tjøtta, J. N., & Tjøtta, S. (1984). Distortion and harmonic generation in the nearfield of a finite amplitude sound beam. *Journal of the Acoustical Society of America*, Vol. 75, pp. 749-768

Bjørnø, L. (2002). Forty years of nonlinear ultrasound. *Ultrasonics*, Vol. 40, pp. 11-17

Bouakaz, A., Merks, E., Lancée, C., & Bom, N. (2004). Noninvasive bladder volume measurements based on nonlinear wave distortion. *Ultrasound in Medicine and Biology*, Vol. 30, pp. 469-476

Cobb, W. N. (1983). Finite amplitude method for the determination of the acoustic nonlinearity parameter B/A. *Journal of the Acoustical Society of America*, Vol. 73, pp. 1525-1531

D'hooge, J., Bijnens, B., Nuyts, J., Gorce, J. M., Friboulet, D., Thoen, J., Van de Werf, F., & Suetens, P. (1999). Nonlinear Propagation effects on broadband attenuation measurements and its implications for ultrasonic tissue characterization. *Journal of the Acoustical Society of America*, Vol. 106, pp. 1126-1133

Enflo, B. O., & Hedberg, C. M. (2002). *Theory of Nonlinear Acoustics in Fluids*, Kluwer Academic Publishers, ISBN 1-4020-0572-5, the Netherlands

Fenlon, F. H. (1971). A recursive procedure for computing the nonlinear spectral interactions of progressive finite-amplitude waves in nondispersive fluids. *Journal of the Acoustical Society of America*, Vol. 50, pp. 1299-1312

Germain, L., Jacques, R., & Cheeke, J. D. N. (1989). Acoustics microscopy applied to nonlinear characterization of biological media. *Journal of the Acoustical Society of America*, Vol. 86, pp. 1560-1565

Goldberg, Z. A. (1957). On the propagation of plane waves of finite amplitude. *Soviet Physics Acoustics*, Vol. 3, pp. 340-347

Gong, X. F., Zhu, Z. M., Shi, T., & Huang, J. H. (1989). Determination of the acoustic nonlinearity parameter in biological media using FAIS and ITD methods. *Journal of the Acoustical Society of America*, Vol. 86, pp. 1-5

Hamilton, M. F., Tjøtta, J. N., & Tjøtta, S. (1985). Nonlinear effects in the farfield of a directive sound source. *Journal of the Acoustical Society of America*, Vol. 78, pp. 202-216

Hamilton, M. F., & Blackstock, D. T. (1988). On the coefficient of nonlinearity β in nonlinear acoustics. *Journal of the Acoustical Society of America*, Vol. 83, pp. 74-77

Haran, M. E., & Cook, B. D. (1983). Distortion of finite amplitude ultrasound in lossy media. *Journal of the Acoustical Society of America*, Vol. 73, pp. 774-779

Hedberg, C. (1994). Nonlinear propagation through a fluid of waves originating from a biharmonic sound source. *Journal of the Acoustical Society of America*, Vol. 96. pp. 1821-1828

Hedberg, C. (1999). Multifrequency plane, nonlinear, and dissipative waves at arbitrary distances. *Journal of the Acoustical Society of America*, Vol. 106, pp. 3150-3155

Labat, V., Remenieras, J. P., Bou Matar, O., Ouahabi, A., & Patat, F. (2000). Harmonic propagation of finite amplitude sound beams: experimental determination of the nonlinearity parameter B/A. *Ultrasonics*, Vol. 38, pp. 292-296

Khelladi H., Plantier F., Daridon J. L., & Djelouah H. (2007). An experimental Study about the Combined Effects of the Temperature and the Static Pressure on the Nonlinearity Parameter B/A in a Weakly Dissipative Liquid, *Proceedings International Congress on Ultrasonics*, Vienna, April 9 – 13, 2007

Khelladi H., Plantier F., Daridon J. L., & Djelouah H. (2009). Measurement under High Pressure of the Nonlinearity Parameter B/A in Glycerol at Various Temperatures. *Ultrasonics*, Vol. 49, pp. 668-675

Krassilnikov, V. A., Shklovskaya-Kordy, V. V., & Zarembo, L. K. (1957). On the propagation of ultrasonic waves of finite amplitude in liquids. *Journal of the Acoustical Society of America*, Vol. 29, pp. 642-647

Law, W. K., Frizzell, L. A. & Dunn, F. (1983). Comparison of thermodynamic and finite amplitude methods of B/A measurement in biological materials. *Journal of the Acoustical Society of America*, Vol. 74, pp.1295-1297

Menounou, P., & Blackstock, D. T. (2004). A New method to predict the evolution of the power spectral density for a finite amplitude sound wave. *Journal of the Acoustical Society of America*, Vol. 115, pp. 567-580

Muir, T. G., & Carstensen, E. L. (1980). Prediction of nonlinear acoustic effects at biomedical frequencies and intensities. *Ultrasound in Medicine and Biology*, Vol. 6, pp. 345-357.

Naugolnykh, K., & Ostrovosky, L. (1998). *Nonlinear Wave Processes in Acoustics*, Cambridge University Press, ISBN 0-521-39984-X, United States of America

Ngoc, D. K., King, K. R., & Mayer W. G. (1987). A numerical model for nonlinear and attenuative propagation and reflection of an ultrasonic bounded beam. *Journal of the Acoustical Society of America*, Vol. 81, pp. 874-880

Nightingale, K. R., Kornguth, P. J., & Trahey, G. E. (1999). The use of acoustic streaming in breast lesion diagnosis: clinical study. *Ultrasound in Medicine and Biology*, Vol. 25, pp. 75-87.

Plantier, F., Daridon, J. L., Lagourette, B. (2002). Measurement of the B/A nonlinearity parameter under high pressure: application to water. *Journal of the Acoustical Society of America*, Vol. 111, pp. 707-715

Sehgal, C. M., & Greenleaf, J. F. (1982). Ultrasonic absorption and dispersion in biological media: a postulated model. *Journal of the Acoustical Society of America*, Vol. 72, pp. 1711-1718.

Sehgal, C. M., Bahn, R. C., Greenleaf, J. F. (1984). Measurement of the acoustic nonlinearity parameter B/A in human tissues by a thermodynamic method. *Journal of the Acoustical Society of America*, Vol. 76, pp. 1023-1029

Smith, M. C., & Beyer, R. T. (1948). Ultrasonic absorption in water in the temperature range 0°-80°C. *Journal of the Acoustical Society of America*, Vol. 20, pp. 608-610

Thuras, A. L., Jenkins, R. T., & O'Neil, H. T. (1935). Extraneous frequency generated in air carrying intense sound waves. *Journal of the Acoustical Society of America*, Vol. 6, pp. 173-180

Trivett, D. H., & Van Buren, A. L. (1981). Propagation of plane, cylindrical and spherical finite amplitude waves. *Journal of the Acoustical Society of America*, Vol. 69, pp. 943-949

Willard, G. W. (1941). Ultrasonic absorption and velocity measurement in numerous liquids. *Journal of the Acoustical Society of America*, Vol. 12, pp. 438-448

Woodsum, H. C. (1981). Author's reply. *Journal of Sound and Vibration*, Vol. 76, pp. 297-298

Zhang, J., Dunn, F. (1991). A small volume thermodynamic system for B/A measurement. *Journal of the Acoustical Society of America*, Vol. 89, pp. 73-79

Zhang, J., Kuhlenschimdt, M., & Dunn, F. (1991). Influence of structural factors of biological media on the acoustic nonlinearity parameter B/A. *Journal of the Acoustical Society of America*, Vol. 89, pp. 80-91

5

Ultrasonic Waves on Gas Hydrates Experiments

Gaowei Hu and Yuguang Ye
Qingdao Institute of Marine Geology
China

1. Introduction

In this chapter, the acoustic properties of gas hydrate-bearing sediments are investigated experimentally. The flat-plate transducers and a new kind of bender elements are developed to measure both compressional wave velocity (Vp) and shear wave velocity (Vs) of hydrated consolidated sediments and hydrated unconsolidated sediments, respectively. The main purpose is to construct a relation between gas hydrate saturation and acoustic velocities of the hydrate-bearing sediments, with which we can give suggestions on the usage of various velocity-models in field gas hydrate explorations.

Gas hydrates, or clathrates, are ice-like crystalline solids composed of water molecules surrounding gas molecules (usually methane) under certain pressure and temperature conditions [Sloan, 1998]. In recent years, gas hydrates have been widely studied because of their potential as a future energy resource [Kvenvolden, 1998; Milkov and Sassen, 2003], their important role in the global carbon cycle and global warming [Dickens, 2003; Dickens, 2004], and their potential as a geotechnical hazard [Brown et al., 2006; Pecher et al., 2008]. To assess the impact of gas hydrates within these areas of interest, an understanding of their distribution within the seabed and their relationship with the host sediment is essential and helpful.

Seismic techniques have been widely used for mapping and quantifying gas hydrates in oceanic sediments [Shipley et al., 1979; Holbrook et al., 1996; Carcione and Gei, 2004]. In general, gas hydrates exhibit relatively high elastic velocities (both Vp and Vs), compared to the pore-filling fluids; therefore, the velocity of gas hydrate-bearing sediments is usually elevated [Stoll, 1974; Tucholke et al., 1977]. To quantify the amount of gas hydrate or to infer the physical properties of gas hydrate-bearing sediments, an understanding of the relationship between the amount of gas hydrate in the pore space of sediments and the elastic velocities is needed.

Two different approaches were used to relate the hydrate saturation and velocity in oceanic sediments: (1) empirical methods including Wyllie's time average [Wyllie et al., 1958; Pearson et al., 1983], Wood's equation [Wood, 1941] and weighted combinations of the Wyllie's time average and Wood's equation [Lee et al., 1996], and (2) physics-based models, such as the effective medium theory (EMT) [Helgerud et al., 1999; Dvorkin and Prasad, 1999] and the Biot-Gassmann theory modified by Lee (BGTL) [Lee, 2002a, 2002b, 2003]. However, the gas hydrate volumes within sediments estimated with these approaches are quite different. Chand et al. [2004] made a comparison of four current models, i.e., the WE (Weighted Equation) [Lee et al., 1996], the EMT, the three-phase Biot theory (TPB) [Carcione and Tinivella, 2000; Gei and Carcione, 2003] and the differential effective medium theory (DEM) [Jakobsen et al., 2000], in predicting hydrate saturation with field data sets obtained from ODP Leg 164 on Blake Ridge, and from the Mallik 2L-38 well, Mackenzie Delta, Canada. The results show that three of the models predict consistent hydrate saturation of 60-80% for the Mallik 2L-38 well, but the EMT model predicts 20 per cent higher. For the clay-rich sediments of Blake Ridge, the DEM, EMT and WE models predict 10-20% hydrate saturation, which is similar to the result inferred from resistivity data, but lower than the result predicted by the TPB model. Ojha and Sain [2008] estimated the saturation of gas hydrate at Makran accretionary prism using the BGTL and EMT models. The BGTL model shows hydrate saturation of 7-9%, but the EMT model predicts the saturation of gas hydrate as 14-33%. Apparently, it is important to validate these elastic velocity models with data obtained from synthesized hydrate-bearing sediment in the laboratory.

The relationship between acoustic velocities and hydrate saturation has been studied in laboratory by several researchers. Although the experimental results may not automatically be applied to the seismic frequencies of field data, however, as the general variability of acoustic velocities with the variable pore fluid is similar with that of seismic velocities [Sothcott et al., 2000], the experiments can provide some basic geophysical parameters for the exploration of gas hydrate reservoirs.

2. Ultrasonic waves in hydrate-bearing consolidated sediments

In this section, acoustic properties of gas hydrate-bearing consolidated sediments are investigated experimentally. Gas hydrate was formed and subsequently dissociated in consolidated sediments. In the whole process, ultrasonic methods and Time Domain Reflectometry (TDR) are simultaneously used to measure the acoustic properties and hydrate saturations of the host sediments, respectively. The whole experimental processes and results are presented here.

2.1 Methods

It's needed to keep a suitable circumstance, e.g. a certain high pressure and low temperature, for measuring parameters of hydrate-bearing sediments. Thus, ultrasonic method and TDR technique used in this study are a little different from their conventional ways because they must sustain pressure during the measurements. A characteristic of our methods is that real-time measurements of both hydrate saturation and acoustic velocities were conducted in one system.

2.1.1 Ultrasonic method

P wave and S wave velocities were measured by transmission using two transducers (0.5 MHz frequency) placed at each end of the cylindrical core (Fig. 1). Signals were digitized by a CompuScope card from GaGe Applied Technologies. Because the CompuScope 14100 is a 14 bit, 50 million samples per second dual-channel waveform digitizer card and data transfer rates from the Compu-Scope memory to PC memory run as high as 80 Mb s^{-1}, it is thought that the CompuScope card caused few errors in velocity estimation. However, the errors in velocity estimation resulted mainly from picking t_1 and t_2, which are the travel times of the compressional and shear waves, respectively (Fig. 2). The velocities were calculated by Vp =

$L/(t_1 - t_0)$ and $Vs = L/(t_2 - t_0)$, where L is the sample length and t_0 is the inherent travel time of the transducers. Two different lengths of standardized cylindrical aluminum rods were used to calibrate the t_0 of the transducers. The final errors in estimating compressional wave velocity and shear wave velocity were ±1.2% (±50 m s^{-1}) and ±1.6% (±40 m s^{-1}), respectively. The amplitudes of compressional and shear waves can be read directly (Fig. 2).

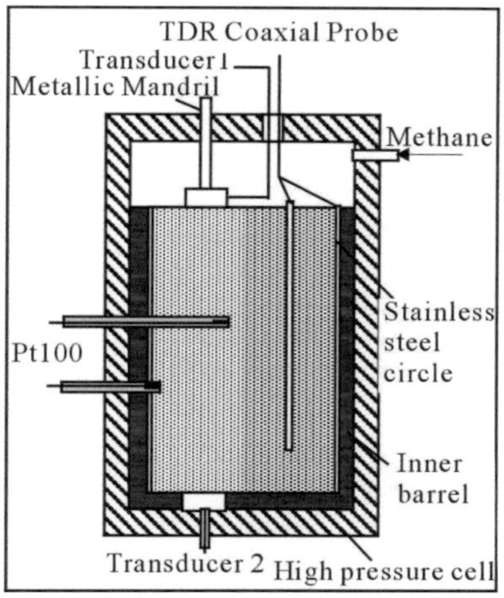

Fig. 1. Cross section through the high-pressure vessel. A single probe and a stainless steel circle around the cylindrical core are two poles of the TDR coaxial probe.

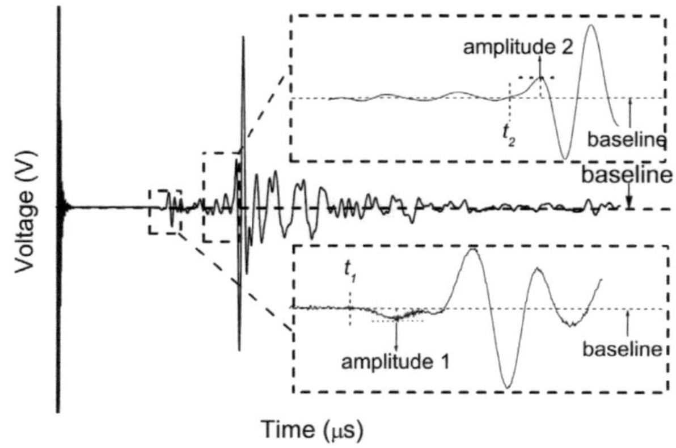

Fig. 2. Ultrasonic waveform measured by flat-plate tranducers. Arrival times of the first arrival wave of compressional and shear waves are t_1 and t_2, respectively. Amplitude 1 and amplitude 2 represent the amplitudes of compressional and shear waves, respectively.

2.1.2 TDR technique

TDR was initially used for detecting the position of breaks in transmission line cables. The technique was introduced to measure water contents of soil samples in 1980s, and then it developed rapidly [Topp et al., 1980; Dalton et al., 1986]. Topp et al. [1980] was probably the first one who measured water contents of soil samples with TDR technique. They found a practical relation between dielectric constants and water contents of the soil samples based on various types of experimental results. With regard to some particular substance, special calibration is needed before measurements. For example, Regalado et al. [2003] proposed a empirical equation for calculating water contents of volcanic soils; Wright et al. [2002] found the relationship between dielectric constants and water contents of hydrate-bearing sediments. Thereafter, TDR was effectively used to measure hydrate pore saturations of the hydrated sediments [Ye et al., 2008; Hu et al., 2010].

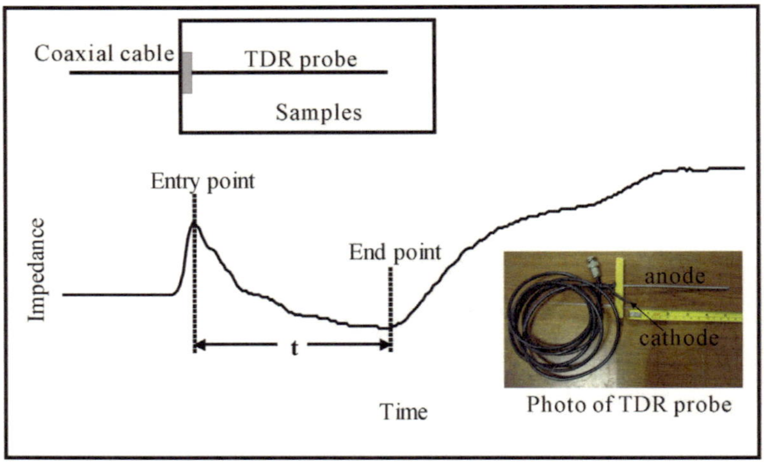

Fig. 3. TDR waveform of low-salty sediments with traditional TDR probe

TDR waveform of a soil or sediment sample is shown in Fig 3. The electromagnetic wave is generated by TDR instruments, and transmitted along the coaxial cable and the TDR probe (Fig 1 & Fig 3). Because there is loss current during electromagnetic wave transmitting in the samples, the characteristics of the entry point and the end point are obviously. The velocity of electromagnetic wave transmitting in the samples can be calculated with:

$$V = l / t \tag{1}$$

Where l is the length of the TDR probe, t is the time-interval between entry point and end point (Fig 3). At the same time, the velocity of electromagnetic wave in the samples can be also related to dielectric constants:

$$V = c / Ka^{1/2} \tag{2}$$

Where c is the velocity of propagation in free space (approximately 3×10^8 m/s). From equation 1 and 2 it solves:

$$Ka = [ct / l]^2 \tag{3}$$

When the Ka is calculated, with the relationship between Ka and water contents we can obtain the water contents of the sample. For soils [Topp et al.,1980]:

$$\theta v = -5.3\times10^{-2} + 2.92\times10^{-2}Ka - 5.5\times10^{-4}Ka^2 + 4.3\times10^{-6}Ka^3 \tag{4}$$

For hydrate-bearing sediments, it's effective to use Wright et al. [2002] 's empirical equation:

$$\theta v = -11.9677 + 4.506072566Ka - 0.14615Ka^2 + 0.0021399Ka^3 \quad (5)$$

As a result, the hydrate pore saturation of the hydrated sediments can be calculated with:

$$Sh = (\varphi - \theta v) / \varphi \times 100\% \quad (6)$$

Where φ is the porosity of the sample.

2.1.3 Synthesized method of hydrated sediments

Gas hydrates could be formed in sediments by at least four different methods in many types of apparatus. Gas can be introduced to specimens containing (1) partially water-saturated sediment [Waite et al., 2004], (2) water-saturated sediment [Winters et al., 2007], (3) seed ice-sediment [Priest et al., 2005] and (4) continually feeding gas-saturated water into the specimen, which is rarely used because of much difficult and time-consuming. The "gas + water-saturated sediment" system was used to synthesize hydrate-bearing sediments in this study.

The geophysical experimental apparatus in the Gas Hydrate Laboratory of Qingdao Institute of Marine Geology (GHL-QIMG) [Ye et al., 2005, 2008; Hu et al., 2010] can simulate in situ pressure and temperature conditions conducive to hydrate formation (Fig. 4). The apparatus is composed of five functioning units: (1) A high-pressure vessel with a plastic inner barrel for simulating in situ pressure and temperature, in which there are two platinum (Pt100) resistance thermometers with precision of ±0.1°C used for measuring the temperature of inner and surface of the sample. (2) a vessel used for making gas-saturated water, (3) a gas compressor and a pressure transducer (precision, ±0.1MPa) responsible for gas pressure control, (4) a cooling system and a bathing through for temperature control , and (5) a computer system for measuring and logging data.

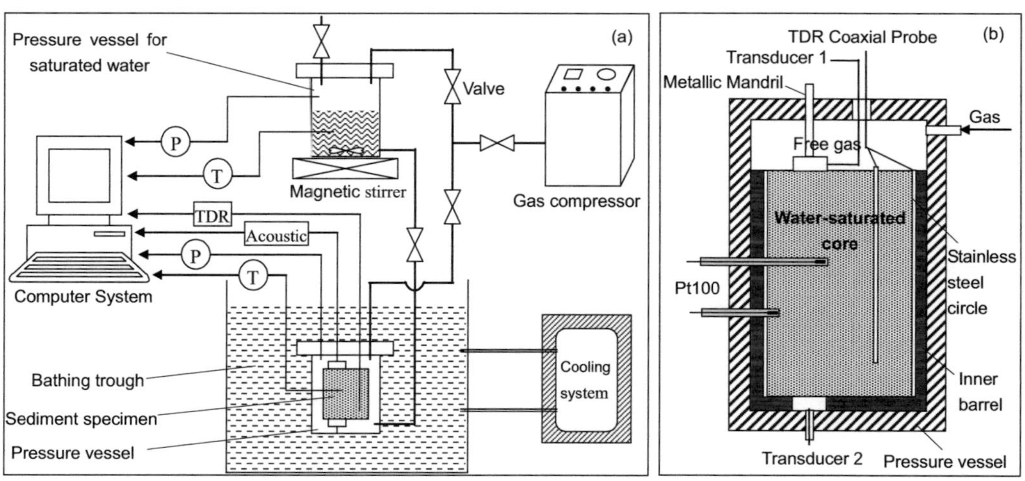

Fig. 4. Schematic diagram of experimental apparatus for geophysical research on gas hydrate-bearing sediments.

The experimental processes are as follows: (1) the artificial core was immersed by pure water or 300ppm SDS solution to gravimetric water content of about 40%, and then loaded into the high-pressure vessel; (2) Methane gas was introduced into the vessel to a scheduled pressure after vacuum, and more than 24 hours are allowed for methane dissolving into the fluid; (3) Temperature in the high pressure-vessel was controlled to ~2°C for hydrate formation; and (4) Temperature was increased naturally to room temperature for hydrate dissociation.

During the whole process of hydrate formation and dissociation, real-time measurements on temperature, pressure, the ultrasonic waveform, and the TDR waveform were recorded by the computer system.

2.2 Acoustic properties of hydrate-bearing consolidated sediments

During gas hydrate formation and subsequent dissociation in the consolidated sediments, the acoustic properties of the sample were measured and the phenomenon is described below. Also, with the results we obtain the relationship between hydrate saturations and acoustic properties of the hydrate-bearing consolidated sediments.

2.2.1 Hydrate formation and dissociation processes

Methane gas was charged into the specimen until the pressure reached to ~5MPa. Then the temperature was decreased gradually and finally at 5°C hydrates began to form (Fig. 5). A temperature-anomaly could be detected in the sample due to the exothermic reaction when hydrate forms. At the same time, the decrease of water content and pressure also indicated that hydrates began to form. In order to get more hydrates, we kept the temperature-pressure condition of the high-pressure vessel for 1~2d. The hydrate saturations range from 0% to 65.5% (water content 40.18% to 13.85%). Gas hydrate dissociation was induced by increasing temperature and the hydrate-dissociation process typically lasts about 8~10 hours.

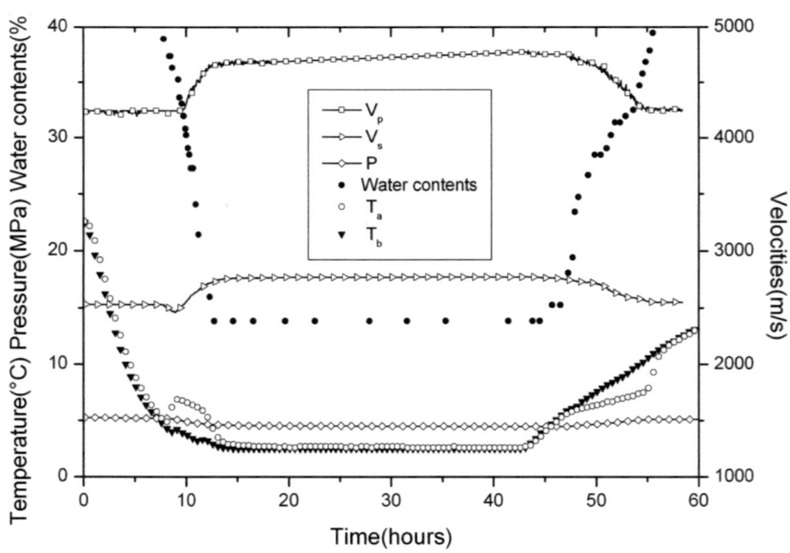

Fig. 5. Variation in temperature (T), pressure (P), water content and acoustic velocities (Vp&Vs) during gas hydrate formation and subsequent dissociation in the sediment core. Ta and Tb are temperatures of the inner and surface of the sediment core, respectively.

2.2.2 Acoustic properties

Figure 6 shows the changes of acoustic velocities (Vp, Vs) and hydrate saturations in hydrate-formation process. The Vp and Vs of the water-saturated sediment are 4242m/s and 2530m/s, respectively. During the first stage of hydrate-formation process, Vp changes hardly (time: 7.7h-~9.7h ; Sh: 0-~20%). Later, it begins to increase and gets to 4643m/s when the hydrate saturation is up to ~65.5% (time: 12.75h). The Vs of the sediment decreases slightly to 2470m/s at the beginning of hydrate-formation process (time: 7.7h-8.95h; Sh: 0-~10%). After that it begins to increase and gets to 2725m/s when the hydrate saturation is up to ~65.5%. Although the pressure-temperature condition was maintained 1~2d after hydrate saturation up to 65.5%, the hydrate saturation didn't increase. However, both Vp and Vs increase slightly in hydrate-maintained process. Vp and Vs increase to 4770m/s and 2770m/s, respectively. During the hydrate-dissociation process, Vp and Vs of the sediments decrease with the increasing water contents (decreasing hydrate saturation). And they decrease to 4250m/s and 2550m/s respectively when gas hydrate was completely dissociated (Fig. 5).

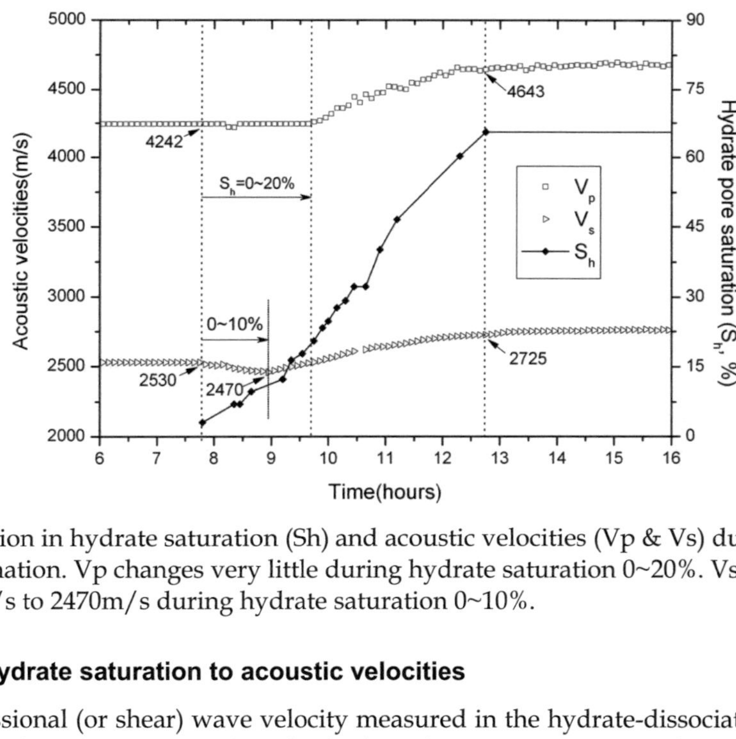

Fig. 6. Variation in hydrate saturation (Sh) and acoustic velocities (Vp & Vs) during gas hydrate formation. Vp changes very little during hydrate saturation 0~20%. Vs decrease from 2530m/s to 2470m/s during hydrate saturation 0~10%.

2.3 Relate hydrate saturation to acoustic velocities

The compressional (or shear) wave velocity measured in the hydrate-dissociation process is much higher than that measured in the hydrate-formation process at the same saturation degree. It may be caused by the hydrate morphology. As indicated in Yoslim and Englezos [2008], when surfactant (SDS) is present in the system, porous hydrate is believed to form at the gas-water interface. In addition, hydrate formation may occur in two stages: first the formation of a water-hydrate slurry, and then a very slow solidification stage [Beltrán and Servio, 2008]. Thus, in the hydrate formation process, the hydrates are porous and soft; as time lapses, the hydrates become rigid solids and consequently lead to an increase of the velocity. Because it's difficult to judge whether in situ gas hydrates are in the process of formation or dissociation during gas hydrate exploration, we use the average Vp (or Vs) of the compressional (or shear) wave velocities obtained in the two processes as the measured velocity to relate with gas hydrate saturations in this paper (Fig. 7). The result shows that acoustic velocities are insensitive to low hydrate saturations (0-~10%). However, the velocities increase rapidly with hydrate saturation when saturation is higher than 10%, especially in the range of 10-30%.

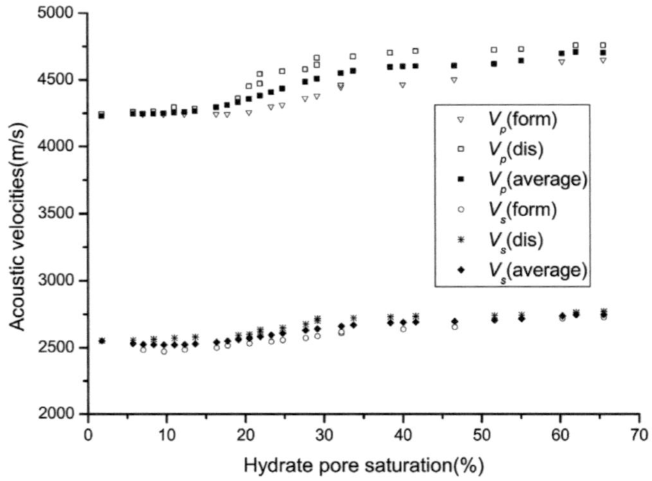

Fig. 7. Variation in Vp during hydrate formation (Vp(form)) and hydrate dissociation (Vp(dis)), Vs during hydrate formation (Vs(form)) and hydrate dissociation (Vs(dis)), the average Vp of Vp(form) and Vp(dis), and the average Vs of Vs(form) and Vs(dis).

3. Ultrasonic waves in hydrate-bearing unconsolidated sediments

The attenuation of ultrasonic wave in unconsolidated sediments is usually much higher than that in consolidated sediments. In order to obtain both Vp and Vs of the hydrate-bearing unconsolidated sediments, various techniques including bender elements, resonant column, etc, are developed to measure acoustic properties of the hydrated samples. In this section, the bender elements are successfully used in measuring both Vp and Vs of the hydrate-bearing unconsolidated sediments.

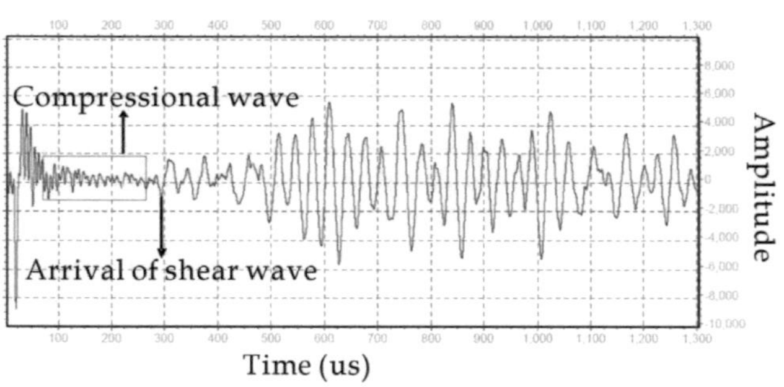

Fig. 8. Waveform of the hydrated unconsolidated sediments measured by bender elements

A waveform of the hydrated sediments measured by bender elements is shown in Fig. 8. From the waveform, it's easy to read the arrival time of shear wave. However, the arrival time of compressional wave is hard to get because of the noise. Thus, we combine the FFT transformation and wavelet-transformation (we called FFT-WT method hereafter) to interpret the compressional wave and obtain the Vp data. Calibration has been made and the method is considered to be correct.

3.1 Bender elements technique

Bender elements are commonly used to measure shear wave velocity of unconsolidated sediments. In order to obtain both Vp and Vs of the hydrate-bearing unconsolidated sediments, a new kind of bender element transducers are developed. With the FFT-WT method, the new transducers are used successfully in this study.

3.1.1 Preparation of bender element transducers

Bender elements consist of two sheets of piezoceramic plates rigidly bonded to a center shim of brass or stainless steel plate (Fig. 9a). When the "cantilever beam" of the transducer is excited by an input voltage, it changes its shape and generates a mechanical excitation (Fig. 9b), and then the signal transmits to the receiver bender element. The in-plane directivity of bender elements was explored by Lee and Santamarina (2005). The results show that amplitude of the signal is more pronounced when the installations of bender elements are parallel (Fig. 10). The amplitude in the transverse configuration is about 75% of the amplitude at 0° in the parallel axes configuration, which suggest the potential use of bender elements in a wide range of in-plane configurations besides the standard tip-to-tip alignment. In order to obtain good signal, we use the parallel installations in our experiments.

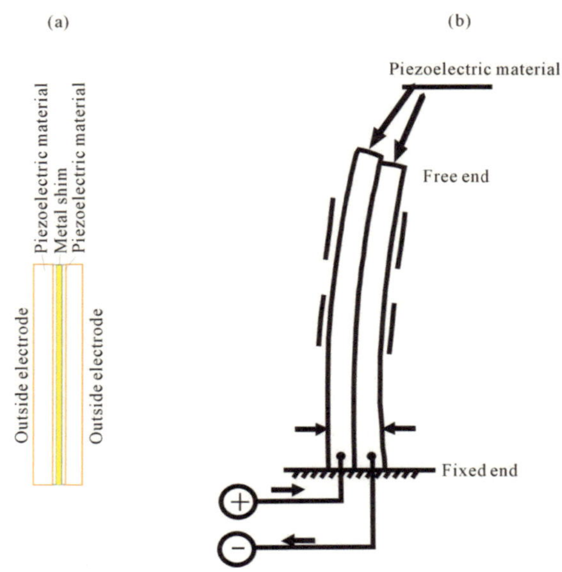

Fig. 9. (a) Schematic representation of bender element; (b) Mechanical excitation of bender element.

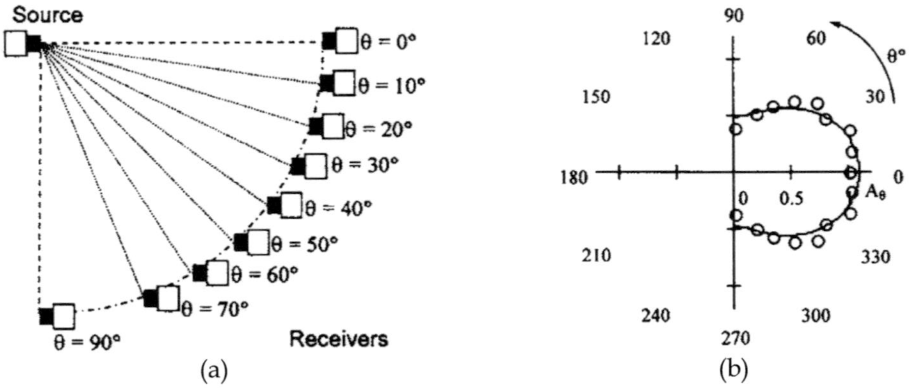

Fig. 10. Source-receiver directivity: (a) test setup, and (b) polar plot of peak amplitudes [Lee and Santamarina, 2005]

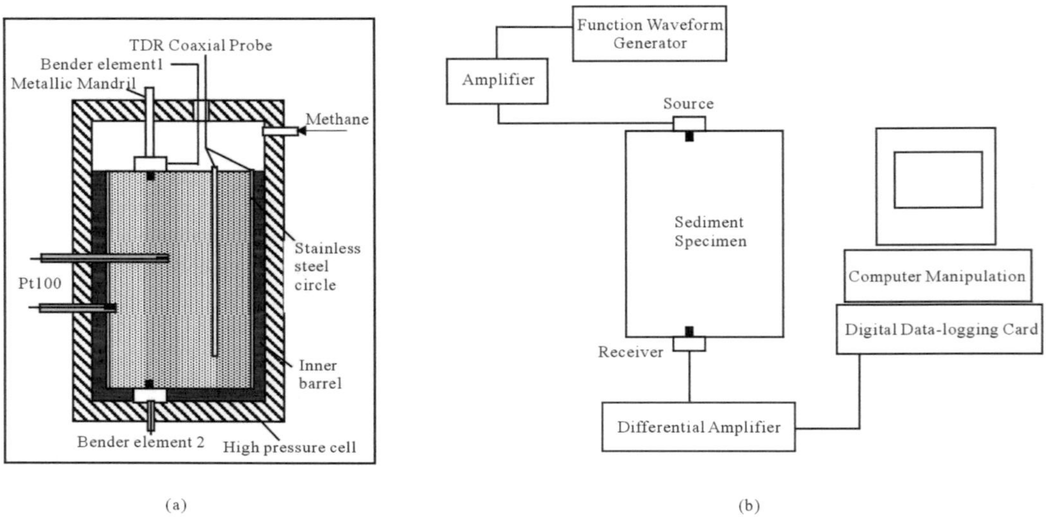

Fig. 11. (a) Schematic map of the high pressure cell for hydrate formation and acoustic measurements; (b) Measuring system of bender elements

Simultaneous measurements of compressional and shear wave velocity of methane hydrate bearing sediments using bender elements are explored. The apparatus is shown in Fig. 11. In the acoustic measuring system (Fig. 11b), signal is generated, amplified and then transmitted by the source bender element. Because the mechanical excitation of bender element is transverse, the waveform received by the receiver bender element is mainly shear wave. In order to obtain both shear wave and compressional wave, a new kind of bender element is developed (Fig. 12).

To overcome a high pressure environment, the bender elements are filled with phenolic resin and protected by stainless steel shell. The mechanical excitation of the new bender element is shown in Fig 12b. The cantilever beam is driven by two piezoelectric circles. When the excitation is generated, the cantilever beam will be distorted and the torsional vibration is occurred. At the same time, a small longitudinal movement is also occurred on the cantilever beam. In order to magnify the longitudinal movement, we add a longitudinal piezoelectric slice clinging to the bender elements. Therefore, there is also compressional wave in the integrative waveform (Fig. 8). From the waveform, it's easy to read the first arrival of shear wave. However, as the noise is largely, we develop the FFT-WT method to analysis the first arrival time of compressional wave.

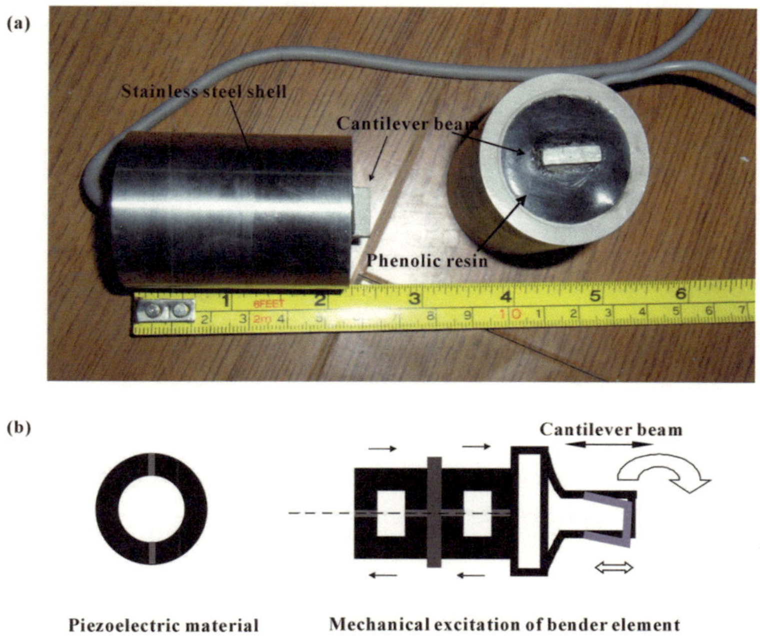

Fig. 12. (a) Photography of the bender element transducers; (b) mechanical excitation of the new bender element.

3.1.2 Calibration

Compressional and shear wave velocities are calculated with: $Vp=L_{tt}/(t_p-t_{0p})$, $Vs=L_{tt}/(t_s-t_{0s})$, where L_{tt} is the tip-to-tip distance of two bender elements, t_p and t_s are travel times of compressional wave and shear wave in the sediments respectively, t_{0p} and t_{0s} are the measured travel times of compressional wave and shear wave in the bender element transducers respectively. Four different lengths of cylindrical Polyoxy-methylene (POM) columns were used to calibrate t_{0p} and t_{0s} of the bender elements (Table 1). The diameter of the POM columns are about 6cm, which is close to the diameter of samples (6.8cm). The waveform of the POM column is shown in Fig. 13. It shows that it's easy to read both arrival times of the P-wave and S-wave using the new type of bender elements. According to lengths and wave-arrival times of the four POM columns, we obtained the t_{0p} and t_{0s}, which are 7.017us and 18.63us respectively. And the P-wave and S-wave velocities of the POM material are 2294.5m/s and 933.9m/s respectively [Fig. 14], which is very close to the reported values [Choy et al., 1983]. The P-wave velocity is also close to the measured results by our flat-plate transducers, which is 2280m/s for POM-I and 2319m/s for POM-II.

Number	Diameter (mm)	Length (mm)	Trough depth (mm)
POM-I	60.3	120	3.9~4.0
POM-II	60.3	150	3.9~4.0
POM-III	60.4	204	3.94
POM-IV	60.4	250	3.84

Table 1. Parameters of the POM columns

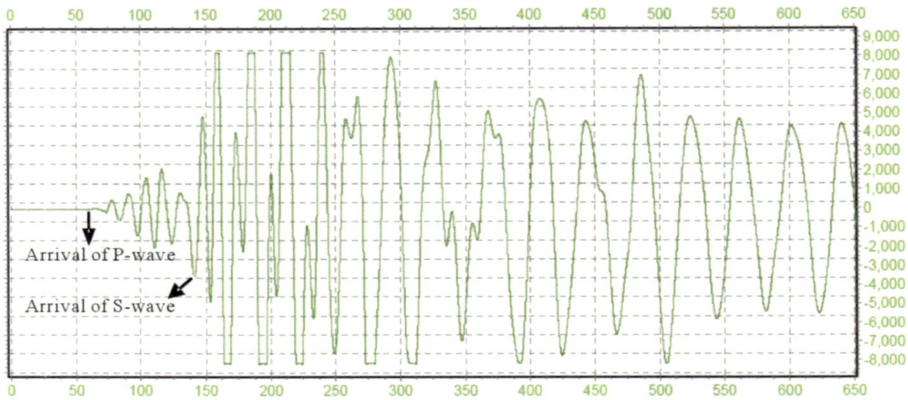

Fig. 13. Waveform of POM-I by bender element transducers

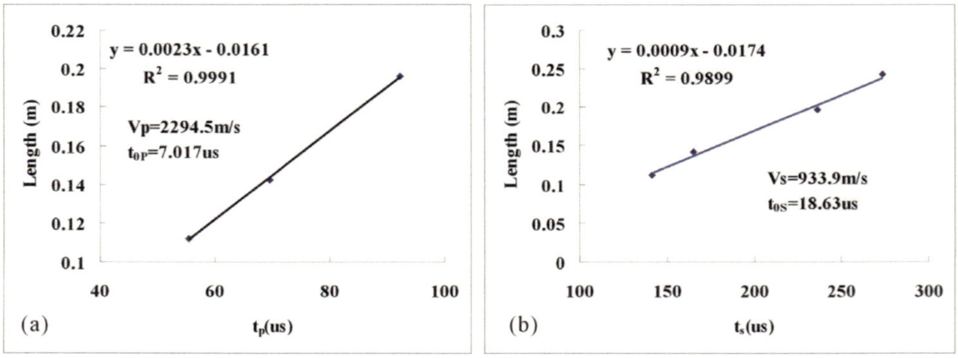

Fig. 14. Calibrating results of the new bender elements

3.1.3 FFT-WT method

Usually, there are two approaches to obtain the travel time of shear wave when using bender elements. In the first approach, the travel time can be directly read from the waveform of the receiver bender element. The characteristic point of wave's arrival must be very markedly when using this approach. The second approach is based on detailed analysis of the waveform, such as the Dynamic Finite Element Analysis, Cross-Correction Analysis, Phase Velocity Analysis, Phase Sensitive Detection, etc. Because it's easy to read the characteristic point of shear wave's arrival in our experiments, we used the first approach to determine the travel time of shear wave. However, as compressional wave is significantly influenced by the acoustic noise of the samples, the FFT-WT method is developed to determine the travel time of compressional wave.

The analysis process is as follows: (1) measuring the main frequencies of the bender element transducers; (2) choosing the compressional waveform, make a Fast Fourier Transform (FFT) on the waveform to obtain the main frequency of compressional wave; (3) making Wavelet Transform (WT) on the chosen compressional waveform to obtain frequencies versus arrival time, from which the arrival time of compressional wave can be obtained. An example of the analysis process is given below.

Firstly, the frequencies of the bender element transducers are determined by admittance curves. The results indicate that the main shear frequency is 30kHz, while the main compressional frequencies are 75kHz, 125kHz, and 140kHz. Secondly, the frequency of compressional wave is analyzed by FFT (Fig. 15). It shows that the frequencies of compressional wave mainly consist of 122kHz and 73kHz. Thirdly, the chosen compressional waveform is analyzed by WT (Fig. 16). With the frequency versus arrival time by WT, it shows that at about 96.1μs the frequency characteristics are the same with that analyzed by FFT. Thus, the travel time of compressional wave is 96.1μs with the above FFT-WT analysis.

Using above FFT-WT method, the travel time of shear wave can be also obtained. The results of Vs obtained using FFT-WT method are comparable with that measured by the first approach (in which the travel time of shear wave is read directly) (Fig. 17), which indicates that the FFT-WT method is credible.

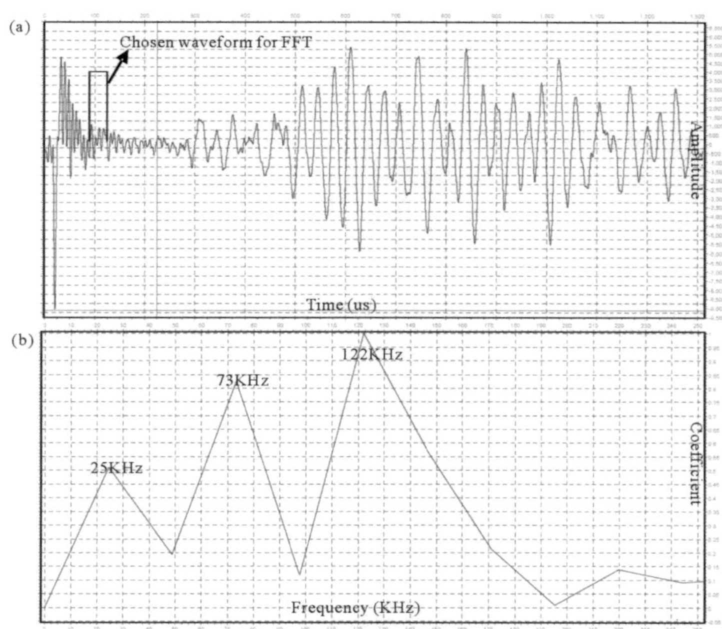

Fig. 15. (a) chosen compressional waveform for FFT analysis; (b) main frequencies of compressional wave

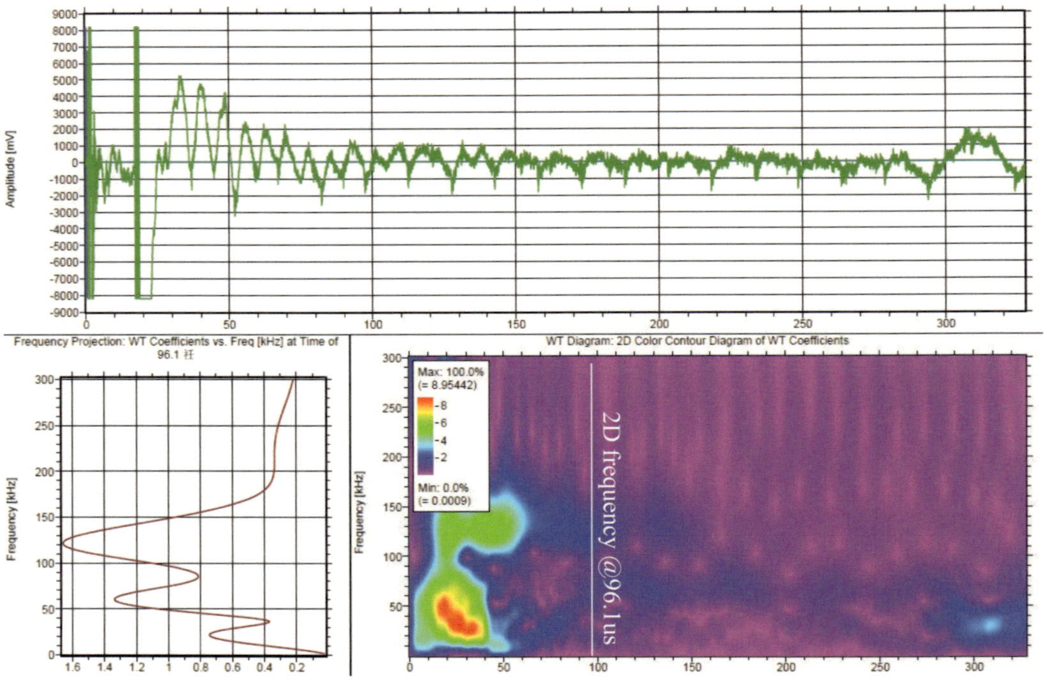

Fig. 16. WT analysis of waveform by bender element measurement

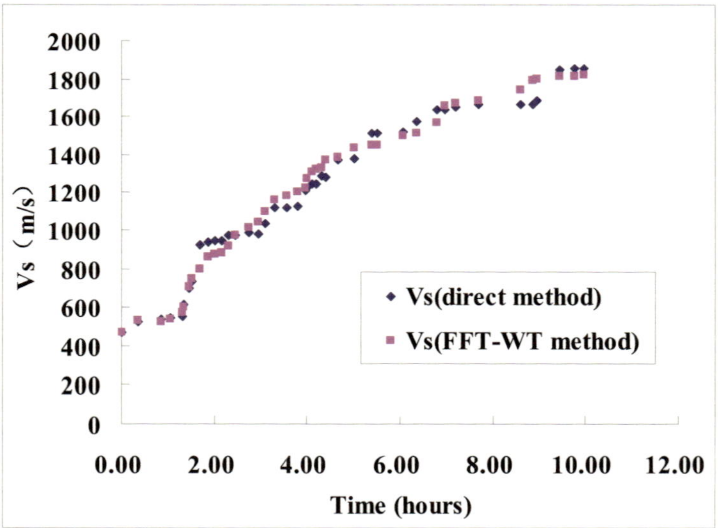

Fig. 17. Comparison of Vs determined by the direct method and the FFT-WT method

3.2 Acoustic properties of hydrate-bearing unconsolidated sediments

Methane hydrate was formed and then dissociated in the 0.09~0.125mm sands (with saturated water), during the process the acoustic velocities (Vp and Vs) of the samples are measured simultaneously with the new type of bender element transducers and analyzed with the FFT-WT method. Also, the water content, temperature, pressure of the porous media are measured (Fig. 18). The results show that the time point of gas hydrates begin to form (or dissociate) detected by the acoustic velocities is the same with that detected by the temperature-pressure method, which indicates that the bender element technique is very sensitive with gas hydrate formation and dissociation. Thus, it is effective for using the new type of bender elements in measuring both Vp and Vs of hydrate-bearing unconsolidated sediments under high pressure conditions.

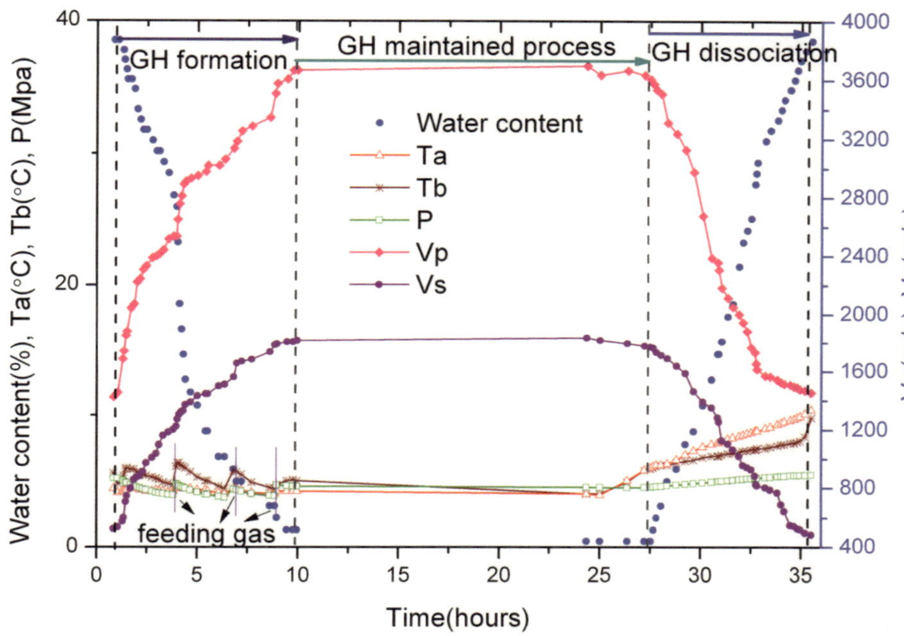

Fig. 18. Changes of parameters during gas hydrate formation and subsequent dissociation

The experimental results also show that the compressional (or shear) wave velocity measured in the hydrate-dissociation process is much lower than that measured in the hydrate-formation process at the same saturation degree (Fig. 19). This may be caused by the influence of gas hydrates on the sediment frame. In the unconsolidated sediments, gas hydrates may act as a kind of cement. A small amount of gas hydrates may dramatically affects the acoustic velocities in this condition (Priest et al., 2005). During gas hydrate formation and dissociation in the unconsolidated sediments, the influences of gas hydrates on the sediment frame became smaller as time lapse. As a result, compressional (or shear) wave velocity of the hydrated unconsolidated sediments in the hydrate-dissociation process is lower than that in the hydrate-formation process. With the average Vp (or Vs) of the compressional (or shear) wave velocities obtained in the two processes, we obtained the relationship between gas hydrate saturation and acoustic velocities of hydrate-bearing unconsolidated sediments. The result shows that Vp and Vs increase rapidly with hydrate saturations, although they increase relatively slow in the range of saturation 25%~60%. It indicates that gas hydrate may first cement grain particles of the unconsolidated sediments, when hydrate saturation is higher, gas hydrate may contact with the sediment frame, or continue cementing sediment particles.

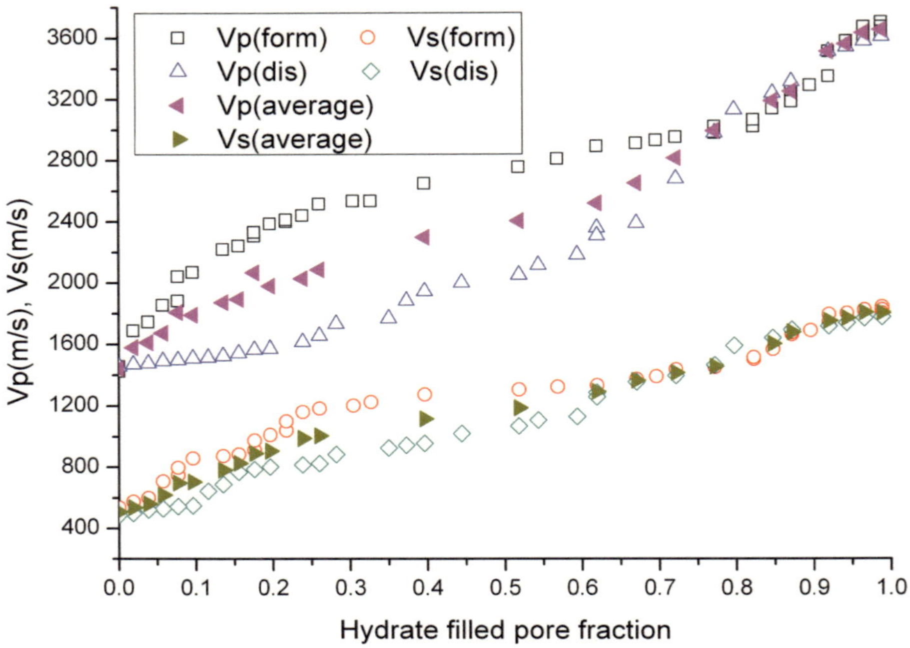

Fig. 19. For unconsolidated sediments: variation in Vp during hydrate formation (Vp(form)) and hydrate dissociation (Vp(dis)), Vs during hydrate formation (Vs(form)) and hydrate dissociation (Vs(dis)), the average Vp of Vp(form) and Vp(dis), and the average Vs of Vs(form) and Vs(dis).

4. Discussions and conclusions

Some interesting acoustic phenomena have been observed in the above experiments. For example, we noticed that compressional (or shear) wave velocities are different at the same saturation degree during hydrate-formation process and hydrate-dissociation process. The morphology of gas hydrates may be a possible factor. As discussed above, hydrate formation may occur in two stages: first the formation of a water-hydrate slurry and then a very slow solidification stage. In the consolidated sediments, gas hydrates became more and more rigid as time lapse during hydrate formation and dissociation. Thus, although the amounts of gas hydrates is the same, compressional (or shear) wave velocity in the hydrate-dissociation process is much higher than that in the hydrate-formation process. However, in the unconsolidated sediments, gas hydrates may act as a kind of cement which mainly affects the sediment frame. A small amount of gas hydrates may dramatically affects the acoustic velocities in this condition (Priest et al., 2005). During gas hydrate formation and dissociation in the unconsolidated sediments, the influences of gas hydrates on the sediment frame became smaller as time lapse. As a result, compressional (or shear) wave velocity of the hydrated unconsolidated sediments in the hydrate-dissociation process is lower than that in the hydrate-formation process. The experimental results may provide basic knowledge for field geophysical interpretations.

In this chapter, two kinds of ultrasonic methods, namely, the flat-plate transducers and a new kind of bender elements have been successfully used in measuring the acoustic properties of gas hydrate bearing sediments. The results show that it's an effective way to use classic flat-plate transducers to measure both Vp and Vs of the consolidated sediments. However, in unconsolidated sediments the bender element technique is much appropriate because the bender elements can sustain larger attenuation. Thus, although significant attenuation was occurred during the unconsolidated experimental process, a developed FFT-WT method is capable of reading the time of the first arrival of P-wave, while the S-wave can be read directly.

Both methods have shown sensitivity in detecting the formation and dissociation of gas hydrate in sediments. With the flat-plate transducers, the acoustic properties of the hydrate bearing consolidated sediments were obtained, which shows that acoustic velocities are insensitive to low hydrate saturations but they increase rapidly with hydrate saturation when saturation is higher than 10%, especially in the range of 10-30%. The measurements by bender element technique figure out that Vp and Vs of the unconsolidated sediments increase rapidly with hydrate saturations, although they increase relatively slow in the range of saturation 25%~60%. It indicates that gas hydrate may first cement grain particles of the unconsolidated sediments, and then contact with the sediment frame.

5. Acknowledgment

This work was financially supported by National Natural Science Foundation of China (40576028, 41104086), Natural Gas Hydrate in China Sea Exploration and Evaluation Project (GZH200200202), Gas Hydrate Reservoir Mechanism Research Project (GZH201100306) and Key Laboratory of Marine Hydrocarbon Resources and Environmental Geology, Ministry of Land and Resources (MRE201113).

6. References

Sloan, E. D., Jr. (1998), *Clathrate Hydrates of Natural Gases*, CRC Press, Boca Raton, Fla.

Kvenvolden, K. A. (1998), *A primer on the geological occurrence of gas hydrate*, Geological Society, London, Special publications, 137, 9-30.

Milkov, A. V., and R. Sassen (2003), Preliminary assessment of resources and economic potential of individual gas hydrate accumulations in the Gulf of Mexico continental slope, *Marine and Petroleum Geology*, 20, 111–128.

Dickens, G. R. (2003), A methane trigger for rapid warming?, *Science*, 299, 1017.

Dickens, G. R. (2004), Hydrocarbon-driven warming, *Nature*, 429, 513-515.

Brown, H. E., W. S. Holbrook, M. J. Hornbach and J. Nealon (2006), Slide structure and role of gas hydrate at the northern boundary of the Storegga Slide, offshore Norway, *Marine Geology*, 229, 179-186.

Pecher, I., R. F. Ayoub, and B. Clennell (2008), Seismic time-lapse monitoring of potential gas hydrate dissociation around boreholes - could it be feasible? A conceptual 2D study linking geomechanical and seismic FD models, *paper presented at Sixth International Conference on Gas Hydrate*, Brithish Columbia, Canada.

Shipley, T. H., M. H. Houston, R. T. Buffler, F. J. Shaub, K. J. McMillen, J. W. Ladd, and J. L. Worzel (1979), Seismic evidence for widespread occurrence of possible gas hydrate horizons on continental slopes and rises, *AAPG Bull.*, 63(12), 2204-2213.

Holbrook, W. S., H. Hoskins, W. T. Wood, R. A. Stephen, and D. Lizarralde (1996), Methane hydrate and free gas on the Blake Ridge from Vertical Seismic Profiling, *Science*, 273, 1840-1843.

Carcione, J. M., and D. Gei (2004), Gas-hydrate concentration estimated from P- and S-wave velocities at the Mallik 2L-38 research well, Mackenzie Delta, Canada, *Journal of Applied Geophysics*, 56, 73-78.

Stoll, R. D.(1974), Effects of gas hydrate in sediments, in *Natural Gases in Marine Sediment*, edited by I. Kaplan, pp. 235-248, Springer, New York.

Tucholke, B. E., G. M. Bryan, and J. I. Ewing (1977), Gas hydrate horizons detected in seismic-profile data from the western North Atlantic, *Am. Assoc. Petrol. Geol. Bull.*, 61, 698-707.

Wyllie, M. R. J., A. R. Gregory, and G. H. F. Gardner (1958), An experimental investigation of factors affecting elastic wave velocities in porous media, *Geophysics*, 23, 459-493.

Pearson, C. F., P. M. Halleck, P. L. McGulre, R. Hermes, and M. Mathews (1983), Natural gas hydrate; A review of in situ properties, *J. Phys. Chem.*, 87, 4180-4185.

Wood, A. B. (1941), *A text book of sound*, Macmillan, New York, 578 pp

Lee, M. W., D. R. Hutchinson, T. S. Collett and W. P. Dillon (1996), Seismic velocities for hydrate-bearing sediments using weighted equation, *J. Geophys. Res.*, 101, 20347-20358.

Helgerud, M. B., J. Dvorkin, A. Nur, A. Sakai, and T. Collett (1999), Elastic-wave velocity in marine sediments with gas hydrate: Effective medium modeling, *Geophysical Research Letters*, 26, 2021-2024.

Dvorkin, J., and M. Prasad (1999), Elasticity of marine sediments: Rock physics modeling, *Geophysical research letters*, 26(2), 1781-1784.

Lee, M. W. (2002a), Biot-Gassmann theory for velocities of gas hydrate-bearing sediments. *Geophysics*, 67, 1711-1719.

Lee, M. W. (2002b), Modified Biot-Gassmann theory for calculating elastic velocities for unconsolidated and consolidated sediments, *Marine Geophysical Researches*, 23, 403-412.

Lee, M. W. (2003), *Velocity ratio and its application to predicting velocities*, U. S. Geological Survey Bulletin 2197, 19 pp., U. S. Geological Survey.

Chand, S., T. A. Minshull, D. Gei, and J. M. Carcione (2004), Elastic velocity models for gas-hydrate-bearing sediments-a comparison, *Geophys. J. Int.*, 159, 573-590.

Carcione, J. M., and U. Tinivella (2000), Bottom-simulating reflectors: seismic velocities and AVO effects, *Geophysics*, 65(1), 54-67.

Gei, D., and J. M. Carcione (2003), Acoustic properties of sediments saturated with gas hydrate, free gas and water, *Geophys. Prospect.*, 51, 141-157.

Jakobsen, M., J. A. Hudson, T. A. Minshull, and S. C. Singh (2000), Elastic properties of hydrate-bearing sediments using effective medium theory, *J. Geophys. Res.*, 105, 561-577.

Ojha, M., and K. Sain (2008), Appraisal of gas-hydrate/free-gas from Vp/Vs ratio in the Makran accretionary prim, *Marine and Petroleum Geology*, 25, 637-644.

Sothcott, J., C. McCann, and S. G. O'Hara (2000), The influence of two different pore fluids on acoustic properties of reservoir sandstones at sonic and ultrasonic frequencies, *paper presented at 70th Ann. Mtg., SEG*, Calgary, Exp. Abst., 2, 1883-1886.

Topp G C, Davis J L, Annan A P (1986). Electromagnetic determination of soil-water content：Measurement in coaxial transmission line. *Water Resour. Res.*, 1980, 16(3)：574-582.

Dalton F N, van Genuchten M T. The time domain reflectometry for measuring soil water content and salinity. *Geoderma*, 38: 237-250.

Ye Y G, Zhang J, Hu G W, et al (2008). Combined detection technique of ultrasonic and time domain reflectometry in gas hydrate. *Marine Geology & Quaternary Geology (in Chinese with English abstracts)*, 28(5): 101-107.

Hu G W, Ye Y G, Zhang J, et al (2010). Acoustic properties of gas hydrate-bearing consolidated sediments and experimental testing of elastic velocity models. *Journal of Geophysical Research*, 115: B02102. doi: 10.1029/2008JB006160.

Ye Y G, Zhang J, Hu G W, et al (2008). Experimental research on the relationship between gas hydrate saturation and acoustic parameters. *Chinese Journal of Geophysics*, 51(4): 819-828.

Regalado C M (2003). Time domain reflectometry models as a tool to understand the dielectric response of volcanic soils. Geoderma, 117:313-330.

Wright J F, Nixon F M, Dallimore S R, et al (2002). A method for direct measurement of gas hydrate amounts based on the bulk dielectric properties of laboratory test media. *Fourth International Conference on Gas Hydrate*, Yokohama, 745-749.

Waite, W. F., W. J. Winters, and D. H. Mason (2004), Methane hydrate formation in partially water-saturated Ottawa sand, *American Mineralogist*, 89, 1202-1207.

Winters, W. J., W. F. Waite, D. H. Mason, L. Y. Gilbert, and I. A. Pecher (2007), Methane gas hydrate effect on sediment acoustic and strength properties, *Journal of Petroleum Science and Engineering*, 56, 127-135.

Ye, Y. G., C. L. Liu, S. Q. Liu, J. Zhang, and S. B. Diao (2005), Experimental studies on several significant problems related marine gas hydrate, *paper presented at Fifth International Conference on Gas Hydrate*, Trondheim, Norway.

Yoslim, J., and P. Englezos (2008), The effect of surfactant on the morphology of methane/propane clathrate hydrate crystals, *paper presented at Sixth International Conference on Gas Hydrate*, Brithish Columbia, Canada.

Beltrán, J. G., and P. Servio (2008), Morphology studies on gas hydrates interacting with silica gel, *paper presented at Sixth International Conference on Gas Hydrate*, Brithish Columbia, Canada.

Lee J S, Santamarina J C (2005). Bender elements: Performance and signal interpretation. *Journal of Geotechnical and Geoenvironmental Engineering*, 1063-1070.

Choy C L, Leung W P, and Huang C W (1983). Elastic moduli of highly oriented polyoxymethylene. *Polymer Engineering and Science*, 23(16): 910-922.

Priest, J., C. R. I. Best, and R. I. Clayton (2005), A laboratory investigation into the seismic velocities of methane gas hydrate-bearing sand, *J. Geophys. Res.*, 110, B04102, doi:10.1029/2004JB003259.

6

Intense Aerial Ultrasonic Source and Removal of Unnecessary Gas by the Source

Hikaru Miura
Nihon university
Japan

1. Introduction

Ultrasonic energy is widely used with liquids and solids, for example, for cleaning substances in liquids, atomization of liquids (Miura, 2007b), (Ueha et al., 1985), sedimentation of dispersed fine particles (Miura, 2004), deflection of water (Ito, 2005) , removal of liquid in a pore (Ito & Takamura, 2010), and cutting (Asami & Miura, 2010, 2011) and welding solids (Miura, 2003, 2008). In addition, it is also used with gases, for example, to enhance the removal of unnecessary gases (Miura, 2007a), (Kobayashi et al., 1997), the aggregation and removal of airborne substances such as smoke, antifoaming, drying of wet substances containing moisture, thawing of frozen materials, and the decomposition of methane hydrates (Miura et al., 2006). However, extremely intense sound waves (with a sound pressure of at least 160 dB) are required to perform the above processes in air. Therefore, intense aerial ultrasonic sources are required so that ultrasonic waves can be used in air. Unfortunately, such ultrasonic sources have seldom been developed because it is difficult to generate intense ultrasonic waves in air.

In this chapter, I describe the structure of an ultrasonic source with a flexurally vibrating plate that can radiate extremely intense ultrasonic waves in air as well as the vibration distribution and directivity of the radiated ultrasonic waves. Next, I examine the effect of using the above-mentioned intense aerial ultrasonic source on the enhancement of the removal rate of an unnecessary gas in air. In general, some gas components in air may have adverse effects on humans and the environment, causing environmental problems. Currently, gas absorption, in which a gas is dissolved in a liquid for transport, is widely adopted as a means of collecting gas components in a mixed gas and removing unnecessary gas components. General absorption systems are roughly classified into gas-dispersion-type systems, in which a gas is dispersed in a liquid in the form of microbubbles, and liquid-dispersion-type systems, in which a liquid is sprayed into containers filled with the gas. Both systems allow the gas and liquid to come into contact without applying an external force.

I attempted to enhance the effect of gas absorption by applying ultrasonic waves to such systems. In this chapter, I describe a method of allowing liquid mist obtained using aerial ultrasonic waves to absorb a gas.

2. Intense aerial ultrasonic source

2.1 Outline

Ultrasonic sources using a thin metal plate that vibrates flexurally can be used to radiate extremely intense ultrasonic waves in air. Such ultrasonic sources consist of a longitudinal vibration transducer, a horn for increasing the amplitude, and a flexurally vibrating plate attached at the end of the horn. The length and width of the flexurally vibrating plate are sufficiently greater than the wavelength of flexural vibration, and the plate shape may be square, rectangular, or circular. Flexural vibration is usually generated by vibrating the center of the plate (Miura & Honda, 2002). Such ultrasonic sources can efficiently radiate sound waves in air and achieve a high electroacoustic conversion efficiency by generating an appropriate mode of vibration, i.e., a vibration mode in which the vibration nodes are distributed in a lattice pattern for square plates, a striped pattern for rectangular plates, and a concentric circular pattern for circular plates (Onishi & Miura, 2005), (Miura & Ishikawa, 2009).

In this section, I describe a method for designing a square plate that vibrates flexurally in the lattice mode and can be used as an ultrasonic source, and I discuss the vibration mode of the fabricated plate (Miura, 1994). Next, the measured distributions of vibration displacement and sound pressure near the plate surface are described. Moreover, the directivity of the sound waves radiated from the vibrating plate into remote acoustic fields is theoretically and experimentally examined.

2.2 Intense aerial ultrasonic source

Figure 1 shows a schematic diagram of an ultrasonic source that can radiate intense ultrasonic waves in air. As shown in the figure, the ultrasonic source consists of a longitudinal vibration transducer, a horn for increasing the amplitude, a rod for tuning the resonance of the longitudinal vibration, and a flexurally vibrating plate attached at the end of the horn.

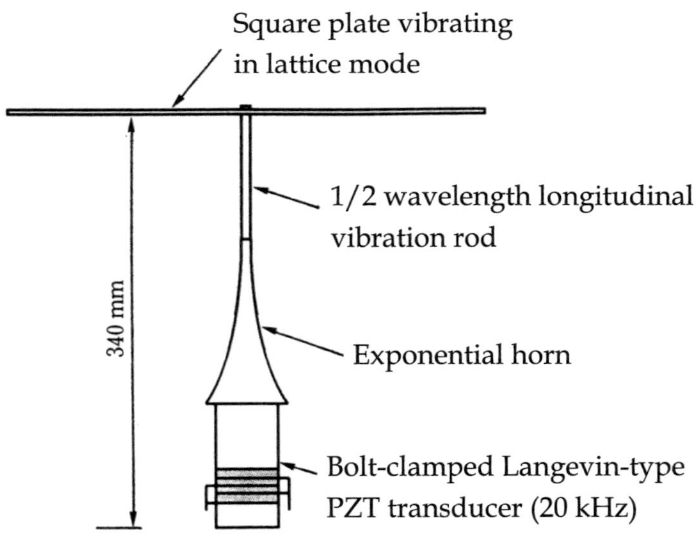

Fig. 1. Schematic diagram of an aerial ultrasonic source.

These parts are joined with screws. In general, transducers are required to generate longitudinal vibration with a frequency of approximately 20-100 kHz; I used a 20 kHz bolt-clamped Langevin-type piezoelectric (PZT) transducer (BLT transducer) to obtain large amplitudes. To increase the amplitude of the vibration of the transducer, an exponential horn (duralumin; diameter of thin end face, 10 mm; amplification rate, approximately was used. A rod (duralumin; diameter, 10 mm; length, 93 mm) was used to tune the resonant frequency of the 20 kHz longitudinal vibration of the transducer to that of the flexural vibration of the plate. The length of this rod should be an integral multiple of the half-wavelength of its longitudinal vibration; however, the rod can be omitted if not required. The flexurally vibrating plate attached at the end of the rod is allowed to vibrate by applying an amplified longitudinal vibration to the plate center as the driving point. Thus, lattice-mode vibration, in which the nodes of the flexural vibration are distributed in a lattice pattern, can be realized.

2.3 Design and fabrication of plate vibrating flexurally in the lattice mode

2.3.1 Design of vibrating plate

To distribute the nodes of a high-order flexural vibration in a lattice pattern by vibrating the center of a square plate, all sides of which undergo free vibration, high-order stripe-mode flexural vibration is induced between each pair of parallel sides of the plate, and the sum of the two flexural vibrations, which orthogonally intersect, is considered to be the lattice-mode flexural vibration. The equation used to design a plate vibrating flexurally in the stripe mode between opposite sides of the plate is given as

$$\lambda_t = \left\{ \frac{2\pi C_p h}{f} \right\}^{1/2} \tag{1}$$

where λ_t is the wavelength of the flexural vibration of the plate, C_P is a constant specific to the plate material, h is the thickness of the plate, and f is the frequency (Yamane et al., 1983). The length of the plate in the direction perpendicular to the striped pattern, L, is given by

$$L = (N - 0.5)\frac{\lambda_t}{2} \tag{2}$$

where N is the number of nodal lines and is even. When the plate is assumed to be a square of side L given by eq. (2), lattice-mode vibration can be obtained.

2.3.2 Vibration mode of plate

Figure 2 shows a schematic of a square flexurally vibrating plate viewed from above. When the center of the plate is assumed to be the origin and the x- and y-axes are set as shown in the figure, the vibration displacement at an arbitrary point (x,y) on the plate, ξ, is approximated using eq. (3) for $N \geq 10$.

$$\xi = \frac{\xi_0}{2}\left\{ \cos\frac{(N-0.5)\pi x}{L} + \cos\frac{(N-0.5)\pi y}{L} \right\} \tag{3}$$

Here, ξ_0 is the maximum vibration displacement.

The displacements of the plate flexurally vibrating in the stripe mode along the x- and y-directions are shown at the bottom and to the left of Fig. 2, respectively, and were calculated using eq. (3) assuming N = 10. Because the vibration mode is obtained by superimposing the vibration displacement in the x-direction (the first term in eq. (3)) onto that in the y-direction (the second term in eq. (3)), the nodal lines of the flexural vibration are represented by the broken lines in the figure. These nodal lines form a lattice pattern having an angle of 45° to each side of the plate. The number of nodal lines in each direction of the lattice mode is equal to N in the stripe mode. The interval between the lattice-mode nodal lines, ds, is given by

$$d_s = \frac{\lambda_t}{\sqrt{2}} \tag{4}$$

The antinodes of vibration displacement are positioned at the center of the regions surrounded by the nodal lines, and the displacements at adjacent antinodes have opposite phases (Miura, 1994).

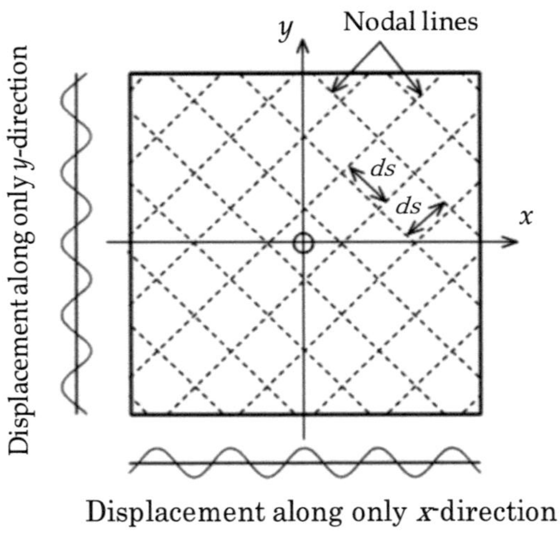

Fig. 2. Outline of a square plate vibrating in a lattice mode. Broken lines are nodal lines. ds is the interval of the lattice mode.

Thickness h [mm]	Nodal number N	Frequency f [kHz]	Wavelength of transverse vibration λ_t [mm]	Width L [mm]	Nodal interval of lattice-mode d_s [mm]
1	10	19.89	22.4	106	15.8
2	10	20.64	30.7	146	21.7
3	10	20.45	37.3	177	26.4
5	10	19.71	48.2	229	34.1

Table 1. Details of the square plate vibrating in the lattice mode.

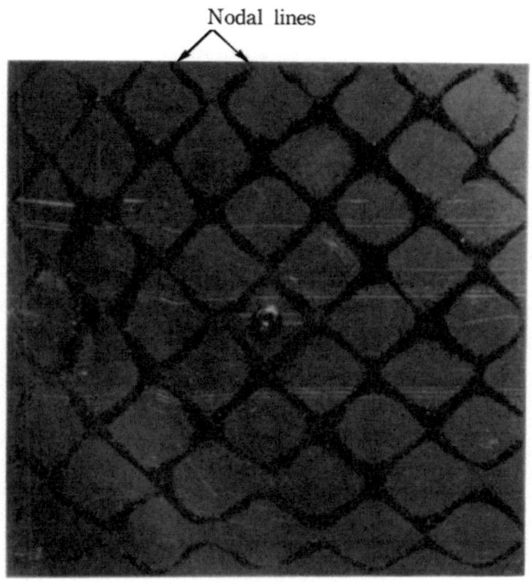

Fig. 3. Chladni sand figure showing nodal pattern of the square vibrating plate.

2.3.3 Fabrication of vibrating plate

Duralumin (JIS A2017P-T3; thickness h, 1-5 mm) was used to fabricate vibrating plates. The resonant frequency of the lattice mode vibration was approximately 20 kHz. The vibrating plates were designed using eqs. (1) and (2). Table 1 summarizes the details of the fabricated plates vibrating in a lattice mode when $N = 10$ and $h = 1, 2, 3$, or 5 mm.

2.3.4 Chladni sand figures

To determine the flexural vibration mode of each plate, Chladni sand figures were observed. Figure 3 shows the result for the plate with $h = 3$ mm (see Table 1). In the figure, the nodal lines on which sand particles (silicon carbide, #100) are concentrated are distributed in a lattice pattern similar to the nodal pattern (dotted lines) shown in Fig. 2. Vibrating plates with N not equal to 10 were also fabricated to observe the Chladni sand figures, and the vibration mode was confirmed to be the lattice mode with a different number of nodal lines.

2.4 Amplitude distribution of vibration displacement of plate

The vibration displacement at each position on the plate surface was measured using a noncontact microdisplacement meter to determine its distribution. The vibrating plate with $h = 3$ mm, which was used to obtain Fig. 3 (see Table 1), was used also for this measurement. The electric power input to the transducer was maintained at 1 W, and the square region indicated as (A) in Fig. 4 was targeted. Figure 5 shows the distribution of the normalized vibration displacement at various positions on the plate surface (using the center of the plate as the reference). The nodal lines of vibration displacement were distributed in a lattice pattern and the antinodes were positioned at the centers of the regions surrounded by the nodal lines, in good agreement with the schematic of the distribution of the vibration displacement within region (A) shown in Fig. 4, where the nodes are indicated by dashed lines.

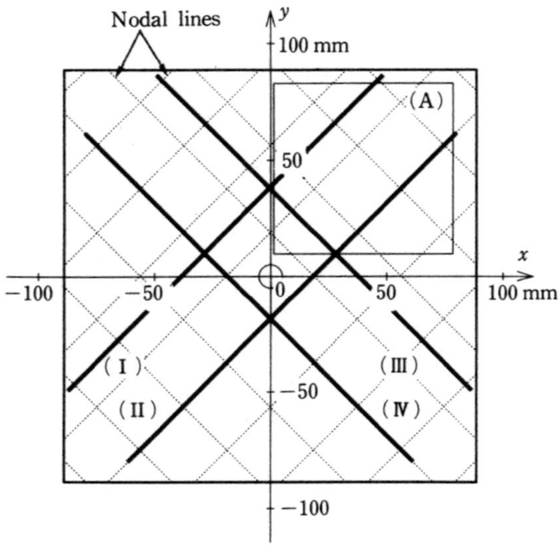

Fig. 4. Measurement positions of vibration displacement and sound pressure.

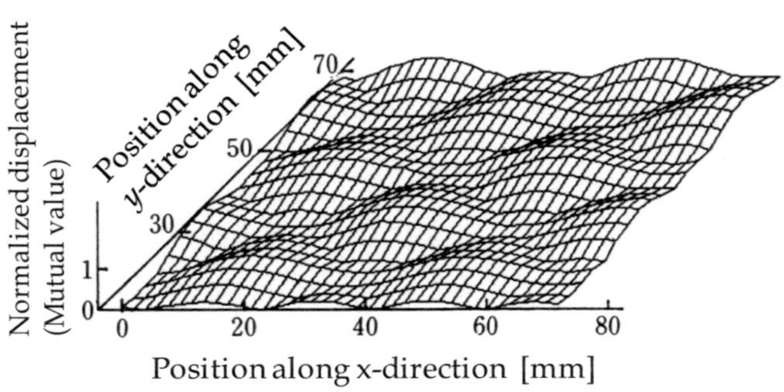

Fig. 5. Distribution of the vibration displacement.

2.5 Distribution of sound pressure near vibrating plate surface

To determine the distribution of sound pressure near the vibrating plate surface, the sound pressure at various positions approximately 1 mm above the plate along measurement lines (I)-(IV), indicated by bold lines in Fig. 4, was measured using condenser microphones (Bruel & Kjaer, 4138) while maintaining the input electric power at 1 W. Figure 6 shows the results, where the ordinate represents the normalized sound pressure and the abscissa represents the distance along each measurement line (using the foot of the perpendicular from the plate center to each measurement line as the reference). The distributions of sound pressure along measurement lines (I) and (III), which passed through antinodes of sound pressure, were similar to a sine wave with a specific amplitude and period. In this case, *ds*, the interval between adjacent nodes of sound pressure, was approximately 26 mm, in good agreement with the value of 26.4 mm calculated using eq. (4). In contrast, sound pressures along measurement lines (II) and (IV), which were located on the nodal lines, were low.

2.6 Directivity of sound waves in remote acoustic fields

In this section, I theoretically and experimentally examine the directivity of the sound waves radiated from the plate vibrating in the lattice mode into remote acoustic fields.

2.6.1 Method of calculating directivity

The directivity of the sound waves radiated from the square plate vibrating in the lattice mode into remote acoustic fields is calculated as follows.

As shown in Fig. 7, it is assumed that the distance from the plate center to a sufficiently remote observation point P is R_0 and that the distance from an arbitrary point (x,y) on the

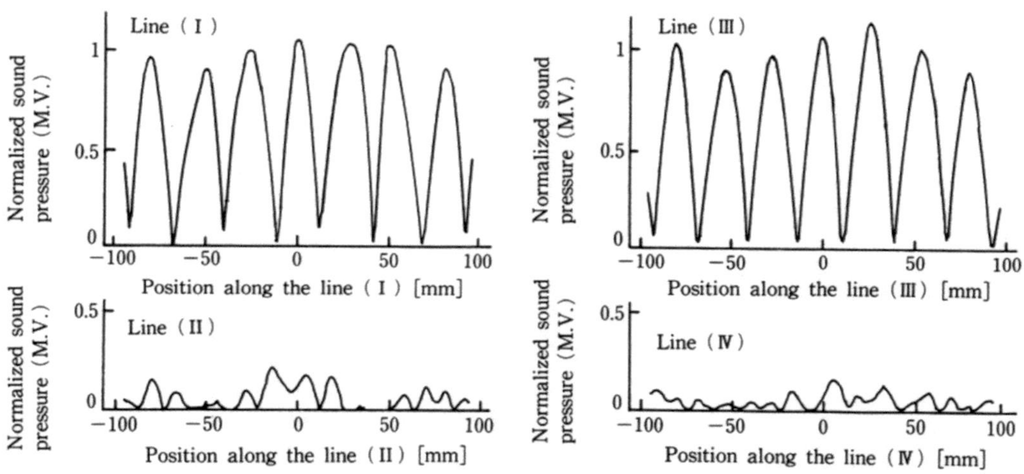

Fig. 6. Distributions of the sound pressure near the vibrating plate. Lines (I) - (IV) are indicated in Fig.4.

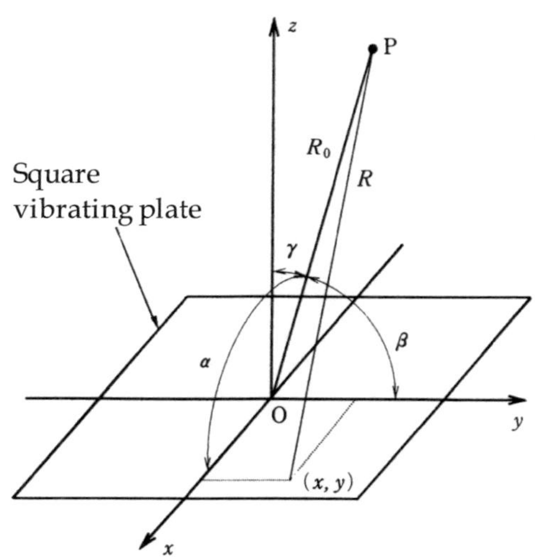

Fig. 7. Coordinate system for the evaluating directivity of radiated sound waves.

plate surface to P is R. When the angles between OP and the x-, y-, and z-axes are assumed to be a, β, and γ, respectively, R is given by

$$R = R_0 - (x\cos\alpha + y\cos\beta) \tag{5}$$

Therefore, the component of the volume velocity over a small area $dxdy$ in the OP direction is $\xi dxdy\cos\gamma$, and the velocity potential $d\Phi$ at P is given by

$$d\Phi = \frac{\xi dxdy}{2\pi R}\cos\gamma\, e^{j(\omega t - kR)} \tag{6}$$

where ω is the angular frequency and k is the wavelength constant of sound waves. When P is sufficiently remote from the plate center, $R_0 \gg x\cos\alpha + y\cos\beta$. When assuming that R only affects the phase difference of the velocity potential Φ, Φ at P is obtained by integrating eq. (6) over the entire plate area.

$$\Phi = \frac{\cos\gamma}{2\pi R_0}\int_{\frac{-L}{2}}^{\frac{L}{2}}\int_{\frac{-L}{2}}^{\frac{L}{2}} \xi e^{j(\omega t - kR)} dxdy \tag{7}$$

The sound pressure p at P is expressed by

$$p = \rho\frac{\partial\Phi}{\partial t} \tag{8}$$

where ρ is the density of air. By substituting eq. (7) into eq. (8),

$$p = \frac{\omega\rho\xi_0\cos\gamma}{2\pi R_0}\left(\frac{A\sin\frac{k_\beta L}{2}}{k_\beta} + \frac{B\sin\frac{k_\alpha L}{2}}{k_\alpha}\right)\bullet e^{j(\omega t - kR_0 + \frac{\pi}{2})} \tag{9}$$

is obtained. Here, A, B, k_a, k_β, and k_N are given as follows.

$$A = \frac{1}{k_\alpha + k_N}\sin\frac{(k_\alpha + k_N)L}{2} + \frac{1}{k_\alpha - k_N}\sin\frac{(k_\alpha - k_N)L}{2}$$

$$B = \frac{1}{k_\beta + k_N}\sin\frac{(k_\beta + k_N)L}{2} + \frac{1}{k_\beta - k_N}\sin\frac{(k_\beta - k_N)L}{2}$$

$$k_\alpha = k\cos\alpha \text{ , } k_\beta = k\cos\beta$$

$$k_N = \frac{(N - 0.5)\pi}{L}$$

2.6.2 Method of measuring directivity

The sound pressure at a distance of 1.8 m from the plate center was experimentally measured using a 6.4-mm-diameter condenser microphone (B&K 4136) while maintaining the electric power input to the acoustic source at 1 W.

The characteristics of the vibrating plates used in the experiment are summarized in Table 1. No baffles were used because the dimensions of the vibrating plates were sufficiently greater than the wavelength of the sound waves and baffles were considered to have little effect. The back of the vibrating plates was covered with glass wool with a thickness of 50 mm to absorb the sound waves radiated from the back. The measurement distance was set at values greater than the Fresnel last maximum (at which the phase difference due to differences in the distance from the vibrating plate surface is negligible and a remote acoustic field is assumed to be formed) on the centerline of a disc piston with a diameter equal to the length of the diagonal of the vibrating plates (David & Cheeke, 2002).

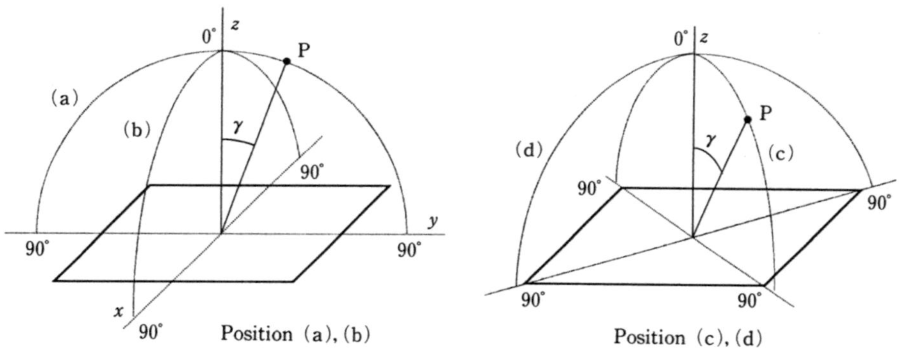

Fig. 8. Measurement positions of directivity.

2.6.3 Directivity in various directions

First, sound pressures in various directions were measured using the vibrating plate with h = 3 mm and N = 10 (see Table 1) for different angles between OP and the z-axis, γ, from -90 (x-y plane), 0 (z-axis), to 90° (x-y plane) to examine the radiation direction of the sound waves. Figure 8 shows the measurement positions: (a) γ = -90 — 90° in the y-z plane (a = 90°), (b) γ = -90 — 90° in the x-z plane (β = 90°), (c) γ = -90 — 90° in the y = x and z-axis plane (a = β), and (d) γ = -90 — 90° in the y = -x and z-axis plane. The measurement results are shown by the solid lines in Figs. 9(a)—9(d). The abscissa represents γ and the ordinate represents the normalized sound pressure. In Figs. 9(a) and 9(b), sharp main lobes of the radiated sound waves are observed in two directions; they have specific z-axis-symmetric angles in the y-z and x-z planes, respectively. However, no sharp main lobes are observed in the y = x and z-axis plane and the y = -x and z-axis plane, as shown in Figs. 9(c) and 9(d), respectively. The dashed lines in Figs. 9(a)—9(d) represent the calculation results obtained using eq. (9), and are in good agreement with the experimental results in all cases.

The directions of the main lobe are symmetric about the z-axis and the angle between the z-axis and the main lobe, γ_m, is given by

$$\gamma_m = \sin^{-1}\frac{\lambda_a}{\lambda_t} \tag{10}$$

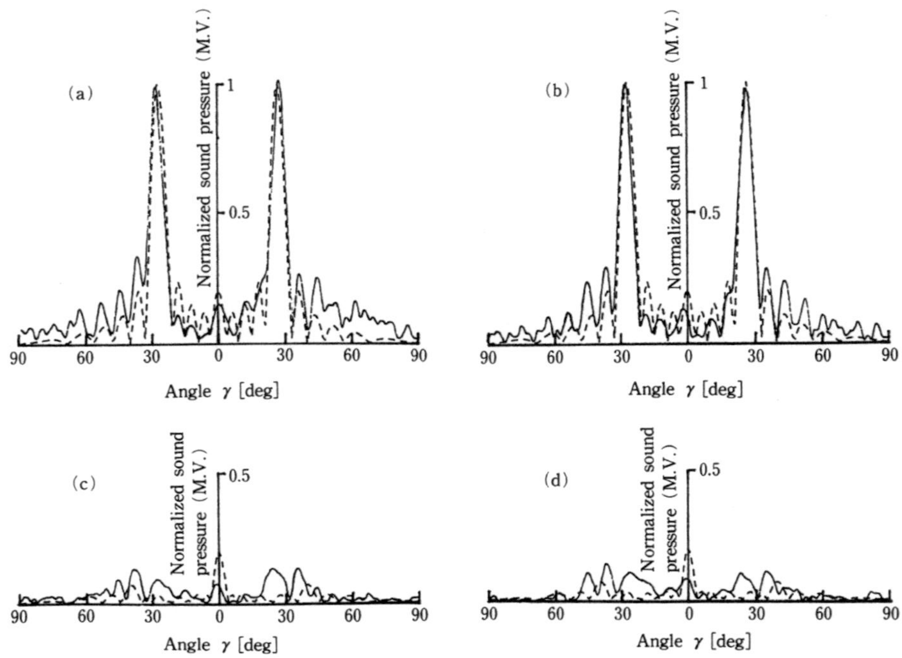

Fig. 9. Directivity patterns in various directions. ------ : Experimental, - - - : calculated. Position (a) is in the y-z axis plane, (b) is in the x-z axis plane, (c) is in the y=x and the z-axis plane, (d) is in the y= - x and the z-axis plane.

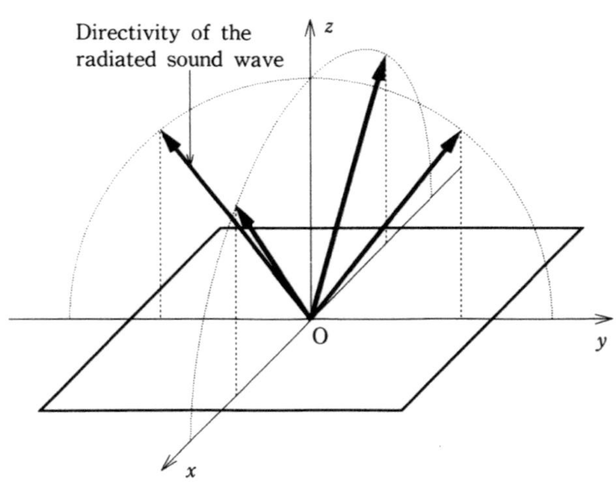

Fig. 10. Outline of the directivity.

where λa is the wavelength of the sound waves in air. The measurement results in Figs. 9(a) and 9(b) reveal that γ_m = 27.2°, which is in good agreement with the value of 27.0° calculated using eq. (10). When similar experiments were carried out for planes other than the above-mentioned planes, the sound pressures were low, similarly to the cases in Figs. 9(c) and 9(d). Therefore, the sound waves radiated from the plate vibrating in the lattice mode have main lobes only in four symmetric directions with a specific angle from the *z*-axis in the *y*-*z* and *x*-*z* planes, as shown in Fig. 10.

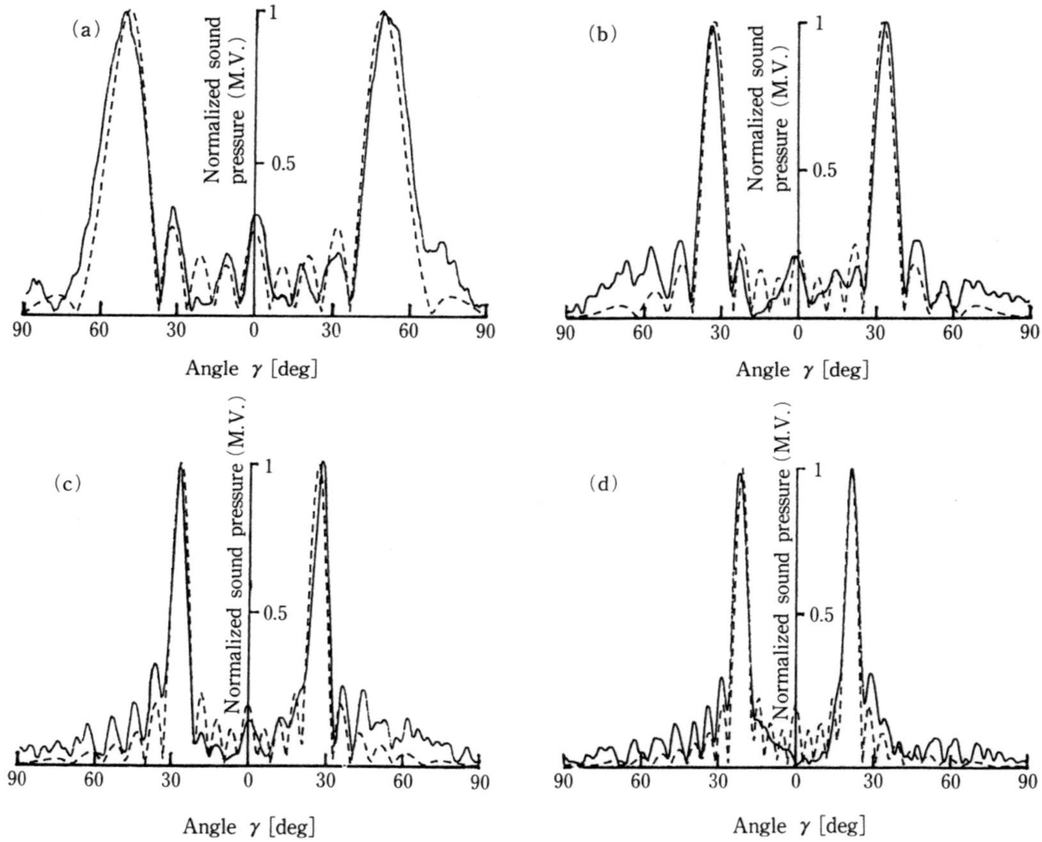

Fig. 11. Directivity patterns for various λ_t/λ_a. ----- : Experimental, - - - : calculated. (a) changing $\lambda_t/\lambda_a = 1.29$ (h=1 mm), (b) $\lambda_t/\lambda_a = 1.83$ (h=2 mm), (c) $\lambda_t/\lambda_a = 2.20$ (h=3 mm), (d) $\lambda_t/\lambda_a = 2.74$ (h=5 mm).

Thickness h [mm]	Nodal interval of lattice-mode d_s [mm]	$\frac{\lambda_t}{\lambda_a}$	Angle from z-axis to main lobe γ_m [deg]		Angle width of half value [deg]	
			Calculated value	Experimental value	Calculated value	Experimental value
1	15.8	1.29	51.1	50.2	19.2	19.6
2	21.7	1.83	33.2	33.7	10.0	9.8
3	26.4	2.20	27.0	27.2	7.8	7.8
5	34.1	2.74	21.4	21.4	6.0	6.4

Table 2. Directivity of the square plate vibrating in the lattice mode.

2.6.4 Directivities for different wavelengths of flexural vibration of plate

Next, to examine the directivity for different wavelengths (λ_t) of the flexural vibration of the plate, the sound pressure on the *y-z* plane (a = 90°, the case of (a) in Fig. 8), on which main lobes were observed, was measured by changing γ from -90 to 90° for the vibrating plates with a constant frequency with N = 10 and h in the range of 1-5 mm (see Table 1). The measurement results are shown by the solid lines in Fig. 11: (a) λ_t/λ_a = 1.29 (h = 1 mm), (b) λ_t/λ_a = 1.83 (h = 2 mm), (c) λ_t/λ_a = 2.20 (h = 3 mm), and (d) λ_t/λ_a = 2.74 (h = 5 mm). The ordinate and abscissa are the same as those in Fig. 9(a). From Figs. 11(a)—11(d), it is found that the radiated sound waves have main lobes in two directions and that their directions are symmetric about the *z*-axis with an angle depending on λ_t/λ_a, similarly to the cases in Figs. 9(a) and 9(b). The dashed lines in Fig. 11 represent the calculation results obtained using eq. (9), and are in good agreement with the experimental results.

As shown in Table 2, γ_m decreased when λ_t/λ_a increased, i.e., when h, and thereby the wavelength of flexural vibration, increased. The experimental and calculated values of γ_m were in good agreement.

Moreover, the angle range of full-width half-maximum of the sound pressure was compared among the main lobes, as shown in the rightmost column in Table 2. This angle range decreased with increasing λ_t/λ_a, indicating that the directivity became sharp.

2.7 Summary

The details of the aerial ultrasonic source with a square plate vibrating in the lattice mode described in this section are summarized as follows.

1. A practical method for designing the flexurally vibrating plate was proposed.
2. The angle between the nodal lines of the lattice-mode vibration and the sides of the plate was 45°.
3. The calculation results for the distributions of the vibration displacement of the plate were confirmed by measuring the distributions of vibration displacement and sound pressure near the plate surface.
4. Regarding the directivity of the plate vibrating in a lattice mode, the theoretical values calculated using eq. (9) and the experimental values for the fabricated plates were in good agreement.
5. The sound waves that were radiated from the vibrating plate had sharp main lobes in four symmetric directions on the *y-z* and *x-z* planes, with a specific angle from the *z*-axis.
6. The angle between the *z*-axis and the main lobe, γ_m, decreased when λ_t/λ_a increased, i.e., when h, and thereby the wavelength of flexural vibration, increased.
7. The angle range of the main lobe in which the sound pressure was half the maximum value decreased with increasing λ_t/λ_a, indicating that the directivity became sharp.

3. Removal of unnecessary gas by intense aerial ultrasonic source

3.1 Outline

In this section, I discuss the effect of the intense aerial ultrasonic source described in section 2 on enhancing the removal of an unnecessary gas from air. In previous studies, We examined the effect of the combined use of aerial ultrasonic waves with water mist on the removal of a gas, with the aim of applying it to the removal of an unnecessary gas. As a result, we found that both aerial ultrasonic waves and water mist are necessary to remove a low-hydrophilicity lemon-odor gas and that the maximum removal percentage reached approximately 40%, (Miura, 2007a).

In this method, intense ultrasonic waves are propagated into air from the acoustic source using the square plate flexurally vibrating in the lattice mode, and water is added dropwise onto the positions of the antinodes of flexural vibration on the plate to generate a mist of water microparticles. Ultrasonic waves are irradiated onto the water microparticles and a gas to enhance the absorption of the gas by the water microparticles. The flexurally vibrating plate has two roles: the radiation of sound waves and the formation of water microparticles. The frequency of collisions between microparticles and gas molecules increases because of differences in the velocity of particles in the sound waves and in their momentum owing to differences in their size, enhancing the aggregation and settlement of water microparticles and gas molecules.

To further increase the removal efficiency for a low-hydrophilicity gas, water mist of small particles (average particle diameter, approximately 3 μm) generated using a water spray system was used, as well as water mist of large particles (average particle diameter, approximately 60 μm) formed by the flexural vibration of the plate. The gas removal efficiency achieved when aerial ultrasonic waves are irradiated onto these water particles is discussed. Here, the particle diameter was calculated using Lang's equation (Lang, 1962).

In this section, I describe the process of the removal of a gas by applying water mist of small particles when intense standing-wave acoustic fields were formed within a gas removal chamber. Also, I discuss the gas removal efficiency achieved upon changing the amounts of the water mist of large and small particles, the electric power input to the transducer, and the initial gas concentration.

Although a toxic gas such as dioxin could be used as the gas to be removed, a gas harmless to humans should be used because the experiments were carried out in a simple laboratory. The gas also was required to have low hydrophilicity so as not to easily aggregate in water and be usable at normal temperatures and pressures, and its gas concentration was required to be easily measurable. To satisfy these conditions, a gas evaporated from lemon oil (Sigma-Aldrich Corporation; Product No., W262528; chemical formula, $C_{10}H_{16}O$; specific gravity, 0.855) was used.

3.2 Gas removal apparatus

In this section, I give an outline of the gas removal apparatus used in the experiment.

3.2.1 Outline of apparatus

Figure 12 shows a schematic of the experimental setup. The gas removal apparatus used in this experiment consists of a supply fan with an activated carbon filter (amount, 92 cm^3/s), a unit for generating the lemon-odor gas, a digital constant-rate pump for supplying water to the vibrating plate, a water spray system for generating water mist of small particles, an acoustic source mainly comprising a 20 kHz BLT transducer and a square plate vibrating flexurally in the lattice mode, an acrylic chamber as the main body, and an exhaust fan for removing air passing through the chamber. The flexurally vibrating plate of the acoustic source, 36 narrow hypodermic needles that supply water to the vibrating plate to form water mist of large particles, and reflective boards used to form standing-wave acoustic fields are included in the acrylic chamber. The digital constant-rate pump is used to supply water at a constant rate to the vibrating plate within the acrylic chamber. In addition, a 2.4 MHz ultrasonic humidifier equipped with a water spray system is used to generate water mist of small particles.

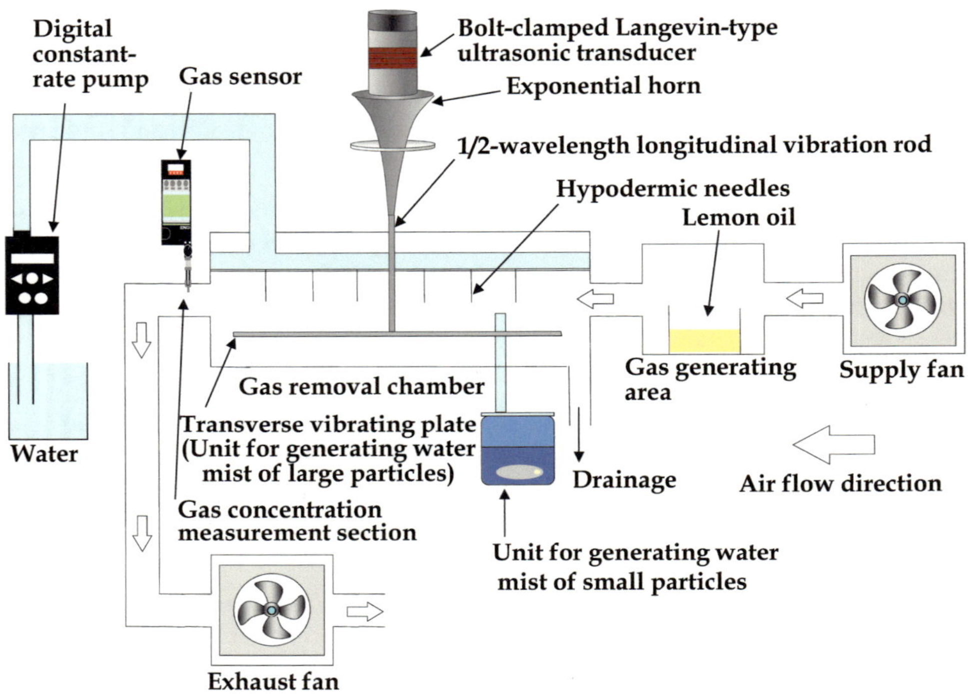

Fig. 12. Schematic of gas removal apparatus.

To measure the gas concentration, a hot-wire semiconductor-type gas sensor is attached to the pipe that connects the acrylic chamber to the exhaust fan.

3.2.2 Aerial ultrasonic source used in experiment

Figure 13 shows a schematic of the ultrasonic source used in the experiment. As shown in the figure, an exponential horn (diameter of thick end face, 70 mm; diameter of thin end face, 10 mm; length, 150 mm; amplification rate, 7.0) and a half-wavelength longitudinal vibration rod (diameter, 10 mm; length, 116 mm) were connected to a 20 kHz BLT transducer (D45520). A duralumin square plate vibrating flexurally in the lattice mode (plate constant, $C_P = 1.51 \times 10^6$ Hz·mm; $N = 12$; $h = 3$ mm; $L = 217$ mm; resonant frequency, 19.8 kHz) was screwed onto the end of the rod.

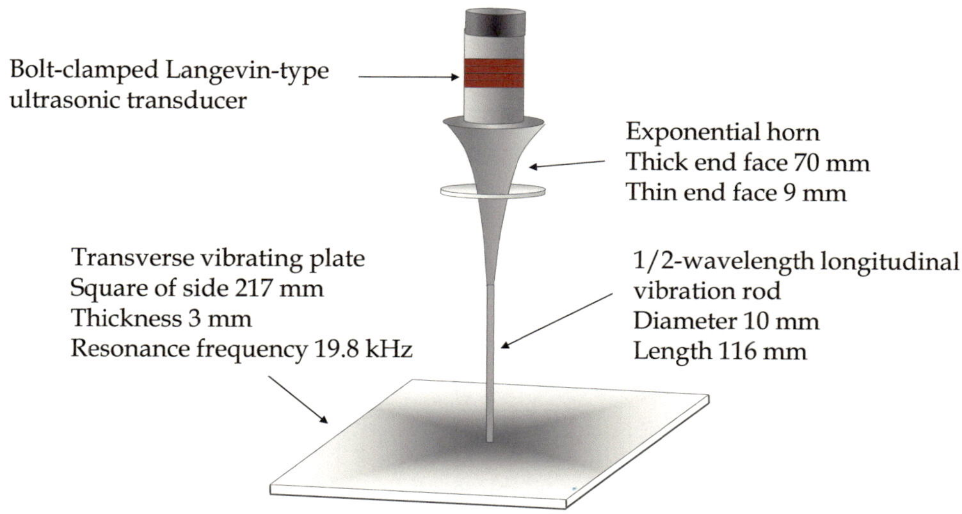

Fig. 13. Schematic of ultrasonic transducer.

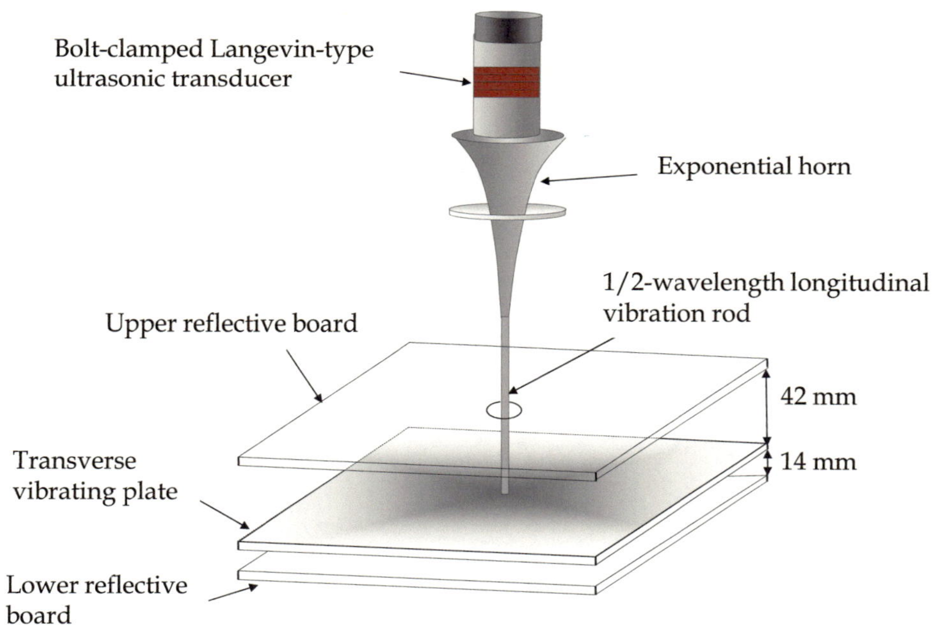

Fig. 14. Positions of reflective boards.

3.2.3 Distances between flexurally vibrating plate and reflective boards

Planar reflective boards were installed above and below the vibrating plate to form an intense standing-wave acoustic field within the acrylic chamber, as shown in Fig. 14.

The relationship between the velocity c and wavelength λ_a of the sound waves radiated from the vibrating plate into free space is as follows.

$$\lambda_a = \frac{c}{f} \tag{11}$$

Here, $c = 331.5 + 0.6t$ (temperature (t) = 25 °C) and f is the frequency. The wavelength of the sound waves transmitted in the vertical direction, λ_y, is given by

$$\lambda_y = \frac{\lambda_a}{\sin\theta} \tag{12}$$

where θ is the radiation angle. Assuming the wavelength of the flexural vibration of the plate to be λ_t, θ is expressed by

$$\theta = \cos^{-1}\left(\frac{\lambda_a}{\lambda_t}\right) \tag{13}$$

To form standing waves between the vibrating plate and the upper reflective board, the distance between them, W, must be set at an integral multiple of the half-wavelength of the sound waves transmitted in the vertical direction and is given as follows.

$$W = \frac{\lambda_y}{2} n \ \ (n \text{ is an arbitrary integer}) \tag{14}$$

Here, W is set to 42 mm, which is double the wavelength of the sound waves transmitted in the vertical direction. In this case, the electric impedance of the transducer is maximized. The distance between the vibrating plate and the lower reflective board is set to 14 mm, at which the electric impedance of the transducer is minimized.

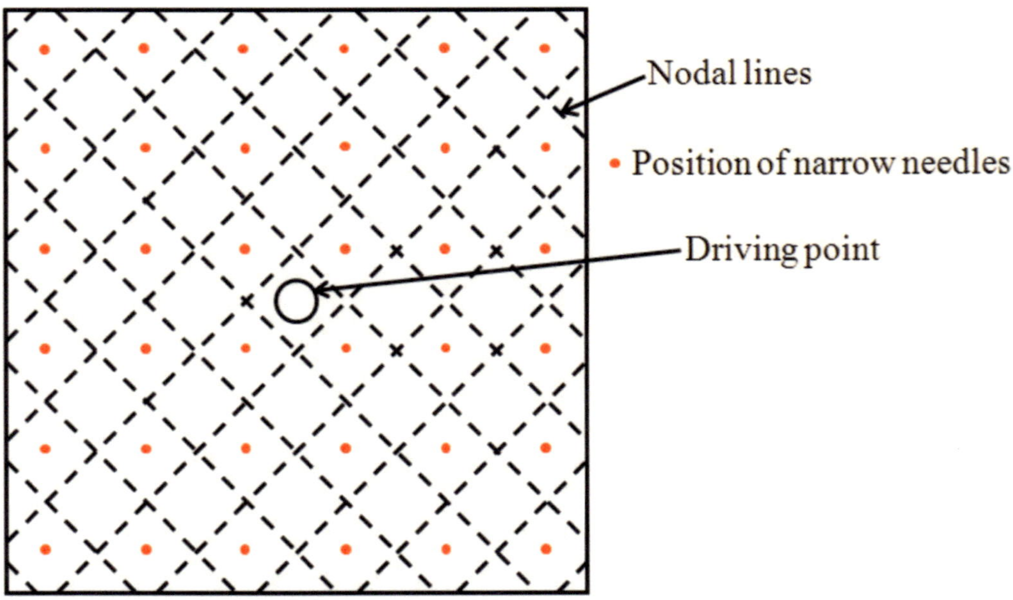

Fig. 15. Positions of narrow needles.

3.2.4 Position of narrow needles

In the experiment, water mist to absorb the gas was generated by supplying water to the vibrating plate. Figure 15 shows the positions of the narrow hypodermic needles installed to supply water to the vibrating plate. The dashed lines in the figure represent nodal lines of vibration. Thirty-six narrow needles are positioned exactly above the antinodes at the centers of the regions surrounded by the nodal lines.

3.3 Effect of combined use of aerial ultrasonic waves and water mist on removal of unnecessary gas

In this section, on the basis of experimental results, We examine the effects of using intense aerial ultrasonic waves and two types of water mist of particles with different diameters on the removal of an unnecessary gas.

3.3.1 Gas removal process

The process for removing the lemon-odor gas by irradiating aerial ultrasonic waves was examined in the presence and absence of water mist of large and small particles.

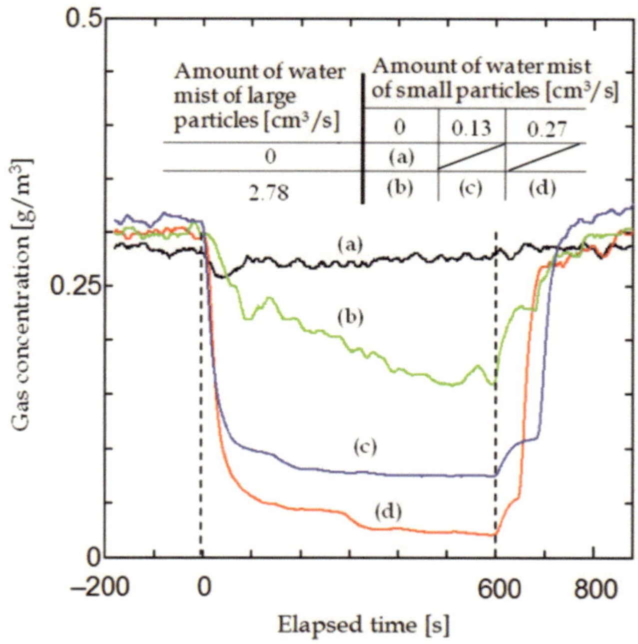

Fig. 16. Gas removal process. The input electric power is 50 W and the driving frequency is 19.8 kHz.

Because the rate of vaporization of the generated gas was unstable when the fan started operating, the apparatus was left to stand for approximately 180 s to stabilize the rate of generation of the gas. The time at which the vaporization rate was stabilized was assigned a time of -180 s, and the apparatus was left to stand for another 180 s. The electric power was input to the ultrasonic source at a time of 0 s to generate ultrasonic waves in air. Simultaneously, water was supplied to the vibrating plate through the narrow needles attached to the upper reflective board, and water mist was formed by vibrating the plate. Moreover, water mist of small particles was also generated by the water spray system. This condition was maintained for 600 s. Then, the generation of the ultrasonic waves and water mist was stopped, and the state within the acrylic chamber was observed.

The experimental conditions were as follows: the input electric power was maintained at 50 W, the driving frequency was 19.8 kHz, the amount of water mist of large particles was 0 or 2.78 cm^3/s, and the amount of water mist of small particles was 0, 0.13, or 0.27 cm^3/s. Figure 16 shows the experimental results, where the ordinate represents the concentration of lemon-odor gas and the abscissa represents the elapsed time. When no mist was generated, case of (a), the gas concentration negligibly decreased during the measurement time from 0 to 600 s (with ultrasonic irradiation). When the amounts of water mist of large and small particles were 2.78 and 0 cm^3/s, respectively, case of (b), the gas concentration decreased slightly. However, when the amount of water mist of small particles was increased to 0.13, case of (c), and 0.27 cm3/s, case of (d), the gas concentration sharply decreased after the start of ultrasonic irradiation, maintained an almost constant value, then increased to the initial value shortly after the end of ultrasonic irradiation (600 s). The results for water mist of small particles in the absence of water mist of large particles are not show, since aggregation and drainage did not occurred in this case. Therefore, the gas was effectively removed by the combined use of ultrasonic waves and the water mist of both large and small particles.

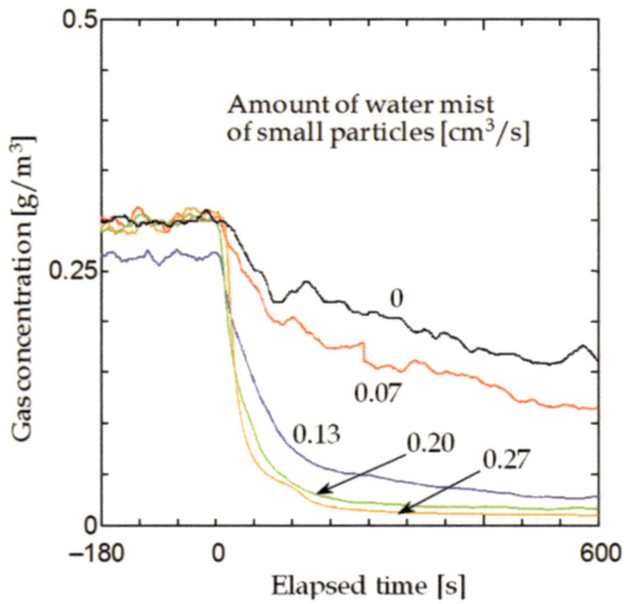

Fig. 17. Relationship between elapsed time and gas concentration. The input electric power is 50 W, the amount of water mist of large particles is 2.78 cm^3/s, and the driving frequency is 19.8 kHz.

3.3.2 Gas removal efficiency with different amounts of water mist of small particles

To examine the gas removal efficiency with different amounts of the water mist of small particles, a gas removal experiment was carried out with the electric power input to the transducer fixed at 50 W and the amount of water mist of large particles maintained at 2.78 cm^3/s while the amount of water mist of small particles was set at 0, 0.07, 0.13, 0.20, or 0.27 cm^3/s. The experimental procedure followed was that described in sec. 3.3.1.

Figure 17 shows the experimental results, where the ordinate represents the concentration of lemon-odor gas and the abscissa represents the elapsed time, with the amount of water mist of small particles as a parameter. In all cases, the gas concentration changed negligibly between -180 and 0 s, during which no ultrasonic waves were irradiated. When ultrasonic waves and the water mist were generated at 0 s, however, the gas concentration sharply decreased.

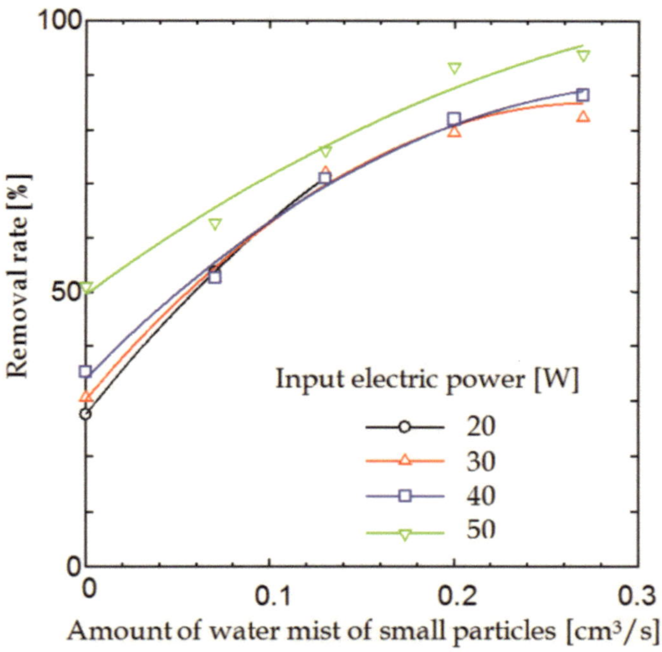

Fig. 18. Relationship between amount of water mist of small particles and removal rate. The amount of water mist of large particles is 2.78 cm^3/s, and the driving frequency is 19.8 kHz.

3.3.3 Gas removal rate with different amounts of water mist of small particles

Similar experiments were also performed for input electric powers of 20, 30, and 40 W as well as above 50 W. The gas removal rate Px was defined as the index describing the removal rate of gas.

$$Px = \frac{G_a - G_b}{G_a} \tag{15}$$

Here, G_a and G_b are the average concentrations of lemon-odor gas [g/m^3] between -180 and 0 s and between 500 and 600 s, respectively. Equation (15) indicates that the greater the value of Px, the higher the gas removal rate.

Figure 18 shows the experimental results, where the ordinate represents the gas removal rate and the abscissa represents the amount of water mist of small particles, with the electric power input to the transducer as a parameter. When the amount of water mist of large particles was constant, the gas removal rate increased with the amount of water mist of small particles; however, no marked difference in the gas removal rate was observed within the range of input electric power examined in this experiment. The gas removal rate reached approximately 90% when the amount of water mist of small particles was 0.27 cm^3/s and the input electric power was 50 W. Moreover, the gas was removed when the amount of water mist of small particles was 0 cm^3/s, i.e., when only water mist of large particles was used; in this case, the gas removal rate reached approximately 60% when the input electric power was 50 W. These results indicate that most of the gas can be removed by using water mist of both large and small particles.

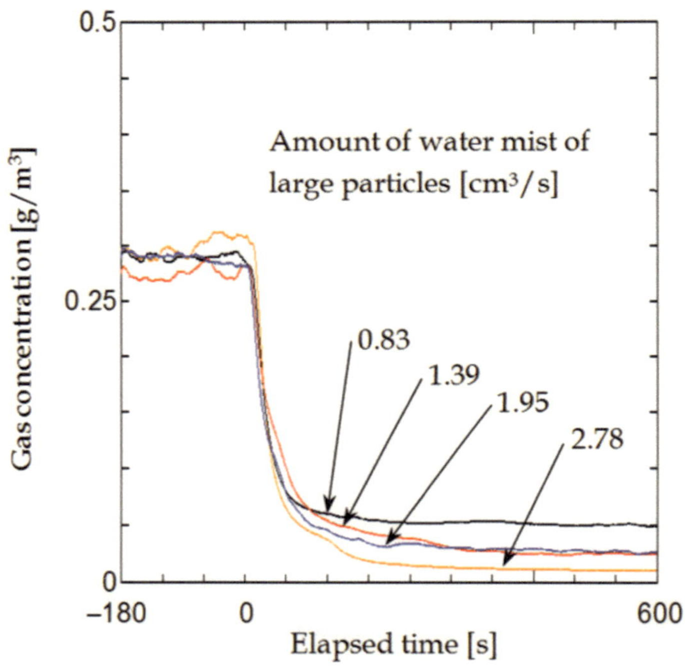

Fig. 19. Relationship between elapsed time and gas concentration. The input electric power is 50 W, the amount of water mist of small particles is 0.27 cm^3/s, and the driving frequency is 19.8 kHz.

3.3.4 Gas removal rate with different amounts of water mist of large particles

To examine the gas removal efficiency with different amounts of water mist of large particles, a gas removal experiment was carried out with the electric power input to the transducer fixed at 50 W and the amount of water mist of small particles maintained at 0.27 cm^3/s while the amount of water mist of large particles was set at 0.83, 1.39, 1.95, or 2.78 cm^3/s. The experimental procedure followed was that described in sec. 3.3.1.

Figure 19 shows the experimental results, where the ordinate represents the concentration of lemon-odor gas and the abscissa represents the elapsed time, with the amount of water mist of large particles as a parameter. In all cases, the gas concentration changed negligibly between -180 and 0 s, during which no ultrasonic waves were irradiated. When ultrasonic waves and the water mist were generated at 0 s, however, the gas concentration sharply decreased and reached a constant value after approximately 200 s.

Next, similar experiments were performed for input electric powers of 30 and 40 as well as 50 W, to obtain the gas removal rate. The gas removal rate was then calculated using eq. (15).

Figure 20 shows the experimental results, where the ordinate represents the gas removal rate and the abscissa represents the amount of water mist of large particles, with the input electric power as a parameter. When the input electric power was in the range examined in this experiment and the amount of water mist of small particles was constant, the gas removal rate tended to increase as the amount of water mist of large particles and the input electric power increased.

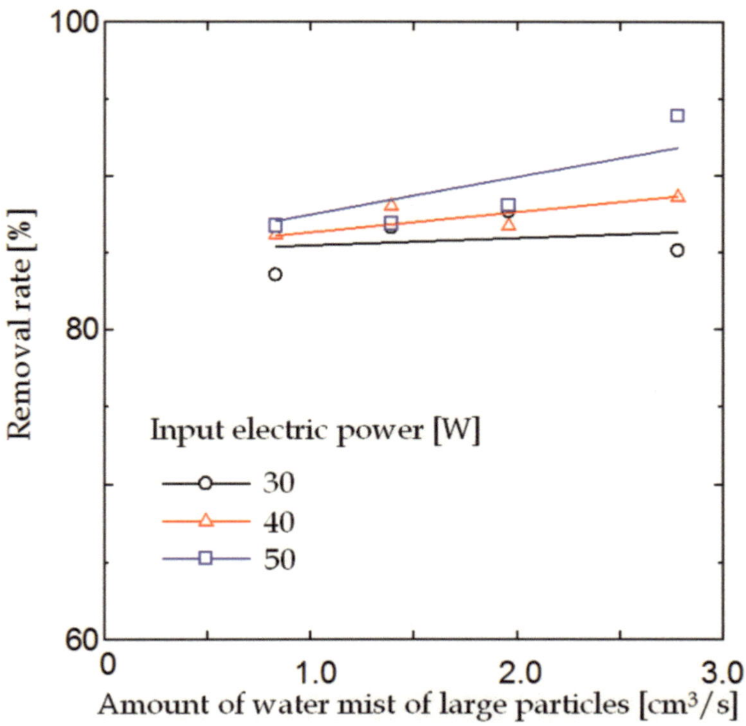

Fig. 20. Relationship between amount of water mist of large particles and removal rate. The Amount of water mist of small particles is 0.27 cm^3/s, and the driving frequency is 19.8 kHz.

3.3.5 Examination of gas removal efficiency with different initial gas concentrations

To examine the gas removal efficiency with different initial concentrations of lemon-odor gas, a gas removal experiment was performed. In this experiment, the area of the base of the petri dish into which the lemon oil was placed was changed to vary the amount of generated gas. The experimental procedure was the same as before.

Figure 21 shows the experimental results, where the ordinate represents the concentration of lemon-odor gas and the abscissa represents the elapsed time, with the initial gas concentration as a parameter. In all cases, the gas concentration remained almost constant between -180 and 0 s, during which no ultrasonic waves were irradiated. When ultrasonic waves and the water mist were generated at 0 s, however, the gas concentration sharply decreased.

Next, a similar experiment was carried out to obtain the gas removal rate with different initial gas concentrations. The gas removal rate was calculated from the experimental results using eq. (15). Figure 22 shows the results, where the ordinate represents the gas removal rate and the abscissa represents the initial gas concentration. When the amounts of water mist of small and large particles and the input electric power were constant, the gas removal rate tended to increase with the initial gas concentration in the range examined in this experiment. This was considered to be because the higher the gas concentration, the greater the number of opportunities for the water mist to absorb gas particles.

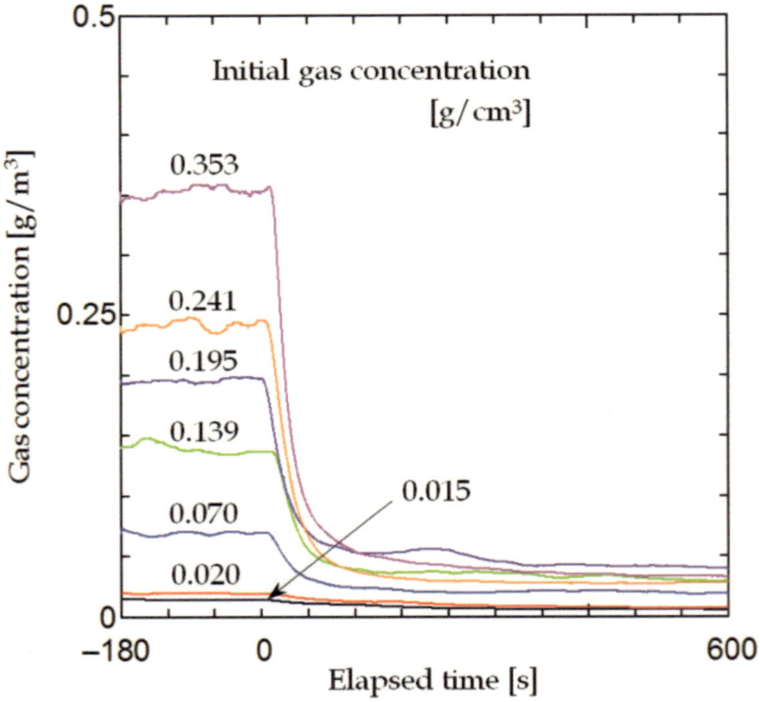

Fig. 21. Relationship between elapsed time and gas concentration. The input electric power is 50 W, the amount of water mist of large particles is 2.78 cm^3/s, the amount of water mist of small particles is 0.27 cm^3/s, and the driving frequency is 19.8 kHz.

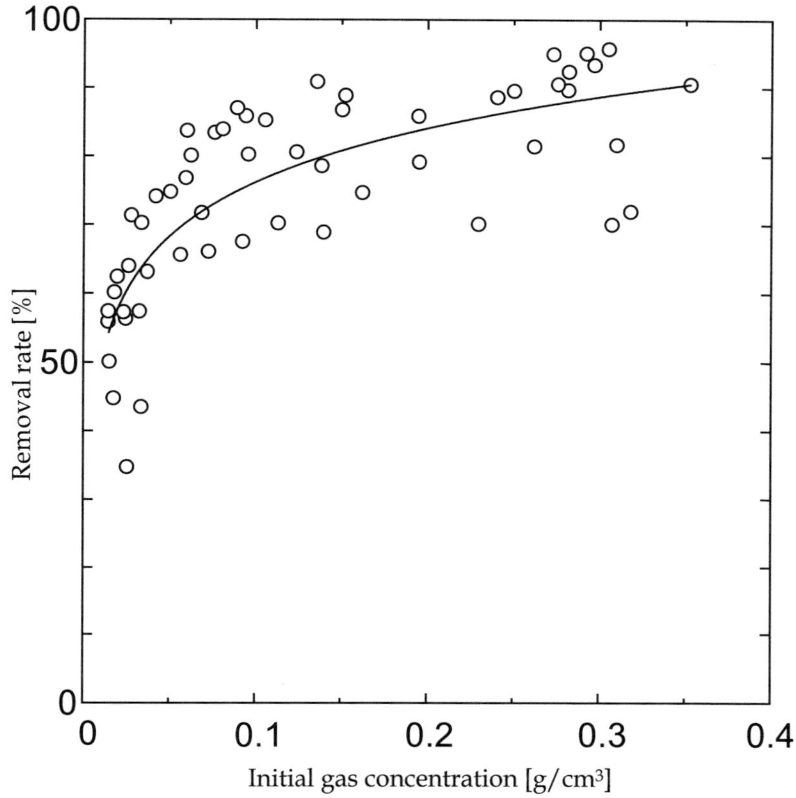

Fig. 22. Relationship between initial gas concentration and removal rate. The input electric power is 50 W, the amount of water mist of large particles is 2.78 cm^3/s, the amount of water mist of small particles is 0.27 cm^3/s, and driving frequency is 19.8 kHz.

3.4 Summary

In this section, I examined the effect on the increase in gas removal rate obtained when a low-hydrophilicity gas was absorbed and aggregated by irradiating aerial ultrasonic waves onto water mist of large particles (average particle diameter, approximately 60 μm) formed by the vibration of the plate and water mist of small particles (average particle diameter, approximately 3 μm) generated using a water spray system.

The following findings were obtained.

1. Even a low-hydrophilicity gas was effectively removed by the combined use of ultrasonic waves and water mist.
2. The gas removal rate increased when two kinds of water mist of large and small particles were simultaneously used.
3. When the amount of water mist of large particles was constant, the gas removal rate increased with the electric power input to the transducer and the amount of water mist of small particles for the ranges of input power and amount examined in this study.

4. When the amount of water mist of small particles was constant, the gas removal rate tended to increase with the electric power input to the transducer and the amount of water mist of large particles.
5. When the two kinds of water mist of large and small particles were used and subjected to ultrasonic irradiation, the gas removal rate increased with the initial gas concentration.
6. The gas removal rate reached approximately 90% when the amounts of water mist of small and large particles were 0.27 and 2.78 cm^3/s, respectively, and the electric power input to the transducer was 50 W.

From these results, it was found that a gas can be effectively removed by irradiating ultrasonic waves onto water mist of both large and small particles.

4. Conclusion

In this chapter, I described a method of designing square plates vibrating flexurally in the lattice mode that can be used as an intense aerial ultrasonic source, as well as the vibration mode of the fabricated plates. The distributions of the vibration displacement of the fabricated plates and the sound pressure near the plate surface were also described. In addition, the validity of the ultrasonic source was verified by theoretically and experimentally determining the directivity of the sound waves radiated from the vibrating plates into remote acoustic fields.

As an application of intense aerial ultrasonic waves, the enhancement of the removal of an unnecessary gas was examined. It was demonstrated that up to 90% of the gas was removed when aerial ultrasonic waves were irradiated onto water mist of large particles (average particle diameter, approximately 60 μm) formed by the flexural vibration of the plate and water mist of small particles (average particle diameter, approximately 3 μm) separately generated using a water spray system.

Thus, the use of aerial ultrasonic energy to remove unwanted gases appears to be promising; however, many unclear points still remain. Their clarification is expected to lead to an expansion in the range of applications of ultrasonic energy.

5. References

Asami, T. & Miura, H. (2010). Longitudinal Vibration Characteristics Required to Cut a Circle by Ultrasonic Vibration. *Japanese Journal of Applied Physics,* Vol. 49, 07HE23, (July 2010), pp. 07HE23-1-7

Asami, T. & Miura, H. (2011). Vibrator Development for Hole Machining by Ultrasonic Longitudinal and Torsional Vibration. *Japanese Journal of Applied Physics,* Vol. 50, 07HE31, (July 2010), pp. 07HE31-1-9

David, J. & Cheeke, N. (2002). *Fundamentals and applications of ultrasonic waves,* CRC Press, ISBN 0-8493-0130-0, Florida, USA.

Ito, Y. (2005). Experimental Investigation of Deflection of High-Speed Water Current with Aerial Ultrasonic Waves. *Japanese Journal of Applied Physics,* Vol. 44, Part 1, No. 6B, (June 2005), pp. 4669-4673.

Ito, Y. & Takamura, E. (2010). Removal of Liquid in a Long Pore Opened at Both Ends Using High-Intensity Aerial Ultrasonic Waves. *Japanese Journal of Applied Physics*, Vol. 49, 07HE22, (July 2010), pp. 07HE22-1-6.

Kobayashi, M., Kamata, C. & Ito, K. (1997). Cold Model Experiments of Gas Removal from Molten Metal by an Irradiation of Ultrasonic Waves. *The Iron and Steel Institute of Japan (ISIJ) International*, Vol. 37, No. 1, (January 1997), pp. 9-1 5.

Lang, R. J. (1962). Ultrasonic Atomization of Liquids. *The Journal of the Acoustical Society of America*, Vol. 34, No. 1, (January 1962), pp. 6-8.

Miura, H. (1994). Aerial Ultrasonic Vibration Source using a Square Plate Vibrating in a Transverse Lattice-Mode. *The Journal of the Acoustical Society of Japan*, Vol. 50, No. 9, (September 1994), pp. 677-684.[in Japanese]

Miura, H., & Honda, Y. (2002). Aerial Ultrasonic Source Using a Striped Mode Transverse Vibrating Plate with Two Driving Frequencies. *Japanese Journal of Applied Physics*, Vol. 41, Part 1, No. 5B, (May 2002), pp. 3223-3227.

Miura, H. (2003). Eggshell Cutter Using Ultrasonic Vibration. *Japanese Journal of Applied Physics*, Vol. 42, Part 1, No.5B, (May 2003), pp. 2996-2999.

Miura, H. (2004). Promotion of Sedimentation of Dispersed Fine Particles Using Underwater Ultrasonic Wave. *Japanese Journal of Applied Physics*, Vol. 43, Part 1, No. 5B, (May 2004), pp. 2838-2839.

Miura, H., Takata, M., Tajima, D. & Tsuyuki, K. (2006). Promotion of Methane Hydrate Dissociation by Underwater Ultrasonic Wave. *Japanese Journal of Applied Physics*, Vol. 45, No. 5B, (May 2006), pp.4816-4823.

Miura, H. (2007a). Removal of Unnecessary Gas by Spraying Water Particles Formed by Aerial Ultrasonic Waves. *Japanese Journal of Applied Physics*, Vol. 46, No. 7B, (July 2007), pp. 4926-4930.

Miura, H. (2007b). Atomization of High-Viscosity Materials by One Point Convergence of Sound Waves Radiated from an Aerial Ultrasonic Source using a Transverse Vibrating Plate. *The 19th International Congress on Acoustics*, (September 2007), ULT-09-017.

Miura, H. (2008). Vibration Characteristics of Stepped Horn Joined Cutting Tip Employed in Circular Cutting Using Ultrasonic Vibration. *Japanese Journal of Applied Physics*, Vol. 47, No. 5, (May 2008), pp. 4282-4286.

Miura, H. & Ishikawa, H. (2009). Aerial Ultrasonic Source Using Stripes-Mode Transverse Vibrating Plate with Jutting Driving Point. *Japanese Journal of Applied Physics*, Vol. 48, 07GM10, (July 2009), pp. 07GM10-1-4.

Onishi, Y. & Miura, H. (2005). Convergence of Sound Waves Radiated from Aerial Ultrasonic Source Using Square Transverse Vibrating Plate with Several Reflective Boards. *Japanese Journal of Applied Physics*, Vol. 44, No. 6B, (June 2005), pp. 4682-4688.

Ueha, S., Maehara, N. & Mori,E. (1985). Mechanism of Ultrasonic Atomization using a Multipinhole Plate. *Journal of the Acoustical Society of Japan (E)*, Vol. 6, No. 1, (January 1985), pp. 21-26.

Yamane, H., Ito, Y., & Kawamura, M. (1983). Sound Radiation from Rectangular Plate Vibrating in Stripes Mode. *The Journal of the Acoustical Society of Japan*, Vol. 39, No. 6, (June 1983), pp. 380-387.[in Japanese]

7

Application of Pulsed Ultrasonic Doppler Velocimetry to the Simultaneous Measurement of Velocity and Concentration Profiles in Two Phase Flow

N. Sad Chemloul, K. Chaib and K. Mostefa
University Ibn Khaldoun of Tiaret
Algeria

1. Introduction

The improvement of knowledge which governs the transport of the solid-liquid suspensions is the subject of many works having generally led to empirical or semi-empirical models which are valid only for specific conditions. Research carried out on solid-liquid suspensions has investigated the continuous phase and particularly the influence of turbulence modulation. In the works by Elghobashi and Truesdell (1993), Michaelides and Stock (1989), Owen (1969), and Parthasarathy and Faeth (1990a, 1990b), dissipation or production of turbulent kinetic energy in the continuous phase were reported. The effect of particles on the carrier flow turbulence was investigated numerically by Varaksin and Zaichik (2000) and, Lei (2000). Recently PUDV (pulsed ultrasonic Doppler velocimetry) was applied to study fluid flow alone or with solid particles by Takeda (1995), Aritomi et al. (1996), Nakamura (1996), Rolland and Lemmin (1996, 1997), Cellino and Graf (2000), Brito et al (2001), Eckert and Gerbeth (2002), Kikura et al. (1999), Kikura et al. (2004), Xu (2003), Alfonsi et al. (2003). Note that the measurement techniques cited in the above references are limited by the nature of the suspensions.

The aim of this experimental chapter is the simultaneous measurement of the local parameters which are the velocity and the concentration fields of the solid particles in flows in a horizontal pipe. The difference between our application of PUDV and those cited in the above references resides in the use of a new measurement approach of the local concentration of the solid particles. This approach consists of the representation of the local concentration profile by the ratio of the number of solid particles crossing the measurement volume, to the number of solid particles crossing the control volume. PUDV technique was selected rather than hot wire or film, or Laser, or the PIV technique, because the first would be destroyed by the particles, and with the second the ultrasonic signal is more attenuated when the volumetric concentration C_V of particles increases. The third technique as reported by Jensen (2004) requires many conditions for its application.

2. Experiment

The working principle of pulsed ultrasound Doppler velocimetry is to detect and process many ultrasonic echoes issued from pulses reflected by micro particles contained in a flowing liquid. A single transducer emits the ultrasonic pulses and receives the echoes. By sampling the incoming echoes at the same time relative to the emission of the pulses, the variation of the positions of scatterers are measured and therefore their velocities. The measurement of the time lapse between the emission of ultrasonic bursts and the reception of the pulse (echo generated by particles flowing in the liquid) gives the position of the particles. By measuring the Doppler frequency in the echo as a function of time shifts of these particles, a velocity profile after few ultrasonic emissions is obtained.

In this study, PUDV technique originally applied in the medical field is used only for the emission and the reception of the ultrasonic signal. This technique was combined with a data processor for the flow measurement of solid-liquid suspension. This combination allows the determination of a local velocity and a local concentration of the solid particles larger than the wavelength of the ultrasonic wave.

2.1 Flow circuit

The flow circuit (Fig. 1a.) consists of a closed loop made of glass pipes with an internal diameter D of 20 mm. The flow is driven by a variable speed centrifugal pump (1). The suspension is kept at a constant temperature by the heat exchanger (2) during measurement. The test section (4) (for detail see Fig 1b), realized in a Plexiglas box which is 150 mm long, 100 mm wide, and 50 mm high, was located at 75D downstream of the pump where the flow was fully-developed. Plexiglas was chosen in order to reduce the reflection of the ultrasonic beam when it crosses the wall. The pressure differential along the test pipe given by two differential pressure transducer (3) allows the determination of the wall shear stress.

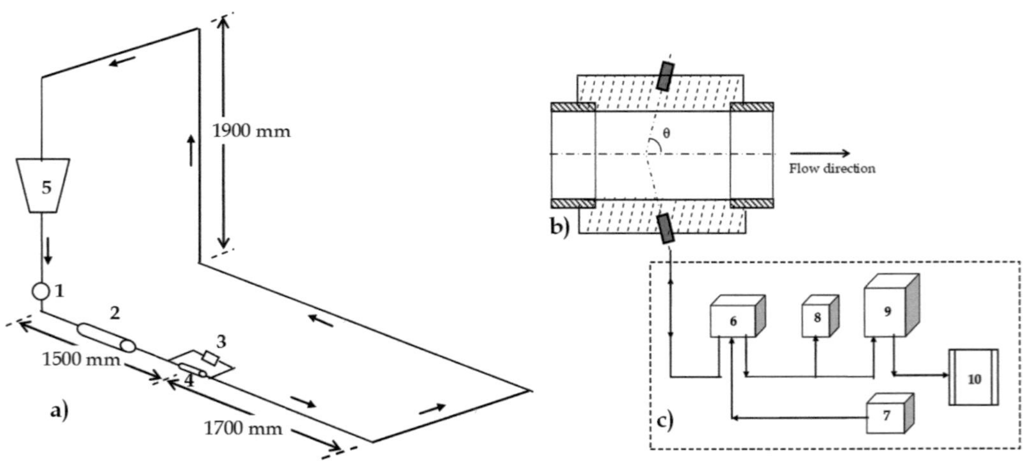

Fig. 1. a) Flow circuit: 1 pump, 2 heat exchanger, 3 differential pressure transducer, 4 test section, 5 tank of suspension; b) Detail of the test section 4: (▬) glass pipe, (▭) Plexiglas box, (▬) Ultrasonic transducer; c) Ultrasonic measurement with acquisition and treatment system: 6 displacement system of the measurement volume, 7 ultrasonic Doppler velocimeter (ECHOVAR CF8, ALVAR), 8 digital storage oscilloscope, 9 data processor (plurimat S), 10 computer

This study was performed using water as the continuous phase and glass beads for solid particles. The particles are spherical with a 5% sphericity defect and density of 2640 kg/m^3. Four samples of different particle size distributions were tested. The volume-averaged mean particle diameters d_p corresponding to these samples are 0.27, 0.3, 0.4, and 0.7 mm. These particles were chosen to be larger than the Kolmogorov length scale η, estimated to 206.6 µm near the wall. The particle diameter is ranging between 1.30 η and 3.38 η. The volumetric concentrations (C_v) of the glass beads used in the suspension are 0.5%, 1%, 1.5% and 2%, and were determined from the volume of the flow circuit. Performed measurements within this work consider a two way coupling, i.e., taking into account the effects of the particles on the carrier fluid and vice versa. The classification of the suspension flow used is made according to Sato (1996), Elghobashi (1994) and Crowe, et al. (1996) who proposed the particle volume fractions as the criterion of classification. The particle volume fractions are ranging between 1.25 x10^{-3} and 5x10^{-3}. Fine starch particles with 6 µm of diameter and density of 1530 kg/m^3 were used as a tracer. The maximum concentration of the starch was fixed at 3% in order to reduce the attenuation of the Doppler signal. The flow mean velocity U_{moy} is ranging from 1 m/s to 2.5 m/s.

2.2 Ultrasonic measurement system of the velocity profiles

The method used for measurement local velocity of the large particles (larger than the wavelength of the ultrasonic wave) is based on a combination of the measurement technique PUDV (7) (type ECHOVAR CF8) used in medical physics, and a data processor (9). The technical specifications of the velocimeter are: emission frequency 8 MHz, pulse durations 0.5 µs, 1 µs, 2 µs, and pulse repetition frequency 64 µs and 32 µs. The maximum measurable distance of these two pulse repetition frequency are respectively 48 mm and 24 mm.

The position adjustment of the measurement volume is done by time step of 0.5 µs between the emission and reception of the ultrasonic wave. This time step corresponds to the penetration depth of the measurement volume of 0.37 mm for an angle θ of 67° (θ angle between the internal pipe wall and the direction of the propagation of the ultrasonic waves). The value of θ is nearly equal to that determined by the calibration of the ultrasonic transducer (67.4°). The dimensions of the cylindrical measurement volume are the same as those of the ultrasonic transducer, diameter of 2 mm and length of 0.8 mm.

In this experimental study, the signal is emitting and receiving by the same ultrasonic transducer, in this case the velocity U (axial velocity component) of the solid particles determined from the Doppler frequency is:

$$U = \frac{cf_D}{2f_e \cos\theta} \tag{1}$$

with f_D the Doppler shift frequency, f_e the emitted frequency by the transducer, c speed of the sound in the water, and θ the angle between the ultrasonic beam and the pipe axis.

The use of the PUDV technique for the solid-liquid suspensions remains however, limited by the concentration of the solid particles. Indeed, the preliminary study results show that the concentration of the solid particles and depth of the measurement volume affect the ultrasonic signal. Figure 2 shows that for a concentration higher than 2.5%, the attenuation of the signal is about 80% (or 20% of coherent signal), and thus the maximum volumetric concentration of solid particles used in this study was C_v= 2%. Because of the non-uniform distribution of the concentration in the test section, the flow is divided in two regions having the horizontal line passing through the position of the maximum concentration as a boundary. To obtain this boundary line, which corresponds to the great number of the Doppler signal visualised by a digital storage oscilloscope (Fig. 1c), we have scanned the test section by the displacement of the ultrasonic transducer along the vertical diameter. Before each measurement, the boundary line is located to be taken as the first measurement point. The first measurements with a two same ultrasonic transducers (one fixed on the pipe top wall and the other on the pipe bottom wall) shown that at the same position of the measurement volume, the velocity measured presents a difference about 3 - 4 %. For the high concentration, this difference is more significant.

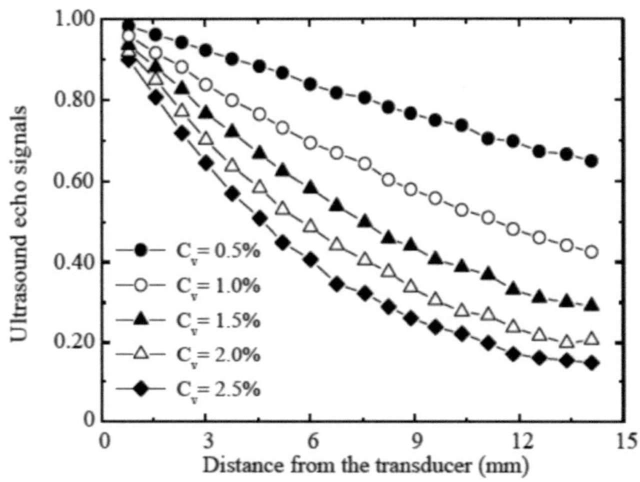

Fig. 2. Signal attenuation function of the distance from the ultrasonic transducer and the volumetric concentration of particle, dp = 0.7 mm

2.2.1 Signal processing

The treatment of the Doppler signal is done using a data processor associated with the ultrasonic velocimeter. The successive Doppler signals received by the ultrasonic transducer are result from either the same particle reached by successive impulses, or various particles crossing the measurement volume. These signals depend on the particles size, the dimensions of the measurement volume, and the velocity of the particles. To avoid the spectrum overlap phenomenon and thus the loss of information, the output signal of the velocimeter is sampled with a sampling frequency about 5 to 10 times the greatest frequency of the signal spectrum. The numerical data of the sampling process are stored in the memory of the data processor to be treated. The signal processing is made by an elaborate software where frequency domain and Fourier transform were used. Figure 3 shows that the power spectral density obtained from different wall distance have the same trend as that of Gauss with a value correlation coefficient close to 0.97. To satisfy the symmetry condition of the power spectrum, a threshold was fixed, only the values greater than 2/3 of the maximum power spectrum were taken into account. The peak amplitude of the power spectra increases with increasing particle diameter; this confirms the difference in energy between the signals coming from the bead glass and those from the starch particles. The separation between the Doppler signals of the continuous phase and large particle is made using two fixed thresholds on the integral of the power spectral density. The higher threshold S_{sup} and the lower threshold S_{inf} are respectively given by the following relations:

$$\begin{cases} S_{sup} = E_{max} - \dfrac{E_{max} - E_{min}}{3} \\ S_{inf} = E_{min} + \dfrac{E_{max} - E_{min}}{3} \end{cases} \tag{2}$$

where E_{max} and E_{min} are respectively, the maximal and the minimal value of the integral of the power spectral density.

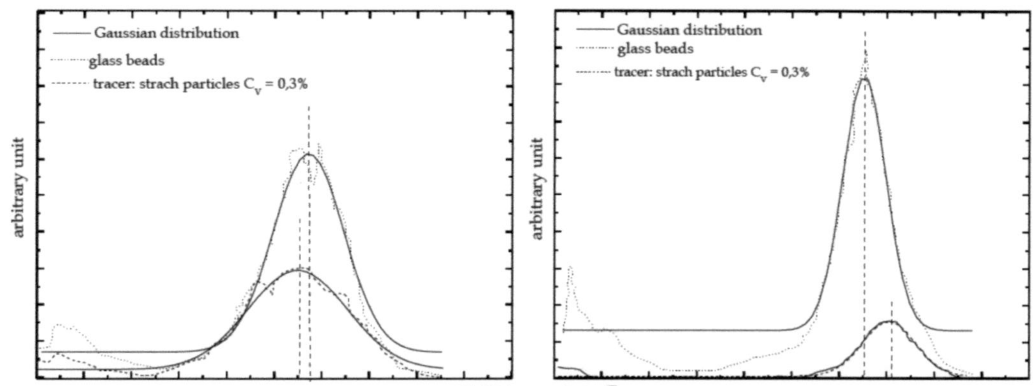

Fig. 3. Doppler power spectrum in the case of glass beads (d_p= 0.7 mm, C_v= 1%) and tracer versus the wall distance

The sampled function $\hat{x}(t)$ of the Doppler signal $x(t)$ is given by:

$$\hat{x}(t) = x(t)\sum \delta(t - \frac{K}{f_e}) \tag{3}$$

where K is a constant and δ the Dirac's impulse. According to the formula of Poisson, we have:

$$\hat{x}(\nu) = X(\nu)\sum \delta(\nu - n f_e) \tag{4}$$

where $X(\nu)$ is the Fourier transform of the Doppler signal $x(t)$. The values treated using the Fourier transform allow to the calculation of the power spectral density P(f) of the Doppler signal from the product of the frequency spectrum and its conjugate. The average frequency Doppler $\overline{f_D}$ is given by the normalized moment of order 1 of the ensemble average spectrum.

$$\overline{f_D} = \frac{\sum_{n=1}^{N} f_n G(f_n)}{\sum_{n=1}^{N} G(f_n)} \tag{5}$$

where $G(f_k) = \sum_{n=1}^{N} P_n(f_k)$ and k = 1, 2, 3,...........,N. Considering that the enlargement of the spectrum is due only to the turbulent velocity fluctuations, we can calculate the normalized moment of second order of the energy spectrum integral associated with the Doppler signals. This moment with a Gaussian trend represents the turbulent intensity given by the relation:

$$\sqrt{\overline{f_D'^2}} = [\sum_{n=1}^{N} (f_n - \overline{f_d})^2 G(f_n) / \sum_{n=1}^{N} G(f_n)]^{1/2} \tag{6}$$

2.2.2 Measurement method of the local concentration profile

This method consists of determining the ratio of the number of particles N_p crossing the measurement volume to the total number of particles N_{pt} crossing the control volume. This control volume is obtained by the displacement of the measurement volume along the vertical diameter of the test section (Fig. 4).

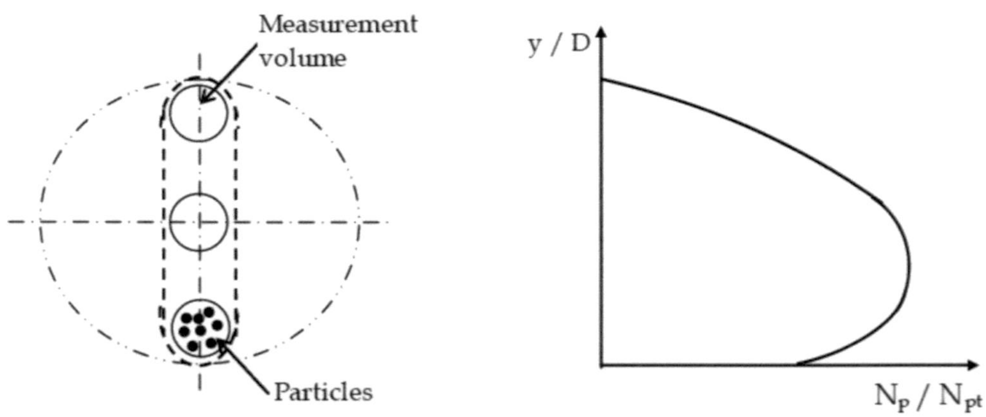

Fig. 4. Determination of the concentration profile: (-) measurement volume Np, (---) control volume N_{pt}, (-·-·-) test section.

The numbers of particles N_p and N_{pt} are obtained by counting the number of the Doppler signal respectively in the volume measurement and in the control volume. For this counting made for each measurement point of the particle velocity, two thresholds are fixed, one on the amplitude of the Doppler signal and the other on the integral of the power spectral density. The total number of particles N_{pt} is given by:

$$N_{pt} = \sum_{i=1}^{n} (N_p)_i \tag{7}$$

with n the number of the measurement point. The local concentration profile is obtained by the plot of the variation of the ratio N_p / N_{pt} versus the depth of the measurement volume along the vertical diameter of the test section (Fig. 4).

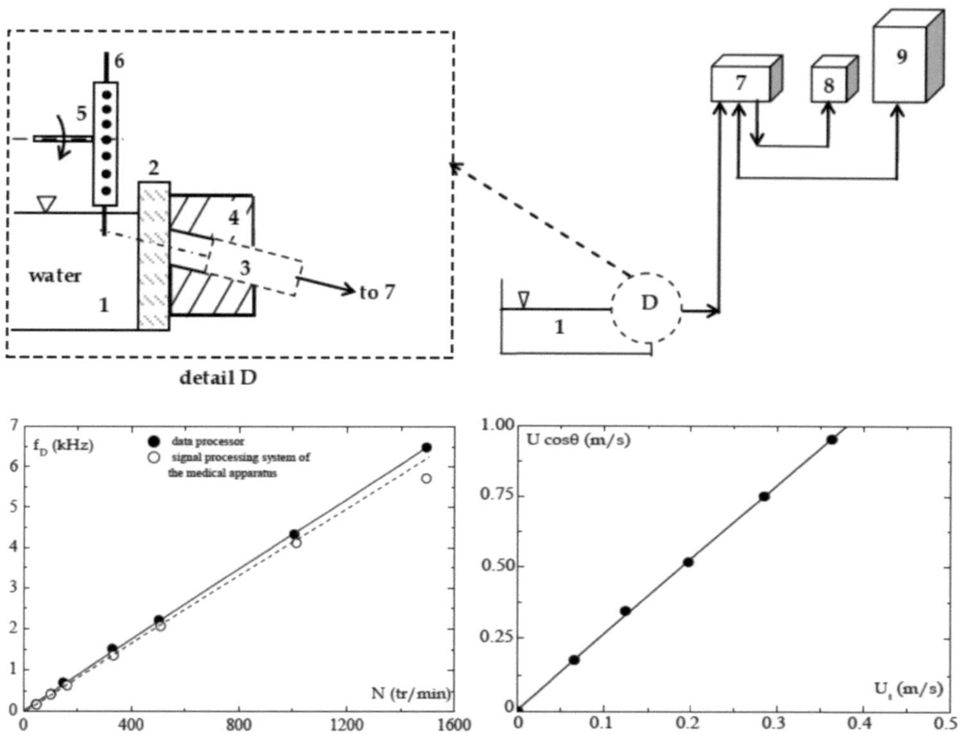

Fig. 5. a) Validation of the method and calibration of the ultrasonic transducer: 1 glass tank of calibration, 2 lateral wall of the tank, 3 ultrasonic transducer, 4 micrometric displacement system of the transducer, 5 plastic disc, 6 Abrasive cloth band or cylindrical metal rods, 7 ultrasonic Doppler velocimeter, 8 digital storage oscilloscope, 9 Data processor (Plurimat S); b) Validation curve of the signal processing method; c) Calibration curve of the ultrasonic transducer

2.2.3 Validation of the processing signal method and calibration of the ultrasonic transducer

The validation of the method applied for the signal processing consists of a comparison between the results obtained by the data processor (9) and those of the processing signal system integrated into the medical apparatus (7). For this validation we developed a device illustrated in Fig. 5a, it consists of the plastic disc (5) of 160 mm diameter turning with a variable speed motor, and on which a abrasive cloth band (6) is stuck (in Figure 5a, only the cylindrical metal rods used for the calibration of the ultrasonic transducer are represented). The fine emery grains act as the centers of diffraction moving with a known tangential velocity in the volume measurement. This intersection between the emery grains and the volume measurement is obtained by a micrometric displacement system of the ultrasonic transducer (4) fixed on one of the lateral walls of the tank calibration (2). The Doppler frequency is determined by the acquisition and processing system (9). Figure 5b shows that, in a large tension range, the results obtained from the signal processing of the data processor have a better linearity than those from the signal processing system integrated into the medical apparatus.

For the calibration of the ultrasonic transducer (3), which consists of determining the exact value of the inclination angle of the ultrasonic transducer, the same device as that of the validation of the signal processing method was used, except for the plastic disc where the abrasive cloth band is replaced by cylindrical metal rods (6) of 1 mm diameter and 15 mm length. Figure 5c shows a better linearity between a known tangential velocity U_t of the rods crossing the measurement volume and the value of $U\cos\theta$ determined from the signal processing. The inclination angle of the ultrasonic transducer (θ= 67.4°) determined from the calibration curve is very close to that used in the medical apparatus (θ= 67°).

3. Experimental results

3.1 Velocity and concentration profiles of fine particles

The rheological study of the starch suspension in water made in a rotating viscometer of coaxial cylinders (Haake RV12), showed that the suspension have a shear thickening behaviour described by the power law rheological model.

$$\tau = k\,\varepsilon^n \tag{8}$$

where τ is the shear stress, k is the flow consistency index and n is the flow behaviour index. These two rheological parameters determined from the flow curves, k = 0.33x10^{-3} kg/ms and n = 1.1, are necessary to determine the velocity profile using the model of Pai (Brodkey, Lee, Chase (1961) and Brodkey (1963)) valid for the Newtonian and non-Newtonian fluids when the Reynolds number Re is lower than 10^5. This model is given by:

$$\frac{U(r)}{U_{max}} = 1 + a_1\left(\frac{r}{R}\right)^{\frac{n+1}{n}} + a_2\left(\frac{r}{R}\right)^{2m} \tag{9}$$

where U_{max} is the maximum velocity generally taken on the pipe axis, R the pipe radius, and r = R - y the variable pipe radius (y is the wall normal distance). The constants a_1, a_2 and m depend on the nature of the fluid, the boundary conditions and the flow mean velocity. Figure 6 shows the velocity and the concentration profiles of the water-starch suspension obtained along the test section diameter. The velocity profile coincides well with the theoretical profile of Pai, and the concentration profile has a uniform distribution.

According to Furuta et al. (1977) we can thus assume that fine particle suspension which represents a tracer behaves as a homogeneous fluid and the slip velocity of the solid - liquid suspension is negligible. In the following results the velocity profile of water alone will be represented by the model of Pai.

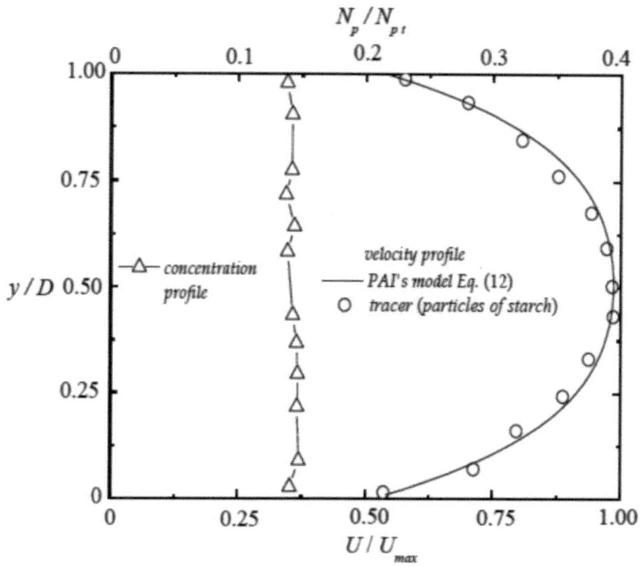

Fig. 6. Velocity and concentration profiles of water-starch suspension, Re= 42000, the volumetric concentration of the starch is C_V = 0.3%.

3.2 Velocity and concentration profiles of large particles

Large particles of high density compared to the carrier fluid, i.e., those where the diameter exceeds the wavelength of the ultrasonic wave, do not follow the flow. In this paper, only the results corresponding to the particles of diameter 0.13 and 0.4 mm are presented.

3.2.1 Effect of the mean flow velocity

A mean flow velocity range of 1 m/s to 2.5 m/s was used. This range corresponds to the heterogeneous and saltation flow regimes. Figure 7 show the influence of the mean flow velocity on the velocity and the concentration profiles of the solid particles of diameter 0.13 and 0.4 mm respectively. The velocity profiles are parabolas with a top around the pipe axis, very near for d_p= 0.13 mm and below for d_p= 0.4 mm. For the other particle diameter of 0.27 mm and 0.7 mm (which are not presented in this paper), the tops are below the pipe axis. The presence of the particles near the top wall was observed only for d_p = 0.13 mm and U_{moy} = 2 m/s.

The concentration profiles are in a better agreement with the observations of the flow made during the experimental tests. Indeed, for a constant volumetric concentration and a mean flow velocity ranging between 1 and 2 m/s, the flow of the particles suspension of diameter 0.27 mm, 0.4 mm and 0.7 mm is a saltation regime. It becomes heterogeneous when the flow mean velocity exceeds 2 m/s. For the particles of diameter 0.13 mm, the flow suspension is heterogeneous regime.

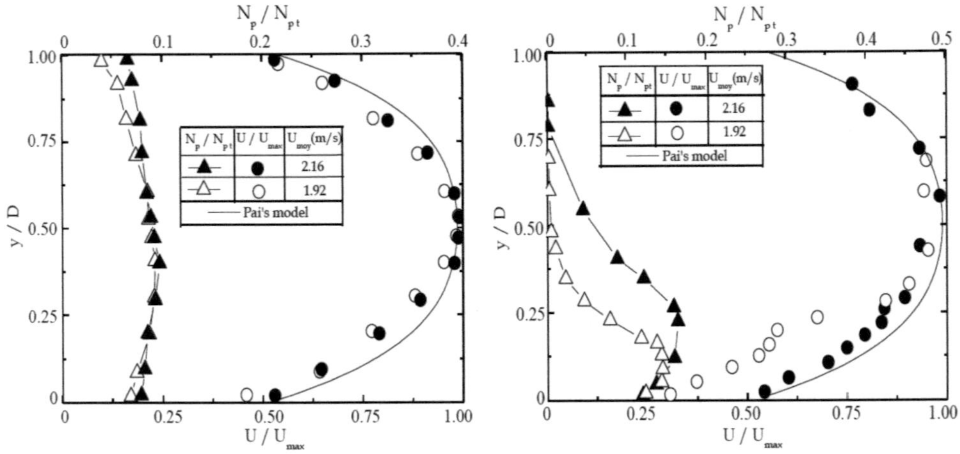

(a) (b)

Fig. 7. Velocity and concentration profiles of glass bead for C_v= 1%: a) d_p= 0.13 mm; b) d_p= 0.40 mm

3.2.2 Effect of volumetric concentration

Figure 8 shows the influence of the volumetric concentration of the solid particles on the velocity and local concentration profiles. Note that the difference between the velocity profile of the solid particles and that of the carrying fluid (Fig. 8a), which represents the solid-liquid slip velocity, increases with the increase in the volumetric concentration. The local concentration profiles (Fig. 8b) present a maximum value, whose position moves away from the internal pipe bottom wall when the volumetric concentration decreases. For the lower volumetric concentration, the local concentration profile tends to that of the carrying fluid (or the fine particles) which is almost uniform. The presence of the particles near the pipe top wall was observed for the volumetric concentration lower than 0.5 %.

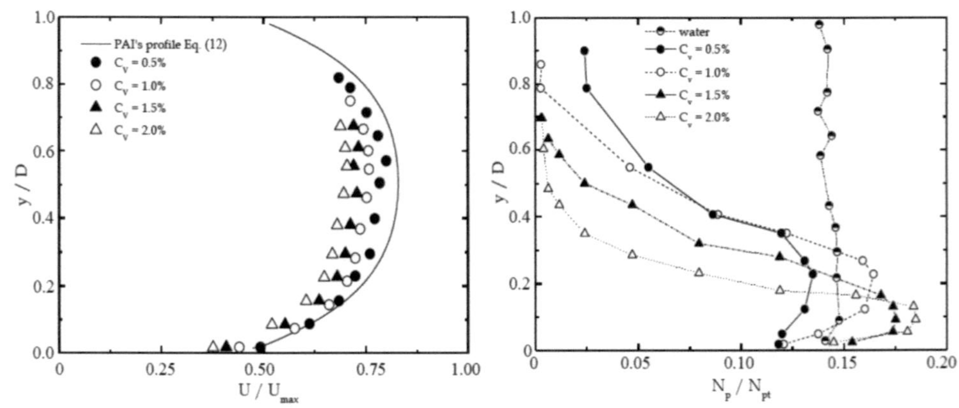

(a) (b)

Fig. 8. Effect of the particle volumetric concentration on: a) the velocity profiles; b) the local concentration profiles for d_p= 0.27 mm and R_e= 44600.

3.2.3 Effect of diameter

The figure 9a shows that for all the particle diameters used, the velocity profiles of the solid particles have the same trend as that of the carrying fluid. The difference between the velocity profile of the continuous phase and that of the solid phase confirms the existence of the solid-liquid slip velocity which, increases with increasing particle diameter. Figure. 9b shows that for the solid particles of diameter 0.13 mm and 0.27 mm, the local concentration profiles tend to that of the carrying fluid.

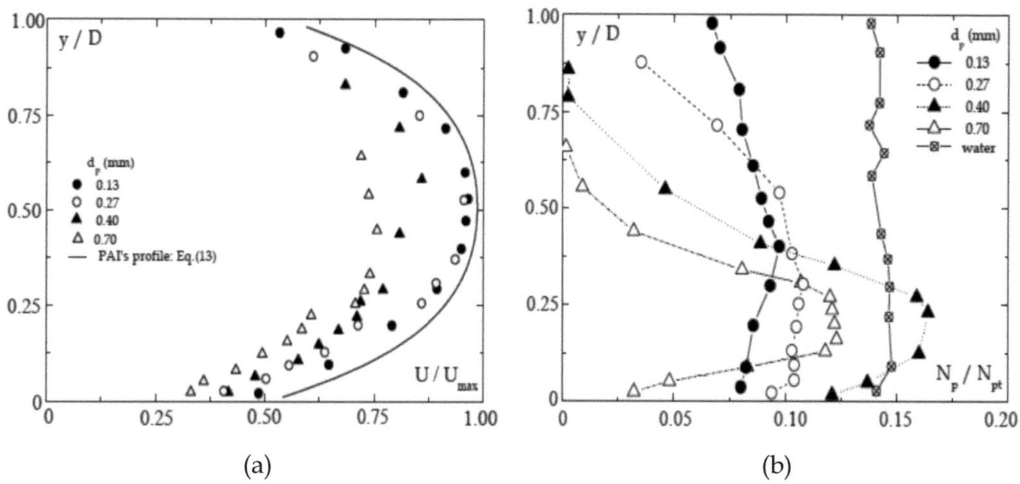

Fig. 9. Influence of the particle diameter on: a) the velocity profiles; b) the concentration profiles for C_V= 1%, R_e = 44600.

The maximum of local concentration profile appears for d_p= 0.4 mm and d_p = 0.7 mm. The solid particles are present near the pipe top wall only for the small diameter (0.13 mm).

4. Conclusion

In this experimental chapter, we have tested a new approach measurement in order to determine simultaneously the velocity profiles and the concentration profiles of the solid particles (glass bead) and the continuous phase (water) of two phase flow in horizontal pipe. The distinction between the Doppler signals coming from the solid phase and the continuous phase was obtained by imposing a threshold on the integral of the power spectral density. The use of this approach measurement is limited to small concentrations lower than 2%. Indeed when a concentration of 2% is exceeded, the ultrasonic signal is attenuated. This approach measurement shows the effects of the particle diameter and volumetric concentration on the local mean velocity and local mean concentration profiles of the suspensions.

The results obtained show that for the fine particles, the suspension behaves like a homogeneous fluid; the velocity profile is in a better agreement with the Pai's model and the concentration profile is almost uniform. For the large particles, saltation and heterogeneous flow regimes were obtained. These two regimes depend on the diameter and volumetric concentration of the particle, and on the flow mean velocity. The slip velocity which is responsible for the fluid-particle interaction depends on the flow regime.

In our next chapter concerning the solid liquid suspension, PUDV and Particle-Tracking Velocimetry (PTV) will be applied together. The most advantage of PTV is the possibility of measurements of large particles, by which the fluid-particle interactions and the particle-response property will be able to be examined. This allows us perhaps the best understanding of the phenomenon caused by the fluid-particle interactions, and also probably by the particle-particle interactions because of high concentration near the wall. It is also necessary to investigate interactions between these phenomenon and particle motion due to two-way and four-way couplings in a wide range of particle diameter, specific density, and sediment concentration, and then to develop reasonable computer simulation models of two phase horizontal pipe flow.

5. Acknowledgment

This work has been supported by Research Laboratory of Industrial Technologies, University Ibn Khaldoun of Tiaret. The first author would like to thank Professor S. Hadj Ziane, Institute of Physics, University Ibn Khaldoun of Tiaret, for useful discussions as well as for providing support to perform this work.

6. References

Alfonsi, G.,Brambilla, S., Chiuch, D. (2003). The Use of an Ultrasonic Doppler Velocimeter in Turbulent Pipe Flow, *Experiments in Fluids*, Vol.35, pp. 553-559.

Aritomi, M., Zhou, S., Nakajima, M., Takeda, Y., Yoshioka, Y. (1996). Measurement System of Bubbly Flow Using Ultrasonic Velocity Profile Monitor and Video Data Processing Unit, *J. Nucl. Sci. Technol*, Vol.33, pp. 915-923.

Brito, D., Nataf, H.C., Cardin, P., Aubert, J., Masson, J.P. (2001). Ultrasonic Doppler Velocimtry in Liquid Gallium, *Experiments in Fluids*, Vol.31, No. 6, pp. 653-663.

Brodkey, R.S., Lee, J., Chasse, R.C. (1961). A Generalized Velocity Distribution for Non-Newtonian Fluids, *A. I. Ch. E. Journal, Ohio.*

Brodkey, R.S. (1963). Limitations on a Generalized Velocity Distribution, *A. I. Ch. E. Journal, Ohio.*

Cellino, M., Graf, W.H. (2000). Experiments on Suspension Flow in Open Channels With Bed Forms, *Journal of Hydraulic Research*, Vol.38, No. 4, pp. 289-298.

Crowe, C.T., Troutt, T.R., Chung, J.N. (1996). Numerical Models for Two-Phase Turbulent Flows, *Annu Rev Fluid Mech.*, Vol.28, pp. 11–43.

Eckert, S., Gerbeth, G. (2002). Velocity Measurements in Liquid Sodium by Means of Ultrasound Doppler Velocimetry, *Experiments in Fluids*, Vol.32, No. 5, pp. 542-546.

Elghobashi, S.E. (1994). On Predicting Particle-Laden Turbulence Flows, *Appl Sci Res.*, Vol.52, pp. 309–329.

Elghobashi, S.E., Truesdell, G.C. (1993). On the Two-Way Interaction Between Homogeneous Turbulence and Dispersed Solid Particles. I: Turbulence Modification, *Phys Fluids A*, Vol.5, No. 7, pp. 1790–1801.

Furuta, T., Tsujimoto, S., Toshima, M., Okasaki, M., Toei, R. (1977). Concentration Distribution of Particles Solid-Liquid in Two Phase Flow Through Vertical Pipe, *Memoirs of the Faculty of Engineering*, Vol. XXXVI, Par 4, Kyoto University.

Kikura, H., Takeda, Y., Sawada, T. (1999). Velocity Profile Measurements of Magnetic Fluid Flow Using Ultrasonic Doppler Method", *Journal of Magnetism and Magnetic Materials*, Vol.201, pp. 276-280.

Kikura, H., Yamanaka, G., Aritomi, M. (2004). Effect of Measurement Volume Size on Turbulent Flow Measurement Using Ultrasonic Doppler Method, *Experiments in Fluids*, Vol.36, No. 1, pp. 187-196.

Lei, K., Taniguchi, N., Kobayashi, T. (2000). LES of Particle-Laden Turbulent Channel Flow Considering SGS Coupling, *Trans JSME B*, Vol.66, No. 651, pp. 2807–2814.

Michaelides, E.E., Stock, D.E. (1989). *Turbulence Modification in Dispersed Multiphase Flows*, FED, Vol.80, ASME, New York.

Nakamura, H. (1996). Simultaneous Measurement of Liquid Velocity and Interface Profiles of Horizontal Duct Wavy Flow by Ultrasonic Velocity Profile Meter, *First International Symposium on Ultrasonic Doppler Methods in Fluid Mechanics and Fluid Engineering*, Villigen PSI, Switzerland.

Parthasarathy, R.N., Faeth, G.M. (1990a).Turbulence Modulation in Homogeneous Dilute Particle-Laden Lows, *J. Fluid Mech.*, Vol.220, pp. 485–514.

Parthasarathy, R.N., Faeth, G.M. (1990b). Turbulent Dispersion of Particles in Self-Generated Homogeneous Turbulence, *J Fluid Mech.*, Vol.220, pp. 515–537.

Rolland, T., Lemmin, U. (1996). Acoustic Doppler Velocity Profilers: Application to Correlation Measurement in Open-Channel Flow, *First International Symposium on Ultrasonic Doppler Methods in Fluid Mechanics and Fluid Engineering*, Villigen PSI, Switzerland.

Rolland, T., Lemmin, U. (1997). A Two-Component Acoustic Velocity Profile for Use in Turbulent Open-Channel Flow, *J. Hydraulic Res.*, Vol.35, pp. 545-561.

Sato, Y. (1996).*Turbulence Structure and Modelling of Dispersed Two-Phase Flows*, Ph.D. Thesis, Keio University, Japan.

Takeda, Y. (1995). Velocity Profile Measurement by Ultrasonic Doppler Method, *Experimental Thermal and Fluid Science*, Vol.10, No. 4, pp. 444-453.

Varaksin, A;Y., Zaichik, L.I. (2000). Effect of Particles on the Carrier Flow Turbulence, *Thermophys. Aeromech.*, Vol.7, pp. 237-248.

Xu, H. (2003). *Measurement of Fiber Suspension Flow and Forming Jet Velocity Profile by Pulsed Ultrasonic Doppler Velocimetry*, Ph.D. Thesis, Atlanta, Institute of Paper Science and Technology.

8

Ultrasonic Thruster

Alfred C. H. Tan and Franz S. Hover
Massachusetts Institute of Technology
United States of America

1. Introduction

Acoustic streaming refers to the bulk net flow of fluid generated as a result of intense, free-field ultrasound. This phenomenon, also known as 'quartz wind' or 'sonic wind', is induced by a loss in mean momentum flux due to sound absorption in the fluid medium, leading to a net flow along the transducer axial direction. The transmission of intense sound energy into the fluid is also associated with a resultant force acting on the transducer surface. This resultant "backthrust" can be exploited in underwater vehicles for propulsion or maneuvering purposes. We refer to the device as ultrasonic thruster (UST), shown in Fig. 1, and define it as an ultrasonic transducer made from piezoelectric material, excited by an alternating high-voltage source in the megahertz. We provide a general comparison between various propulsion technologies in Table 1.

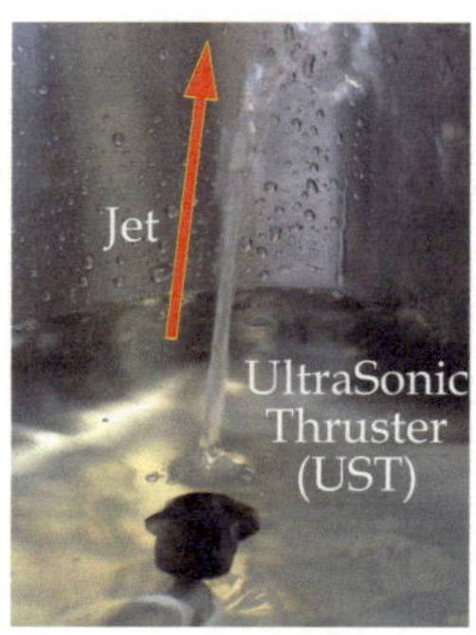

Fig. 1. An ultrasonic thrusters (UST) directing a jet through the free water surface.

The UST we describe is being considered as an alternative small-scale propulsor for underwater robotic devices and systems, because it has no moving parts beyond its membrane and can be mounted flushed with the body of the watercraft. At a sound level of 132dB (re 1μPa) (Panchal, Takahashi, & Avery, 1995; Tan & Tanaka, 2006), it is destructive to biofouling and could perform self-cleaning under prolonged water submersion. These novelties in low maintenance and design robustness, coupled with low cost commercially off-the-shelf (COTS) transducers, are attributes which are not found in rotary or biomimetic propulsors of today. For a small transducer diameter of 1cm, thrust generated is in the order of tens of milli-newtons (mN), suitable for systems operating at low Reynolds numbers.

Propulsors	Advantages	Disadvantages
Fixed-pitch propeller	Very mature technology; cost effective	Load- and speed-dependent performance
Podded drive	Maneuverability; reduced impact on vessel internal layout; noise isolation	High bearing loads; costly and complex
Ducted propeller	High efficiency; directional stability; robustness against line fouling	May be inefficient at off design conditions
Controllable-pitch propeller	High efficiency at different advance speeds and loadings	Complex actuation system and maintenance
Waterjet	Efficient at high vessel speeds; suitable for shallow water	Poor performance at low speeds; vulnerable to ingested debris
Cycloidal propeller	Low-speed maneuverability	Complex mechanical structure and maintenance
Biomimetic – body/caudal fin (BCF) locomotion	Efficient and maneuverable at many speeds; quiet	Complex physical design
Biomimetic – Median and/or paired fin (MPF) locomotion	High maneuverability at low speeds; quiet	Complex physical design
Ultrasonic thruster (this work)	Very small size; no moving parts; short-range acoustic communication	Poor propulsive efficiency; requires a high-voltage supply

Table 1. Broad comparison of propulsor technologies (Carlton, 2007; Sfakiotakis, Lane, & Davies, 1999; Tan & Hover, 2009).

At the same time, large-scale collaborative swarm of small "microrobots" or "pods" systems are also gaining more interest in terms of low cost and practical operation. Small clusters of exploratory underwater vehicles could also acquire true flexibility in formation morphing, and wide spatial/temporal coverage in search-survey work (Trimmer & Jebens, 1989). More importantly, small water submersible is valuable in cluttered or confined environments such as inside a piping network or complex underwater structures (Egeskov, Bech, Bowley, & Aage, 1995). Naval reconnaissance missions could involve the deployment of clusters of expendable (even biodegradable) small underwater robots for hazardous/security missions such as mine-hunting or surveillance mapping (Doty et al., 1998).

In the following sections, we examine the theoretical background of the UST thrust generation, introduce scaled parameters for comparison between various UST devices found in the literature, examine its transit efficiency, identify some thermal anomalies, and discuss some of the UST design considerations. A detailed construction of the UST will be outlined with underlying insights to materials selection and design principles. For practical demonstration, we have also built a small underwater vehicle named *Huygens*, to establish a miniaturized platform for supporting multi-objectives subsea tasks. The main focus will be exclusively on the transducer-only testing (without nozzle appendages) for thrust and wake characterization as these are fundamentals toward understanding the UST technology. Prior works on thrust by ultrasonic means can be found in (Allison, Springer, & Van Dam, 2008; Nobunaga, 2004; Wang et al., 2011; Yu & Kim, 2004); we review and expand upon these. We have also made a comparison among these UST technologies, and provide some insights into some of the design parameters of ultrasonic propulsion.

2. Underwater Ultrasonic Thruster (UST)

The UST is made from a membrane actuator mounted in a specially designed waterproof housing, excited by an electrical source at ultrasonic frequency; the UST is generally applied underwater to generate thrust. For this work, the actuator is made of a thin, circular piezoelectric plate. We establish some fundamental concepts of the UST physics leading to thrust as experienced on the transducer surface, and its resulting jet of acoustic streaming into the farfield. Several nomenclatures are defined and used to describe the experimental results in the preceding sections.

2.1 Thrust generation

As described in the introduction, there are some prior works on the UST found in the literature. The experimental model used in those examples varies in shapes and sizes, and in the following, we provide a basis of comparison among them in terms of thrust density and electrical power density between each UST design. As an introduction, we refer to the mathematical treatment of thrust and acoustic power generation in Eqs. (1) to (4) as originally proposed by (Allison, et al., 2008), and relates thrust to the transducer voltage supplied, E, a parameter reported in most UST-related studies.

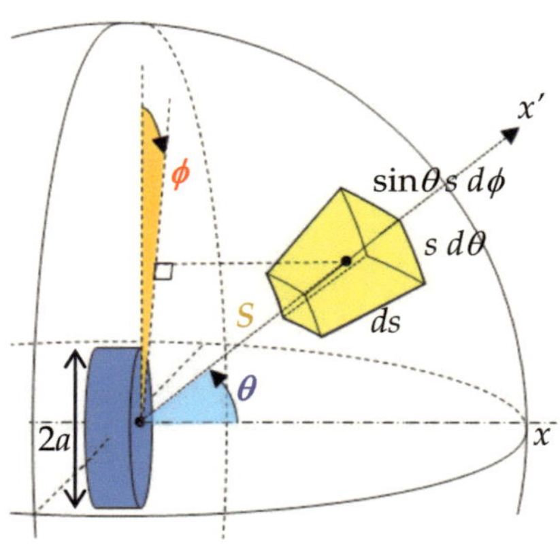

Fig. 2. An elemental control volume in the far-field; the elevation angle, azimuth angle and distance from the center of transducer are θ, ϕ, and s respectively.

In Fig. 2, several variables for the transducer and the acoustic field are introduced. The acoustic energy transmits to the right semi-hemisphere, propagating perpendicularly through an elemental cross-sectional surface area $dS = s^2 \sin\theta d\theta d\phi$ of a control volume $dS = s^2 \sin\theta d\theta d\phi ds$. According to (Allison, et al., 2008), thrust experienced on the surface of the transducer is expressed as

$$T = \rho \upsilon_o^2 a^2 \pi \int_0^{\theta_1} \frac{J_1(ka\sin\theta)^2}{\sin\theta} d\theta \tag{1}$$

where ρ (kg/m^3), υ_o (m/s), a (m), k, β, and $J_1(\cdot)$ denote the fluid density, transducer surface velocity, radius of the transducer, wavenumber, sound absorption coefficient, and Bessel function of the first kind, respectively, and we consider $\theta_1 = \sin^{-1}\frac{3.832}{ka}$ (Blackstock, 2000) as the upper limit of the dominant ultrasonic beamwidth.

The relationship between the acoustic power along the transducer axial direction x and the ultrasonic thrust is given by a simple relationship (Allison, et al., 2008)

$$P_x = {cT}/{\eta} \tag{2}$$

where c (m/s) denotes the sound speed, and η relates to the efficiency which will be elaborated in the next paragraph. As we verify below, this also means the thrust is considerably lower than would a rotary propulsor operating at the same power level.

At s = 0, this acoustic power radiation is associated with the electrical power consumption across the transducer. The electrical power would provide an approximation to the acoustic power loading, which relates to the thrust force, and is reflected in the following considerations through an efficiency constant, η; (i) electrical power lost across the transducer is not wholly transferred into the medium; (ii) acoustic loading at the sharp dominant resonant frequency of the transducer may not be precisely tuned; (iii) it is difficult to consider the equivalent acoustic load in the lumped circuit impedance; (iv) although the acoustic power is mainly generated within the narrow ultrasonic beam, some losses also occur outside the beamwidth.

By relating the thrust production to the root mean square of the transducer voltage supplied, E_{rms}, it can be equated as

$$T = \frac{\eta E_{rms}^2}{c\,\mathrm{Re}(\Re)} \tag{3}$$

where Re($\Re$) is the real component of the transducer impedance, and the electrical power,

$$P = \frac{E_{rms}^2}{\mathrm{Re}(\Re)} \tag{4}$$

corresponds to P_x at s = 0. This thrust scaling with squared voltage in (3) will be used in Section 2.3.4.

2.2 Acoustic streaming

Following Lighthill (Lighthill, 1978), acoustic streaming arises because of acoustic energy absorption along the path of propagation in a viscous, dissipative fluid medium. A simple explanation of the mechanism can be thought of as exit momentum flux from each exposed fluid particle is less than it entered due to sound absorption. The distribution of energy gradient then moves the particle away from the source resulting in an overall streaming effect. In another words, the loss in momentum flux across a control volume (Fig. 2) leads to a net force in the direction of the acoustic path. This net force in turn generates hydrodynamic flow in a steady state incompressible medium, as governed by the Navier-Stokes equation,

$$\rho \bar{u}_x \frac{\partial \bar{u}_x}{\partial x} = -\frac{\partial \bar{p}_0}{\partial x} + \mu \frac{\partial^2 \bar{u}_x}{\partial x^2} + F_x \tag{5}$$

where $\bar{u}_x$ denotes the time-averaged streaming velocity along the x-axis, F_x is the radiation force along the x-axis, p_0 is defined as $p_0 = \rho c \upsilon_0$, and μ (kg/m·s) is the dynamic viscosity of the fluid. Since mass flow is conserved within the control volume, that is, $\rho \frac{\partial \bar{u}_x}{\partial x} = 0$, Eq. (5) can be solved by assuming $F_x = A e^{-\beta x}$ and using boundary conditions $\bar{u}_x(0) = 0$, $\bar{u}_x(\infty) = 0$, to obtain

$$\bar{u}_x = \frac{A}{\nu\left(B^2 - \beta^2\right)}\left[e^{-\beta x} - e^{-Bx}\right] \tag{6}$$

where A and B are characteristic coefficients of the fully developed axial velocity profile ($t \rightarrow \infty$), and ν is the kinematic viscosity $\nu = \mu / \rho$ (m²/s), $\beta = 2\alpha$ for low intensity sound, and α (dB/m) is the absorption coefficient at a particular sound transmission frequency (Rudenko & Soluian, 1977). Accordingly, Lighthill demonstrated that the radiation force can also be represented by the Reynolds stress along the axial direction:

$$\bar{F}' = -\frac{\partial \rho \bar{u}_x^2}{\partial x} \tag{7}$$

In this case, it follows from Eq. (7) that $\bar{F}'$ is proportional to $\bar{u}_x^2$ and knowing that $\bar{F}'$ is also proportional to E^2, $\bar{u}_x$ is seen to scale directly with E.

Next, the total fluid discharge across the lateral section of the flow in Fig. 3 is given by $u_r \delta A = u_r 2\pi r \delta r$ (Fig. 4), and the mass flow rate through the elemental disc is $\rho u_r 2\pi r\, dr$. From (Tan & Hover, 2009) and later in Section 2.3.4, it can be seen that the streaming field across a lateral section can be approximated by a Gaussian distribution of the following form, $u_r = u_x e^{-\frac{r^2}{2C^2}}$, where u_x (m/s) is the axial velocity in unit time, r (m) is the radial distance and C is a constant associated with the standard deviation of the Gaussian distribution. u_x reveals a vital difference from conventional propeller design and biomimetic actuation – that at $x = 0$, the streaming velocity at the surface of the transducer is zero. This observation is further verified in the experimental results in Section 2.3.4. Hence the total kinetic energy of the velocity field in unit time from $x = 0$ to $x = x_f$, where x_f (m) is the focal distance from the flat transducer surface, is given by

$$K.E. = \int_0^{x_f}\int_0^{R} \rho u_x u_r 2\pi r\, dr dx$$
$$= 2\pi\rho \int_0^{x_f}\int_0^{R} u_x^2 r e^{-\frac{r^2}{2C^2}} dr dx \qquad (8)$$

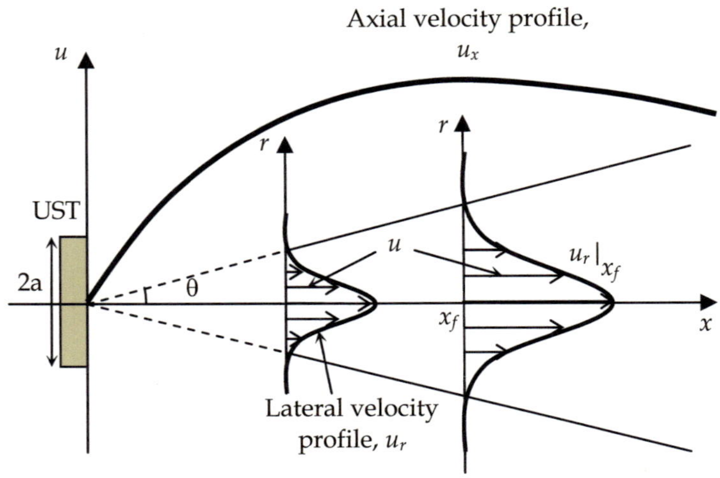

Fig. 3. Velocity distribution of the acoustic streaming. Lateral velocity profile approximates a Gaussian distribution while axial velocity follows a rapid increase in velocity before a gradual decline.

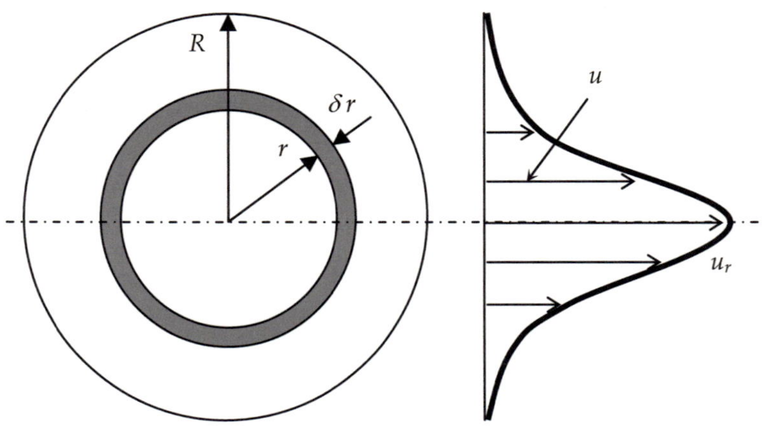

Fig. 4. An elemental disc sectioned laterally from the ultrasonic field. Axis of the transducer passes through the center of the concentric circles. The shaded annulus area, $2\pi r\,\delta r$, is an elemental area through which flow discharges.

where R (m) is the radius of u_r (m/s) profile subtended by θ_1 at x_f, and the overbar is omitted for simplicity. Substituting Eq. (6) into Eq. (8), and solving the double integral, the total kinetic energy in unit time becomes

$$K.E. = \frac{\pi \rho A^2 C^2}{\nu^2 \left(B^2 - \beta^2\right)^2} \left(-\frac{1}{\beta} e^{-2\beta x_f} + \frac{1}{\beta + B} e^{-(\beta + B) x_f} - \right.$$
$$\left. \frac{1}{B} e^{-2Bx_f} + \frac{1}{\beta} - \frac{1}{\beta + B} + \frac{1}{B} \right) \times \left(1 - e^{-\frac{R^2}{2C^2}} \right) \qquad (9)$$

While Eq. (9) sums up the total kinetic energy within the streaming field up to the focal distance, streaming at distances above x_f will evidently slow down, and free turbulence occurs due to the boundary between the stationary ambient water and the insonified flow - a process called entrainment. We consider regime $x > x_f$ to be no longer reliable or valid for *K.E.* calculation. Eq. (9) will be used in Section 2.3.4. Finally, it follows earlier that u_x scales directly with E and knowing that *K.E.* is also proportional to u_x^2 from Eq. (8), *K.E.* is seen to scale directly with E^2. This scaling of *K.E.* with square voltage will be used in Section 2.3.4 to determine the streaming energy and a comparison is made across other UST devices.

2.3 Thrust and wake experimental setup

In this section, we characterize thrust and wake energy for a specific UST design, and then investigate and how these properties can be modified using various source voltages. An underwater vehicle prototype was also constructed to demonstrate the ultrasonic propulsion capability.

2.3.1 UST hardware and methods

We use standard piezoelectric transducer technology for the conversion of electrical to acoustical energy. The transducer (Murata Manufacturing Co. Ltd) is made from a circular PZT plate measuring 7mm in diameter, housed in a 10mm diameter waterproof metallic casing as shown in Fig. 5.

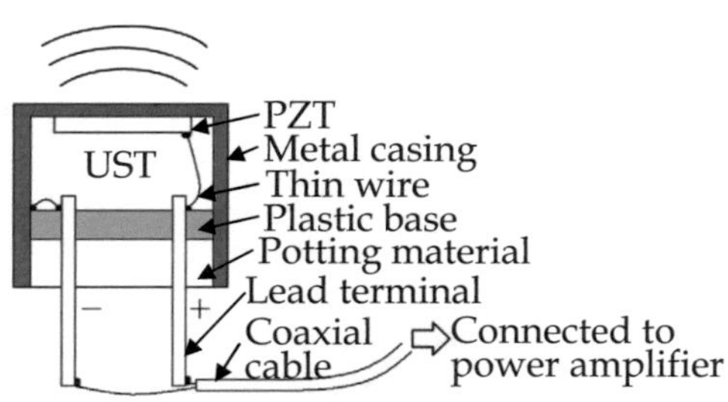

Fig. 5. Construction of the UST (adapted from Murata Manufacturing Co. Ltd).

In our prototype vehicle described below, three USTs are connected to switches for actuation control, and 50Ω coaxial cables connect the three switches to a single power amplifier (ENI 3100L). The ENI unit accepts an oscillatory input up to a maximum of $1V_{rms}$, and amplifies the output voltage by a gain of 50dB for a 50Ω output impedance. It is a Class A amplifier which means it will be unconditionally stable, and maintains linearity even with a combination of mismatched source and load impedance. When the ultrasonic transducer emits intense acoustic energy into the fluid, the thin metal housing provides excellent heat dissipation. Natural convection and acoustic streaming also aid in carrying away heat from the transducer surface (Tan & Hover, 2010a).

2.3.2 Thrust force measurement

It is important to develop a reliable underwater thrust measurement method, especially for small thrust magnitudes as is the case of the UST. Most load cells are either non-submersible or could not provide sufficient sensitivity/resolution required at small driving forces. While thrust can also be inferred from acoustic intensity measurement using a hydrophone, it poses some challenges unique to the UST setup, such as membrane cavitation, heating effect, and reading errors averaged from a finite-size hydrophone. Indeed, (Hariharan et al., 2008) reported that acoustic intensity measurements do not perform well, having an error in excess of 20% with experimental data.

In order to accurately measure the thrust produced by the UST, we developed an approach which allows sensitivity control by sliding the UST along an L-shaped arm. Moment is measured with a high precision torque sensor. Other methods have been proposed for measuring these very fine-scale forces, for example, to attach and submerge a UST on one end of a vertical pendulum, hinged off-center, with the other end flexing a strain gauge (Allison, et al., 2008). Another approach is to attach the UST to a free-hanging wire and take photographs of the displacement as the UST is being actuated (Wang, et al., 2011; Yu & Kim, 2004). We found that the setup in Fig. 6 provides very good accuracy and repeatability, as indicated in the calibration plot of Fig. 7. The torque meter accuracy provided by the manufacturer is ±0.09mN·m, and the absolute error in the calibration thrust force averages about ±0.6mN, when our dial gauge is positioned 0.5m from the torque axis on the L-arm.

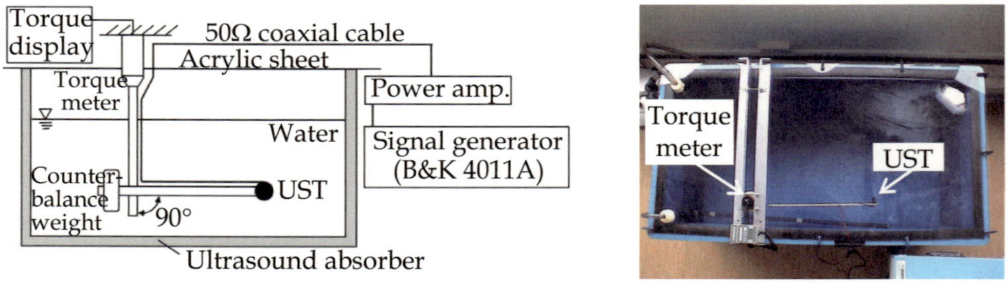

(a) (b)

Fig. 6. (a) The semi-anechoic water tank and torque sensor used to measure UST thrust; thrust is generated in the direction perpendicular to the page. (b) A plan view of the UST thrust measurement setup.

Hence the total force uncertainty is around 0.8mN. Torque calibration is performed prior to all test sets. Thrust stabilizes and is recorded about ten seconds after turning on the power; then the power is turned off. We maintain a minimum of ten-second rest period between all tests, to allow for cooling and for the water to settle. The UST together with the L-shaped arm is submerged in an ultrasound semi-anechoic tank measuring 1.2m×0.6m×0.6m, filled with distilled water and covered with an acrylic sheet.

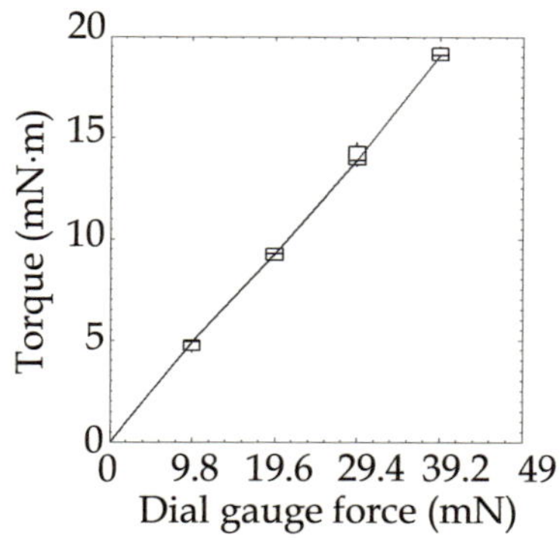

Fig. 7. Calibration of the test rig. Each of the four points shown is an average of four separate force applications using a sensitive dial gauge.

2.3.3 Acoustic streaming measurement

A 2-dimensional Digital Particle Image Velocimetry (DPIV) provides an accurate depiction of the UST wake, and beyond the entrainment boundary in the far-field. In similarly-scaled conditions, it has been reported that a 2% standard deviation at the point of maximum velocity can be expected with DPIV (Myers, Hariharan, & Banerjee, 2008). The DPIV system is set up in a water tank measuring 2.4m×0.7m×0.7m, as shown in Fig. 8. Calculated particle velocities are subject to noise depending predominantly on the interrogation window size, the number of seeded particles, and the sampling rate. We made efforts to tune these for the lowest noise level.

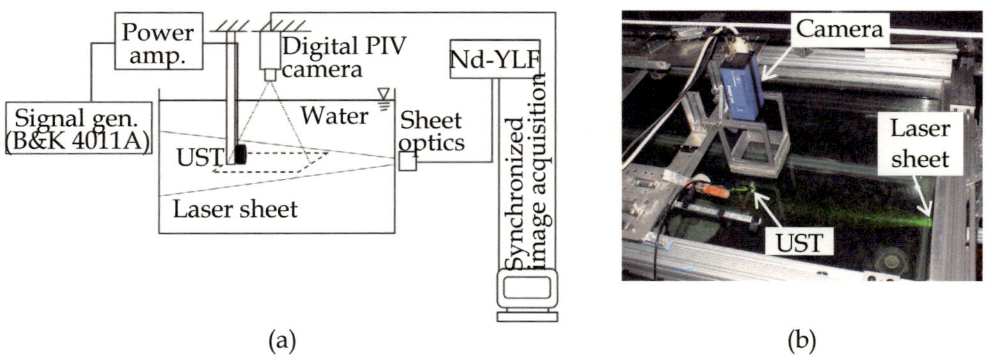

Fig. 8. (a) Schematic diagram of the DPIV tank setup. Acoustic streaming is illuminated by a laser sheet, and images are captured by a digital camera for post-processing. (b) A top view of the UST wake measurement setup.

A pulsed laser sheet, produced by a Quantronix diode pumped Q-switched frequency laser (Darwin-527-30M) and spreads horizontally through the water tank from the outside, coincident with the UST axis. The tank is seeded with 50μm polyamide particles, and a camera viewing the laser sheet perpendicularly traces the streaming particles when the UST is operating. The camera samples 300 timed, paired images at 400Hz, in a 0.2m×0.2m field of view. Post-processing is carried out using DaVis 7.1 software. The UST transducer is positioned at 1.2m from the sheet optics, 0.4m from each adjacent tank wall, and 0.3m below the water surface. The time taken for the stream to become established has been reported variously at about 0.5s (Loh & Lee, 2004) and 20s (Kamakura, Sudo, Matsuda, & Kumamoto, 1996). We allow at least one minute of flow before the camera starts recording. Specific kinetic energy of the flow up to the focal distance x_f is shown in the next section, which averages 150 measurements; recording the frames takes less than one second. Then the power is turned off, and the tank water is allowed to settle for at least one minute. This schedule is not the same for thrust measurements, as the transducer is powered and cooled for a considerably longer time during DPIV tests. A more detailed analysis of the transducer heating is presented in Section 3.

2.3.4 Thrust and streaming results

Fig. 9(a) shows thrust force as a function of frequency, for a sinusoidal waveform. The thrust has obvious peaks near 11mN when operated at 7MHz, which can also be computed using the thickness mode frequency constant, $N_t = f_0 h$, where h (m) is the thickness of the PZT, and $N_t = 1970$ as specified by the manufacturer. We will focus on this frequency in most of the discussion to follow. Fig. 9(b) illustrates the DPIV velocity field for a sinusoidal waveform at 7MHz, with amplified output voltage $54V_{rms}$. From Fig. 10(a) and 10(b), the maximum axial streaming velocity is observed at the point (0mm, 120mm) in the DPIV image, illustrating a fundamental feature of the UST – *that net fluid flow is zero at the transducer face*. Considering the same waveform configurations and frequencies as in Fig. 9(a), a similar peak at 7MHz in the total kinetic energy can be seen in Fig. 9(b).

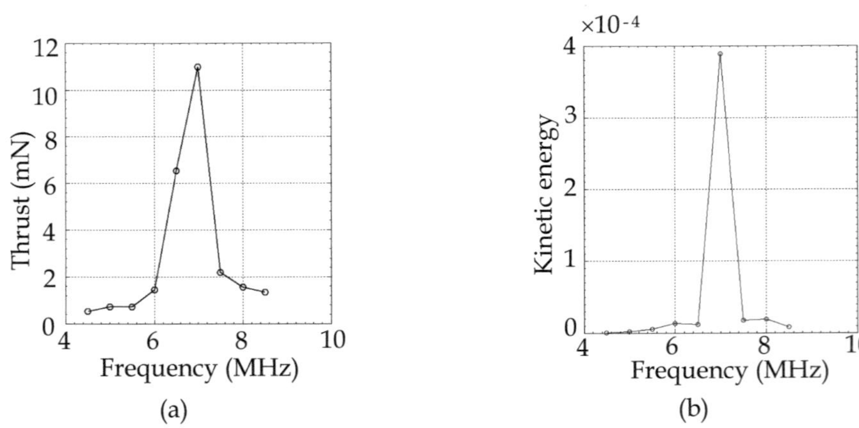

Fig. 9. (a) Measured thrust versus source frequency for sinusoidal waveforms supplied at $59V_{rms}$. (b) Total kinetic energy versus source frequency, from DPIV. The sinusoidal signal is supplied at $54V_{rms}$.

Figs. 11(a) and 11(b) summarize our findings specifically at the 7MHz resonant point. Thrust generally increases with output voltage, but then starts to flatten out above the amplified output voltage of $60V_{rms}$. From Fig. 12 and from Eq. (3), the corresponding scaled thrust level is 3.8×10^{-3}mN/V^2 at $59V_{rms}$. Scaled thrust decreases gradually as the output voltage increases, and the absolute thrust appears to saturate above $60V_{rms}$; see Fig. 12 – a result also reported in another publication (Tan & Hover, 2010a). The kinetic energy data in Fig. 9(b) show similar trend as well. Using the *K.E.*-E^2 proportionality relationship from Section 2.2, the scaled kinetic energy at $54V_{rms}$ is calculated to be $1.5\times10^{-7}W/V^2$ respectively. Both scaling of thrust and *K.E.* is important when we compare various UST devices later in Section 2.5.

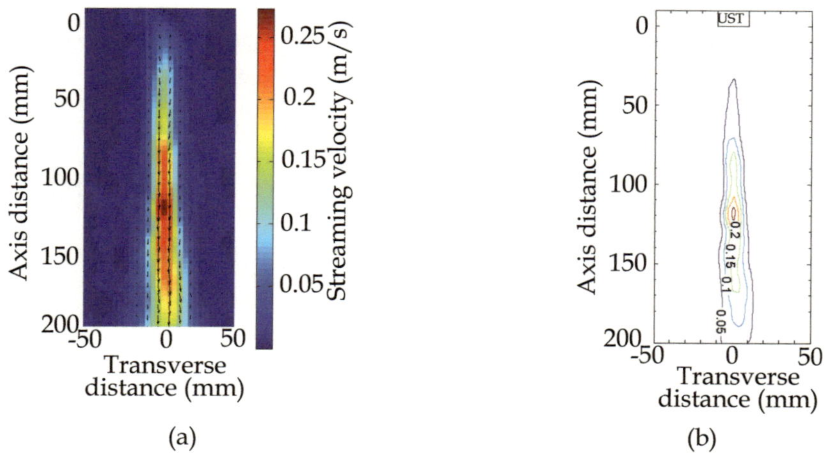

Fig. 10. (a) Digital Particle Image Velocimetry (DPIV) image of a jet at 7MHz, $54V_{rms}$. (b) Processed DPIV velocity field with each velocity contour step at 0.05m/s interval. A focal point can be clearly seen at (0mm, 120mm).

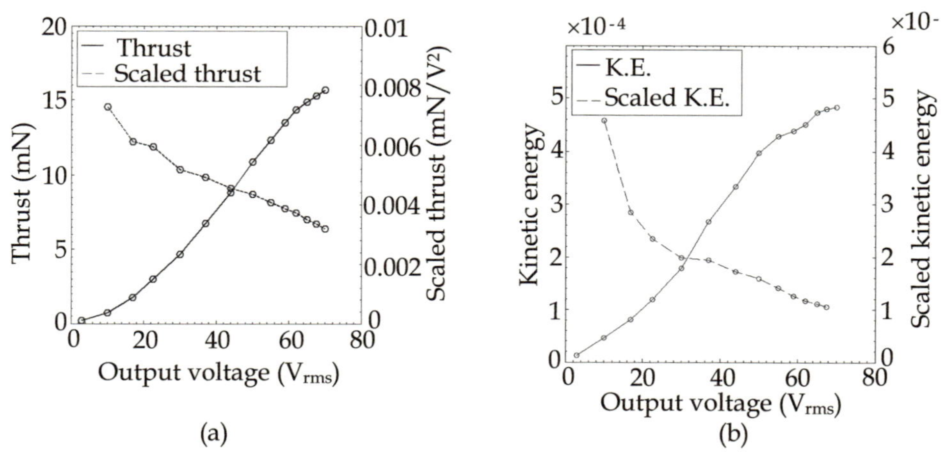

Fig. 11. (a) Measured thrust versus input voltage to the power amplifier using a 7MHz sinusoidal signal. Data appear to saturate above $60V_{rms}$, which implies a disproportionate relationship. (b) Total kinetic energy versus output voltage of the power amplifier using a 7MHz sinusoidal signal. Data appear to saturate above $55V_{rms}$, which also implies a disproportionate relationship.

Although increasing the output voltage always increases thrust, the wake velocity of the sinusoidal source appears to saturate near $55V_{rms}$. The phenomenon of saturation can be explained by distortion in finite-amplitude traveling waves, according to weak shock theory. On the other hand, the fact that thrust in this case increases with input power despite the saturation of velocity highlights an unusual observation - that thrust production mechanism involves the wake only indirectly, and in a manner that is distinct from other propulsors. This fact may offer some interesting avenues for UST design, where the wake and the thrust force could be manipulated independently. This is especially useful in a scenario where larger thrust is desired but the wake has to be weak at the same time, for example, to minimize stirring up particulates near the seabed.

In summary, increasing the output voltage of the power amplifier will no doubt increase the thrust and kinetic energy production of the transducer, but at the same time, introduces an undesirable disproportionate relationship (Tan & Hover, 2010a). This will be further discussed in Section 3.

2.4 Small underwater vehicle, *Huygens*

We designed and constructed a small, streamlined shell with three embedded UST devices for propulsion and steering in the horizontal plane. The shell measures 215mm×160mm×80mm, profiled by a truncated NACA 0054 airfoil in the side view, and a truncated NACA 0025 airfoil in the plan view, as shown in Fig. 12(a). The shell is made from high-strength urethane foam for buoyancy, and coated with polyurethane. A UST is positioned at the rear end of the craft to provide forward thrust, and two USTs subtending 120° are symmetrically located on each side of the front end - at the "fish eyes" position. Together, these provide the right/left steering and backing thrust. Inside the shell, two rectangular cavities are machined, measuring 100mm×100mm×35mm, and 45mm×60mm×30mm. The shell can be opened into two halves via a stepped mid-section opening, lined with double O-rings for a watertight seal. The cylindrical UST seats are also lined with O-rings. The vent shown in the top left corner of Fig. 12(b) allows for a tether or an antenna for shallow-water wireless control.

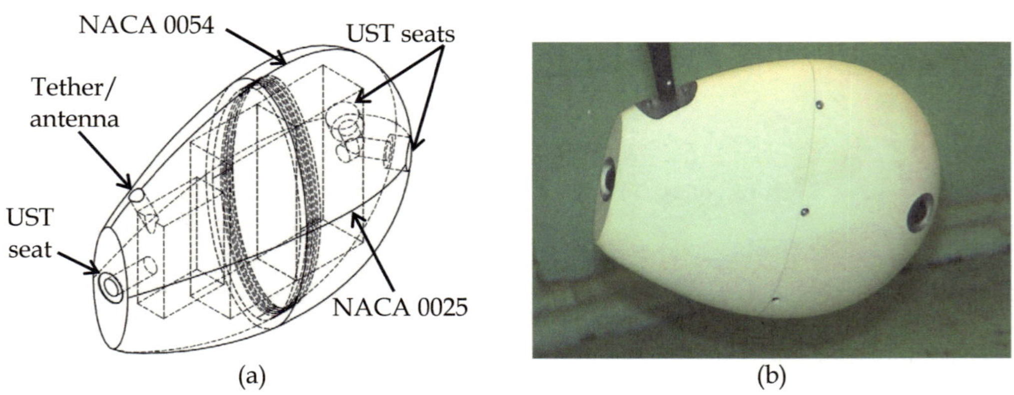

Fig. 12. (a) 3-D wire mesh view of *Huygens*. (b) A side view of *Huygens* prototype. Three transducers are installed on the shell with two at the frontal "eyes" positioned for steering and another in the rear for forward thrust. Tethered signal generates a multiplexed 7MHz sinusoidal input to the three USTs.

2.5 Ultrasonic propulsion design considerations

To our knowledge, there are only three experimental works on ultrasonic propulsors reported from (Allison, et al., 2008), (Yu & Kim, 2004), and (Wang, et al., 2011); we compare them with our UST in terms of thrust density, scaled thrust density and power density, in Table 2. As our UST system was operating at a resonance frequency and high voltage, it is difficult to make a fair and direct comparison of performance. We have employed a much higher power level, resulting in a very high thrust level and higher scaled thrust density.

Properties	Allison et al. (Allison, et al., 2008)	Yu et al. (Yu & Kim, 2004)	Z. Wang et al. (Wang, et al., 2011)	This work
Frequency (MHz)	5.5	10.8	17.8	7.0
Voltage (V)	24.5	46	140	59
Electrical power (W)	5	--	--	69.5
Transducer surface area (mm^2)	$\pi/4\times10^2$	5×5	$\pi/2\times1.28^2$	$\pi/4\times7^2$
Thrust (mN)	2.25	5.6	2.3	13.5
Thrust density (N/m^2)	28.6	224	893.7	350
Scaled thrust density ($N/m^2\cdot V^2$)	0.05	0.11	0.04	0.10
Electrical power density (kW/m^2)	64	--	--	1806

Table 2. Performance of different UST devices.

This latter property is important in applications because it indicates a very compact force source operating with reasonable voltage levels, exploitable to benefit from many of its unique ultrasonic attributes we discussed earlier in the introduction.

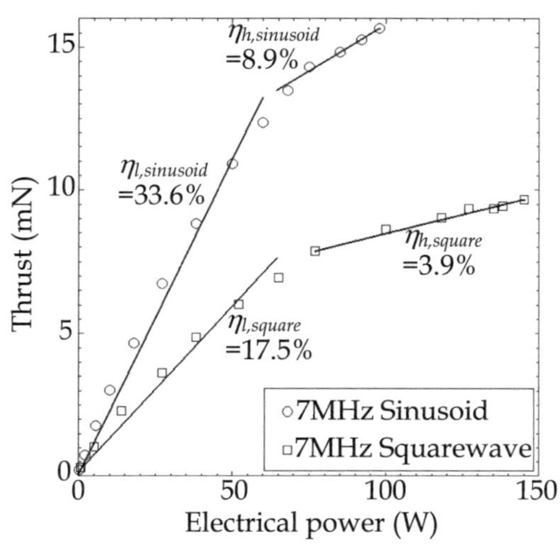

Fig. 13. Thrust force produced by the UST versus driving electrical power. The *l* and *h* subscripts indicate evident low- and high-power regimes. The output voltage to the sinusoid is $59V_{rms}$, and for the square wave is $54V_{rms}$.

Fig. 13 details the acoustic efficiency of the UST transmitting at a sinusoidal 7MHz. From Eq. (2), $T = \eta P/c$, where c is known (1480m/s), and T and P are measured. It can be made out that the sinusoidal waveform has two regimes relating thrust to electrical power – one of lower power with proportionally increasing thrust, and another somewhat saturated thrust at higher power. Below 70W electrical power, the power-thrust curve for sinusoidal input shows an acoustic efficiency of about 34%. Above 70W, the UST efficiency falls to less than 10%. Efficiency is calculated incrementally for each of the lines, that is, using the change in thrust versus the change in power. A square waveform is added in Fig. 13 for the sake of comparison. In general, the sinusoidal excitation is much more efficient than the square waveform. We note that the efficiency numbers given in (Allison, et al., 2008) are somewhat higher than what we show here, in part because of their custom transducer design, but also because they operated at much lower power levels. Below 2mN, we also achieve high efficiency around 70%. In respect to improving acoustic efficiency, other features such as nozzle appendages, shaping of UST, stacked piezoelectric layers could be considered.

2.5.1 Vehicle mission

In this subsection, we infer the expected speed and mission length that could reasonably be achieved from a UST-propelled underwater vehicle similar to *Huygens*. For our vehicle speed measurements, we used a high resolution Vision Research digital camera (Phantom V10) mounted with a wide-angle 20mm lens from Sigma. We sampled the advance speed of *Huygens* at 40samples/second over a straight course of 0.6m; the craft was allowed to accelerate for ten seconds before beginning the velocity measurement. Although a tether was attached to the vehicle for these tests, we maintained a large loop hanging below the vehicle, and moved the top of the tether along with the vehicle, using a sliding car and guiding post. With the recorded images of *Huygens*, the advance speed measured is a constant 0.049m/s.

The vehicle is quite streamlined, with only small holes around the frontal USTs, and a flat trim at the rear. From the top sectional view of *Huygens*, we can approximate the overall profile as an airfoil with a thickness-to-chord ratio of 0.37, and a span of 0.16m. The Reynolds number is $\mathrm{Re} = \frac{\rho U l_c}{\mu}$, where ρ (kg/m^3), U (m/s), l_c (m) and μ (kg/m·s) denote the fluid density, advance speed of the vehicle, chord length of the vehicle, and dynamic viscosity of the fluid respectively. The drag coefficient is $C_d = \frac{2T}{\rho A_w U^2}$, where T (N) and A_w (m^3) denote the thrust force, and wetted surface area respectively. Expressing U on the left hand side of Re and C_d separately, the Reynolds number and drag coefficient is related by

$$\mathrm{Re}^2 C_d = \frac{2\rho l_c^2 T}{A_w \mu^2}. \tag{10}$$

Using the Moody chart for a streamlined strut (Hoerner, 1965), U can then be estimated. To obtain the thrust, we recall that a sinusoidal input at 7MHz and 59V_{rms}, creates a thrust force of T = 13.5mN (Fig. 11(a)). The constants ρ, l_c, A_w, and μ are 1000kg/m^3, 0.215m, 0.053m^2, and 1.002×10-3kg/m·s respectively. The parameter $\mathrm{Re}^2 C_d$, solved using Eq. (10), is 23.4×10[6], and for *Huygens*, a unique point can be identified on the $C_d - \mathrm{Re}$ Moody diagram; Re = 1.17×10[4] and $C_d = 0.17$. It is thus estimated that *Huygens* will advance at a velocity of 0.054m/s - very close to the observed value.

Regarding mission duration and length, a small 11V, 0.75Ah lithium-ion battery would occupy about 10% of the *Huygens* vehicle volume. We assume an average power capacity reduction of 90% in a single discharge cycle, providing about 7.4Wh of energy. The UST consumes 69W of electrical power with an acoustic efficiency of 33.6% (Fig. 13), to produce 13.5mN of thrust with a constant vehicle advance velocity of about 0.05m/s. If we assume the instrumentation and other loads are small compared to the propulsive load, a simple straight-path mission will last around six minutes and travel a distance of about twenty meters. A somewhat larger battery could power the vehicle for perhaps thirty minutes, with a mission length of one hundred meters. While the UST is clearly not competitive with rotary or some biomimetic propulsors in terms of transit efficiency, nonetheless these estimates show that maneuvering a very small-scale vehicle utilizing USTs offers a propulsive force with interesting thrust and wake characteristics.

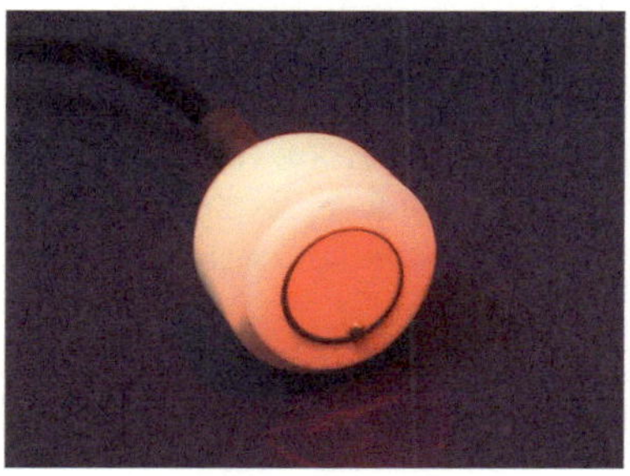

(a)

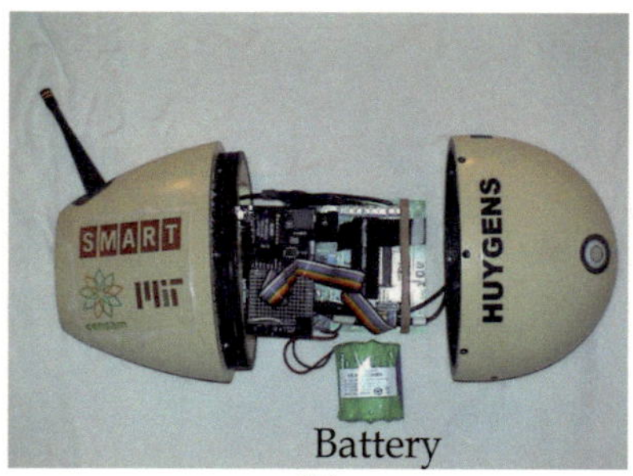

(b)

(c)

Fig. 14. (a) A specialized UST constructed from a one centimeter diameter piezoelectric transducer for underwater operation. (b) An internal view of the components of the wireless underwater vehicle, *Huygens*. (c) Wireless testing of *Huygens* in a laboratory tank.

We have also demonstrated a wireless version of *Huygens* (Fig. 14(c)) powered by a small on board battery (Fig. 14(b)). However, the electronics board could only supply limited power to each UST, and was able to slowly move the vehicle but inadequate to overcome the vehicle drag very well. One solution could be to improve the UST power output through a small size ultrasound amplifier such as reported in (Lewis & Olbricht, 2008). We continue to make improvements to optimize the output thrust density with considerations to on board space budget, using specially constructed underwater UST (Fig. 14(a)), and higher energy density batteries as well.

3. Thermal dissipation of UST

As we observed in our previous work (Tan & Hover, 2009, 2010b), a disproportionate loss in thrust exist under elevated voltage applied across the transducer. As in all piezoelectric transducers, most of the electrical energy is converted into acoustical energy with some lost as superfluous heat through the transducer. In the presence of a large potential voltage, heat dissipation increases significantly, and dependent variables include the dielectric dissipation factor, transducer capacitance, and the presence of heat retardant materials next to the transducer. In instances of high power, localized heating at the soldered points may result in a failure or other undesirable outcome (Zhou & Rogers, 1995).

While temperature studies on PZT have been adequately described and investigated in the literature (Duck, Starritt, ter Haar, & Lunt, 1989; Sherrit et al., 2001), we are not aware of any work that makes a direct connection between transducer temperature rise and the propulsive thrust generated. As the UST is an underwater propulsor, knowledge of the conditions leading it to become a thermal source is important in many applications. In the following, we experimentally quantify the thermal distribution on the surface of the transducer under ultrasonic thrusting conditions, and introduce a dimensionless parameter to relate the thermal loss. In certain strategic applications, knowledge of this heat signature could aid in critical UST and system propulsion designs.

3.1 Heat transfer equations

We designed and constructed a larger UST unit based on (Allison, et al., 2008) but with several new features as shown in Fig. 15(a) and 15(b). Two o-rings are designed to seal the PZT against the water pressure and the air-backed layer provides maximum acoustic power transfer into the water. The screwed-on base holding the transducer is made of polytetrafluoroethylene (PTFE), so is the capping component holding the o-rings. Adhesive heat shrink and silicone potting seal the water from entering the air backing via the coaxial cable. The transducer material and medium will determine the heat transfer profile, rate of heat transfer and its dominant mode of heat transfer. For effective propulsion purpose, we add a layer of epoxy cast to match the acoustic impedance between the PZT and water.

From Fig. 16, in order to calculate the temperature profile through the transducer, $x = 0$ is taken at the centerline through the thickness of the PZT, and the average temperature at the centerline is given by

$$T_{mid} = \frac{Q_p x_p}{2k_p A_p}\left(\frac{1}{\pi^2} + \frac{1}{4}\right) + T_p \tag{11}$$

where x_p (m) is the thickness of the transducer, k_p (W/m·°C) is the thermal conductivity of the transducer, A_p (m^2) is the surface area of the transducer, and T_p (°C) is the temperature at the interface between the transducer and epoxy layer (Sherrit, et al., 2001). Q_p (W) is the average heat transfer rate of the piezoelectric material, which is also the average power dissipation, given by

$$Q_p = 2\pi f C \tan\delta E_{rms}^2 \tag{12}$$

where f (Hz) is the resonance frequency, C (F) is the capacitance of the transducer, tan δ (%) is the dielectric dissipation factor, and E_{rms} (V_{rms}) is the root-mean-square of the applied voltage. It is important to understand that under high power and temperature conditions, the transducer's dielectric dissipation factor will change with the voltage applied, temperature and fluidic load. Consequently, the capacitance and dielectricity also vary nonlinearly as the voltage and temperature increase, incurring significant errors if these are not carefully characterized under elevated settings. The temperature profile of the piezoelectric material takes on a parabolic distribution, peaking at T_{mid} (°C), and either surface of the transducer has the same temperature, T_p; see Fig. 16. In addition, as we verify in Section 3.4 for the Biot number, we consider convection to be more important than the internal conduction which exhibits a nearly uniform temperature gradient within the homogenous solid body (PZT).

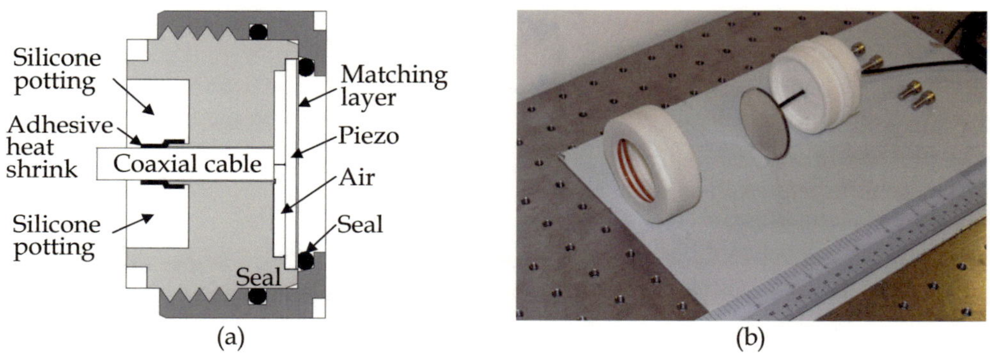

Fig. 15. (a) Sectioned view of the modular UST. A female Teflon cap is screwed onto the base holder which holds the transducer. O-rings provide the water-tight seals, and silicone potting provides flexibility and waterproofing to the coaxial cable connection. (b) An exploded view of the components of the modular UST.

The 1-dimensional Fourier's law is used to describe the thermal conduction of heat through the layer of epoxy cast on the UST water-side surface, and is governed by the heat flux, q (W/m^2), and the heat transfer rate (W) is given by

$$Q_e = qA_e = k_e A_e \frac{(T_p - T_e)}{x_e} \tag{13}$$

where k_e (W/m·°C) is the thermal conductivity, T_e (°C) is the surface temperature of the epoxy cast facing the water, and x_e (m) is the thickness of the epoxy cast. q is positive if heat flows along the positive x-direction, and vice versa.

As heat transfers into the water, convection will be dominant (heat radiation is negligible in our case) and using the steady-state Newton's Law of cooling, the convection governing equation is described as

$$Q_c = \bar{h}A_c\left(T_e - T_\infty\right) \tag{14}$$

where $\bar{h}$ is the averaged convective heat transfer (W/m$^2\cdot$°C), A_c is the area of transducer exposed to the water (m^2), and T_∞ is the ambient water tank temperature (°C).

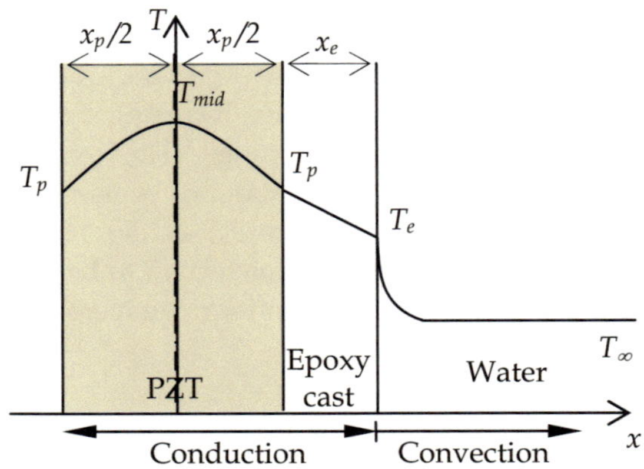

Fig. 16. A schematic diagram of the heat transfer profile. Acoustic transmission is generated by a transducer made from a PZT, through an epoxy cast layer into the water. Temperature gradient in each material outlines their relative heat transmission.

3.2 Temperature experimental setup

In this section, we characterize temperature for a 50mm UST transducer under increasing power, and then investigate in detail how various parameters could provide insights to the overall UST design consideration.

3.2.1 Hardware and methods

Fig. 17(a) shows the experimental setup similar to Fig. 6(a) except with an addition temperature data logger. Voltage to the amplifier is increased incrementally in steps of 0.2V_{pp} up to a maximum of 1V_{pp}. The output power to the transducer depends on the transducer load capacitance and electrical impedance, and is not explicitly controlled.

The PZT (k_p = 1.25W/m·°C) measures 50mm in diameter and 2.1mm thick, and resonates at 1MHz. The epoxy cast layer (k_e = 0.22W/m·°C) is designed to be 0.6mm thick, with the same diameter as the transducer but the o-ring seals part of it, and only 42.2mm of the diameter is exposed to the water. The UST is vertically positioned with the surface facing the side wall of the 1.2m×0.6m×0.6m water tank, submerged at 0.2m from the bottom of the tank. Tap water is filled and stood for at least one day before the experiment begins, and ambient water temperature is regularly maintained at 21.5°C with sufficient cooling. Water is filled up to a depth of 0.45m.

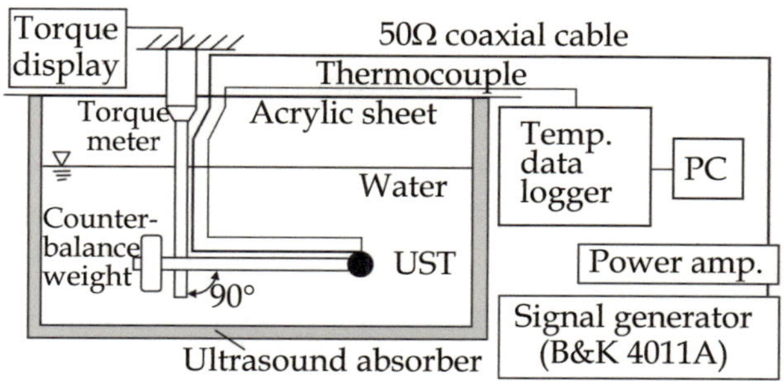

(a)

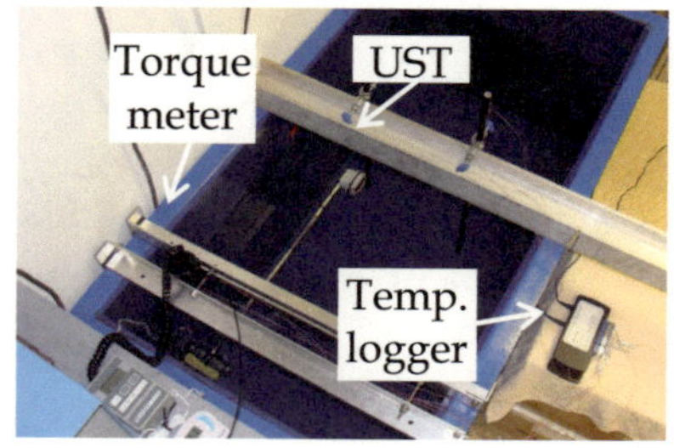

(b)

Fig. 17. (a) The experimental setup used to measure the UST temperature and thrust at the same time. Thrust is generated in the normal direction and out of the page. (b) A top view of the UST temperature measurement setup.

Similar torque calibration is performed prior to each set of test as discussed in Section 2.3.2. The sensing tip of five AWG 36 T-type (copper-constantan) thermocouples (SW-TTC2-F36-CL1) is attached to the transducer surface at five locations; T1 (north), T2 (east), T3 (west), T4 (south), and T5 (center) when viewed face-on as shown in Fig. 18. Thermal grease is applied to the junction of the thermocouples for improved thermal contact and each junction is affixed to the surface with a small adhesive tape. All thermocouples measure 2m in length and each terminates at a type-T miniature plug (CN001-T) is plugged into an Eight Channel Thermocouple Logger (OctTemp, MadgeTech), which has corrected cold-junction compensation internally for improved accuracy and response time. Each of the thermocouple is wrapped with an electromagnetic interference (EMI) tape and grounded to remove any RF interference. All temperature sensors have been calibrated by the manufacturer and are prescribed with an accuracy of ±0.5°C. Temperature data is sampled at 4Hz for all channels. The data logger has a background temperature noise of less than 0.1°C. Overall, temperature measurement uncertainty is estimated to be ±1°C in all cases. We consider a quasi-steady state of heat transfer to be defined by a temperature change of 0.2°C in a minute for 3 minutes.

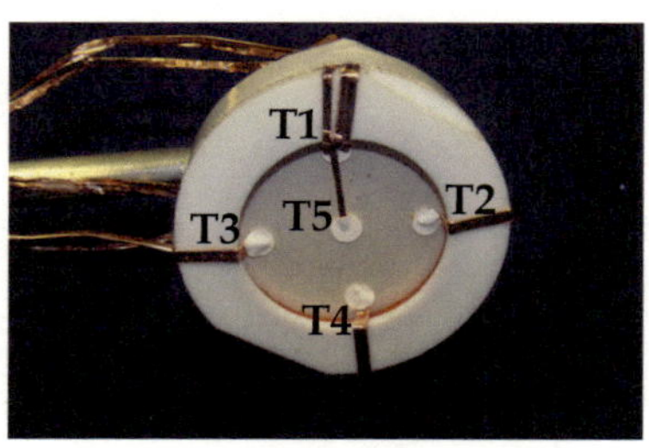

Fig. 18. Locations of thermocouples on the UST surface for average temperature measurement. Thermocouple cables are wrapped with RFI tape to ground RF interference.

3.3 Thermal characteristics results

Each temperature profile at a particular applied power is repeated four times independently under ambient water conditions. The experiment begins by recording 5 seconds of ambient water temperature, after which the power amplifier is switched on for about 9mins. During this time, the transducer is observed to increase its temperature steadily and then stabilizes. 9mins into the actuation, all thermocouple achieve a quasi-steady state and the power amplifier is switched off. Cooling proceeds for another 10mins and the tank is allowed to settle. Ambient temperature of the water tank is monitored separately at the start and end of the experiment, and further cooling is allowed if the ambient water temperature increased significantly.

Generally, increasing the input voltage (in steps of $0.2V_{PP}$ to $1.0V_{PP}$) to the amplifier increases the temperature recorded on the transducer surface. Table 3 summarizes the input voltage to the amplifier, output voltage and output power of the amplifier, the corresponding thrust, and the mean transducer surface temperature. Four sets of data are each tabulated for the output voltage, output power, and thrust, to demonstrate the consistency of the system.

Using the output power and thrust columns of Table 3, a plot of thrust versus power is shown in Fig. 19. From Eq. (2), the acoustic efficiency of the UST can be worked out, which essentially is the gradient at each cluster of points multiply by the speed of sound in the water, *c*. Generally at lower power, the efficiency is higher, however, effective thrust is also lower, which may not be practically useful. As the power increases, efficiency declines rapidly and Fig. 19 illustrates the trend. In Fig. 19, it can be seen that generated thrust starts to decline at higher electrical power level. However, as we discuss below, the acoustic efficiency is observed to vary approximately in a first-order fashion as *P* increases. Ultrasonic thrust appears to begin to saturate near 16mN.

Input voltage (V_{PP})	Output voltage (V_{rms})	Output power (W)	Thrust (mN)	Mean surface temp. (°C)
0.2	13, 13, 12, 13	3.2, 3.2, 2.9, 3.2	4.59, 4.44, 4.59, 4.52	22.9
0.4	30, 29, 28, 29	18.0, 16.8, 15.6, 16.8	7.08, 7.01, 6.93, 6.86	26.7
0.6	48, 47, 47, 48	45.6, 43.6, 43.6, 45.6	10.85, 10.70, 10.70, 10.78	30.6
0.8	67, 66, 66, 67	90.4, 87.2, 87.2, 90.4	14.85, 14.77, 14.62, 14.85	35.0
1.0	86, 84, 86, 86	147.8, 141.8, 147.8, 147.8	16.43, 16.43, 16.5, 16.35, 16.5	39.2

Table 3. Table of output power from the amplifier and the corresponding thrust generated. Four sets of data from each stepped input voltage are recorded. Ambient tank temperature is maintained at 21.5°C.

3.3.1 Thermal losses at high voltage

To calculate the averaged convective heat transfer, we use the lumped-capacity solution for a heated body transferring heat into the water by free convection. The solution can be solved for *T* (*t* = 0) to give (Lienhard IV & Lienhard V, 2002)

$$T = \left(T_e - T_\infty\right)e^{-t/\tau} + T_\infty \tag{15}$$

where T (°C) is the time-variant temperature of the epoxy with respect to time t (s), T_e (°C) is the averaged temperature of the epoxy surface determined as [22.9, 26.7, 30.6, 35.0, 39.2]°C from the stepped input voltage, and T_∞ (°C) is the ambient tank temperature at 21.5°C. $\tau = \frac{\rho_e c_e V_e}{\bar{h} A_e}$ is the time constant of the cooling process, where ρ = 1200 kg/m³ is the density of the epoxy, c_e = 1110 J/kg·°C is the specific heat capacity of epoxy, $V_e = 8.4\times10^{-7}$m³ is the volume of the epoxy through which sound transmits, $\bar{h}$ (W/m²·°C) is the average convective heat transfer coefficient determined as [25, 24, 22, 18, 14]W/m²·°C from the stepped input voltage, and $A_e = 5.6\times10^{-3}$m² is the cross-sectional area of the epoxy through which sound transmits (note $A_e = A_p$).

Most of this transmitted heat energy is convected away into the water as the surface of the transducer heats up. With this knowledge and the acoustic efficiency plot in Fig. 19, where efficiency decline rapidly as output power increases, we can see that the supplied electrical power has been significantly converted into heat energy while thrust increased diminutively - that net thrust begins to saturate above 90W. In the region of the saturated thrust, we also note an increase in heating of the transducer when the power applied is increased further, explaining the fundamental cause of loss of thrust at high electrical power.

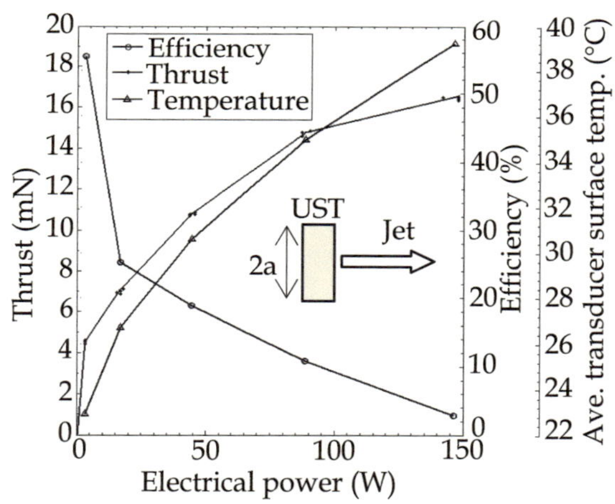

Fig. 19. Thrust force generated, acoustic efficiency, and average surface temperature versus the electrical power driving the UST. Gradients of tangent at each of the clusters of thrust points indicate the UST efficiency. Low power operation generally gives higher efficiency, but at a trivial lower thrust. At high output power, thrust generally saturates.

Next, substitute Eq. (13) into Eq. (14) to determine T_p, which is then substituted into Eq. (11) together with Eq. (12) to determine T_{mid}. With C = [11.26, 11.38, 11.81, 12.53, 13.62]nF and tan $\delta \approx$ 0.4%, the calculated values of T_{mid} are [23.84, 31.12, 42.49, 59.35, 82.59]°C at average output voltage E_{rms} = [12.75, 29, 47.5, 66.5, 85.5]V_{rms}. The superfluous heat, Q_l (W), can be calculated using

$$Q_l = \frac{(T_{mid} - T_e)}{R_{eq}} \tag{16}$$

where the equivalent thermal resistance, $R_{eq} = \frac{x_p\left(\frac{1}{\pi^2}+\frac{1}{4}\right)}{4k_pA_p} + \frac{x_e}{k_eA_e}$, for the transducer distance from $x = 0$ to the epoxy surface. We plot this heat loss against the amplifier output voltage, E_{rms}, in Fig. 20. Introducing a dimensionless parameter, Q_l/P, which is the ratio of the heat loss energy Eq. (16) to the electrical power supplied Eq. (4), as

$$\frac{Q_l}{P} = \frac{(T_{mid} - T_e)\mathrm{Re}(\Re)}{R_{eq}E_{rms}^2}. \tag{17}$$

We refer to this parameter as the lossy ratio. A large lossy ratio means more electrical energy has been converted to superfluous heat and a small ratio indicates most of the electrical energy has been converted to thrust. This ratio is also plotted in Fig. 20. Note that although the minimum of the graph signifies minimal heat loss to the electrical power supplied, it does not indicate the maximum thrust.

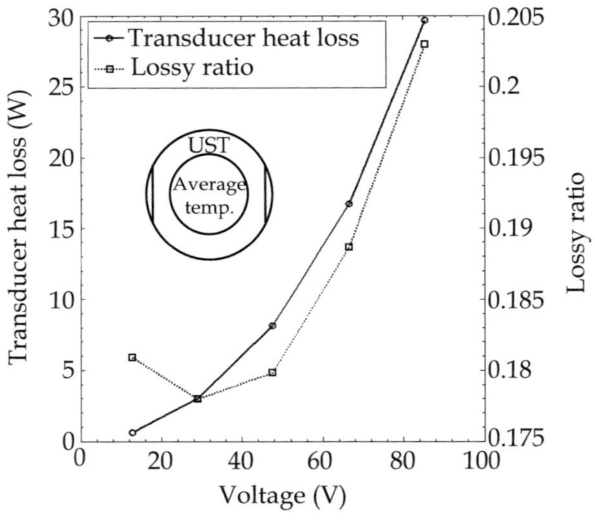

Fig. 20. Superfluous heat loss through the surface of the UST versus output voltage, E_{rms}. An increase in output voltage is generally associated with an increase in heat loss. The minimum of the lossy curve signifies the least heat loss in the system. Higher lossy ratio means more electrical energy is converted into heat losses instead of proportionately thrust.

3.4 Thermal losses design considerations

To determine the maximum thrust without significant loss due to heating, we will consider acoustic efficiency above 10% to be a practical value for this setup. From Fig. 19, the electrical power necessary to generate 10% efficiency would be less than 95W. From (4), and $\Re = 50\Omega$, E_{rms} is found to be 69V$_{rms}$. When this voltage is applied across the transducer, from Fig. 19, a UST thrust of about 15mN can be expected. Obviously increasing the voltage will increase the thrust but not appreciably; instead most added energy will be converted to heat losses, which becomes undesirable in the UST design scheme.

Finally, we verify that T_{mid} is not higher than the Curie temperature of the transducer, specified by the manufacturer at 320°C, to maintain its poled lattice integrity. We also validate the Biot number (Bi), which must be Bi « 1 to justify the temperature within the transducer to be relatively even. From the above parameters, $\mathrm{Bi} = \bar{h}x_e / k_e$ = [0.0682, 0.0655, 0.0600, 0.0491, 0.0382]. More importantly, this condition must also be satisfied for lump-capacity solution in Eq. (15) to be accurate.

4. Conclusion

The ultrasonic thruster technology could bring about interesting and novel attributes to robotic propulsion devices and systems. These include low cost and high robustness when applied at the centimeter scale or smaller. The robustness is due to the fact that the UST effectively has no moving parts, and will not biofoul - these are properties unavailable in the rotary and biomimetic propulsors in use today. Our experiments indicated that frequency, and voltage level can both strongly influence the behavior of the UST, in terms of wake, thrust, and efficiency. We have successfully implemented three sub-centimeter UST devices into a small robot, and made calculations showing short missions can be developed with such craft, despite its inherently low propulsive efficiency.

We have also studied the heating of ultrasonic transducers under conditions of thrust production. In view of practical application of the ultrasonic transducer in the medical field, it has been reported that clinical ultrasonic probes generate considerable heat when driven at off-resonance frequencies (Duck, et al., 1989). Most medical ultrasonic devices have a safety regulation on the level of power that the transducer can produce; for example, the IEC Standard 60606-2-37 limits the surface temperature of ultrasonic transducer to 43°C. In extreme cases, ultrasonic probes could reach a steady-state temperature of 80°C in ambient air at 25°C; obviously this is not suitable for human contact in practice. While the UST is not subject to complying with this standard, a UST device is still limited by extreme heating which may cause physical damage, and also because at high temperature conditions it suffers a saturation in thrust. It may be possible to minimize or harness heat for recycling in the UST system architecture, or even recoup a part of it through specialized nozzle appendages so as to enhance efficiency. For example, the backing layer of the transducer can be ventilated or cooled to remove heat. (Deardorff & Diederich, 2000) demonstrated using a water-cooling system and reported not only it does not reduce the acoustic intensity or beam distribution, but also allows more than 45W additional power supplied to the transducer. Indeed, its thrust assistive quality remains to be investigated.

Clearly UST technology would benefit from further developmental work on application-specific areas. The UST could, for example, complement an existing propulsor system to fine tune maneuvering, or to strategically control or manipulate a flow-field for other purposes. Propulsive efficiency could conceivably be enhanced by developing a waveguide external to the transducer. The use of DPIV for characterizing UST properties is considerably richer than velocity measurement using hot wire method alone, and could also aid new transducer designs traceable to the wake field. New applications, designed to exploit the above UST's unique attributes, will certainly find this technology a valuable solution. With these insights, the UST could uncover its potential as an enabler for very small crafts, and might find uses in other completely separate applications.

5. Acknowledgment

This research was supported by the Singapore National Research Foundation (NRF) through the Singapore-MIT Alliance for Research and Technology (SMART) Centre, Centre for Environmental Sensing and Modeling (CENSAM). The authors are also grateful to M. Triantafyllou and B. Simpson for access to a DPIV system.

6. References

Allison, E. M., Springer, G. S., & Van Dam, J. (2008). Ultrasonic propulsion. *Journal of Propulsion and Power, 24*(Compendex), 547-553.

Blackstock, D. T. (2000). *Fundamentals of physical acoustics*: New York : Wiley.

Carlton, J. (2007). *Marine propellers and propulsion*: Oxford : Butterworth-Heinemann.

Deardorff, D. L., & Diederich, C. J. (2000). Ultrasound applicators with internal water-cooling for high-powered interstitial thermal therapy. *IEEE Transactions on Biomedical Engineering, 47*(Compendex), 1356-1365.

Doty, K. L., Arroyo, A. A., Crane, C., Jantz, S., Novick, D., Pitzer, R., et al. (1998). *An autonomous micro-submarine swarm and miniature submarine delivery system concept.* Paper presented at the Florida Conference on Recent Advances in Robotics, Florida.

Duck, F. A., Starritt, H. C., ter Haar, G. R., & Lunt, M. J. (1989). Surface heating of diagnostic ultrasound transducers. *British Journal of Radiology, 62*(Copyright 1990, IEE), 1005-1013.

Egeskov, P., Bech, M., Bowley, R., & Aage, C. (1995). *Pipeline inspection using an autonomous underwater vehicle.* Paper presented at the Proceedings of the 14th International Conference on Offshore Mechanics and Arctic Engineering. Part 5 (of 5), June 18, 1995 - June 22, 1995, Copenhagen, Den.

Hariharan, P., Myers, M. R., Robinson, R. A., Maruvada, S. H., Sliwa, J., & Banerjee, R. K. (2008). Characterization of high intensity focused ultrasound transducers using acoustic streaming. *Journal of the Acoustical Society of America, 123*(Compendex), 1706-1719.

Hoerner, S. F. (1965). *Fluid-dynamic drag: practical information on aerodynamic drag and hydrodynamic resistance*: Midland Park, N. J.

Kamakura, T., Sudo, T., Matsuda, K., & Kumamoto, Y. (1996). Time evolution of acoustic streaming from a planar ultrasound source. *Journal of the Acoustical Society of America, 100*(Compendex), 132-132.

Lewis, G. K., & Olbricht, W. L. (2008). Development of a portable therapeutic and high intensity ultrasound system for military, medical, and research use. *Review of Scientific Instruments, 79*(Compendex).

Lienhard IV, J. H., & Lienhard V, J. H. (2002). *A heat transfer textbook* Cambridge, Mass. : Phlogiston Press.

Lighthill, J. (1978). Acoustic streaming. *Journal of Sound and Vibration, 61*(Copyright 1979, IEE), 391-418.

Loh, B.-G., & Lee, D.-R. (2004). Heat transfer characteristics of acoustic streaming by longitudinal ultrasonic vibration. *Journal of Thermophysics and Heat Transfer, 18*(Compendex), 94-99.

Myers, M. R., Hariharan, P., & Banerjee, R. K. (2008). Direct methods for characterizing high-intensity focused ultrasound transducers using acoustic streaming. *Journal of the Acoustical Society of America, 124*(Compendex), 1790-1802.

Nobunaga, S. (2004). European Patent Office: H. Electronic.

Panchal, C. B., Takahashi, P. K., & Avery, W. (1995). Biofouling control using ultrasonic and ultraviolet treatments (pp. 13): Department of Energy, Office of Scientific and Technical Information (DOE-OSTI).

Rudenko, O. V., & Soluian, S. I. (1977). Theoretical foundations of nonlinear acoustics. 274.

Sfakiotakis, M., Lane, D. M., & Davies, J. B. C. (1999). Review of fish swimming modes for aquatic locomotion. *IEEE Journal of Oceanic Engineering, 24*(Compendex), 237-252.

Sherrit, S., Bao, X., Sigel, D. A., Gradziel, M. J., Askins, S. A., Dolgin, B. P., et al. (2001). *Characterization of transducers and resonators under high drive levels.* Paper presented at the 2001 IEEE Ultrasonics Symposium. Proceedings. An International Symposium, 7-10 Oct. 2001, Piscataway, NJ, USA.

Tan, A. C. H., & Hover, F. S. (2009). *Correlating the ultrasonic thrust force with acoustic streaming velocity.* Paper presented at the 2009 IEEE International Ultrasonics Symposium, 20-23 Sept. 2009, Piscataway, NJ, USA.

Tan, A. C. H., & Hover, F. S. (2010a). *On the influence of transducer heating in underwater ultrasonic thrusters.* Paper presented at the the 20th International Congress on Acoustics Sydney, Australia.

Tan, A. C. H., & Hover, F. S. (2010b). *Thrust and wake characterization in small, robust ultrasonic thrusters.* Paper presented at the 2010 OCEANS MTS/IEEE SEATTLE, 20-23 Sept. 2010, Piscataway, NJ, USA.

Tan, A. C. H., & Tanaka, N. (2006). *The safety issues of intense airborne ultrasound: Parametric array loudspeaker.* Paper presented at the Proceedings of the 13th International Congress on Sound and Vibration, Vienna, Austria.

Trimmer, W., & Jebens, R. (1989). *Actuators for micro robots.* Paper presented at the Proceedings. 1989 IEEE International Conference on Robotics and Automation (Cat. No.89CH2750-8), 14-19 May 1989, Washington, DC, USA.

Wang, Z., Zhu, J., Qiu, X., Tang, R., Yu, C., Oiler, J., et al. (2011). *Directional acoustic underwater thruster.* Paper presented at the 2011 IEEE 24th International Conference on Micro Electro Mechanical Systems (MEMS 2011), 23-27 Jan. 2011, Piscataway, NJ, USA.

Yu, H., & Kim, E. S. (2004). *Ultrasonic underwater thruster.* Paper presented at the 17th IEEE International Conference on Micro Electro Mechanical Systems (MEMS): Maastricht MEMS 2004 Technical Digest, January 25, 2004 - January 29, 2004, Maastricht, Netherlands.

Zhou, S.-W., & Rogers, C. A. (1995). Heat generation, temperature, and thermal stress of structurally integrated piezo-actuators. *Journal of Intelligent Material Systems and Structures, 6*(Compendex), 372-379.

9

Real-Time Distance Measurement for Indoor Positioning System Using Spread Spectrum Ultrasonic Waves

Akimasa Suzuki, Taketoshi Iyota and Kazuhiro Watanabe
Faculty of Engineering, Soka University
Japan

1. Introduction

The fields of application for position information have been expanded along with developments in the information-driven society. Outdoor position information systems such as car navigation systems that use the global positioning system (GPS) have spread widely. On the other hand, indoor position information is also important for humans and robots for self-location and guided navigation along routes. However, as the signals from satellites seldom reach indoors, it is hard to convert a directly outdoor positioning system to an indoor one. Therefore, systems that use sensors such as pseudolites (Petrovsky et al., 2004), infrared rays (Lee et al., 2004), or ultrasonic waves (Shih et al., 2001) have been investigated for use in indoor positioning.

Compared to other methods, systems using ultrasonic waves have the advantage that they can be built at low cost and have comparatively high accuracy, because the propagation speed of ultrasonic waves is slower than that of electromagnetic waves. However, these systems are generally said to be weak in terms of noise resistance, and to take a longer time to acquire data, because they utilize the time-division multiplexed method, which becomes more taxing as the number of objects to be measured is increased. To overcome these drawbacks, positioning systems with spread spectrum (SS) ultrasonic waves have been investigated (Hazas & Hopper, 2006), (Yamane et al., 2004), (Suzuki et al., 2009); they use code division multiple access (CDMA) methods and are more robust to noise, because they use spread spectrum ultrasonic signals. According to these studies, although the procedure of reception and positioning calculation is computed off-line, the systems are shown to be highly effective.

Correlation calculation is one of the most important procedures when measuring a position using SS ultrasonic waves. This calculation (in air) is difficult to carry out using existing methods, which were developed either for ultrasonic signals traveling through solids and liquids or for radio signals such as GPS because of differences in speed, frequency, and susceptibility to signal loss. The process of correlation calculation in the positioning systems also requires many calculations for long signal sequences if it is to maintain CDMA performance and robustness against noise. Therefore, real-time correlation calculation must be realized with efficient use of limited electronic circuits.

As the positioning system is required to control moving robots, a study of real-time positioning should be carried out. In the positioning process, correlation calculation is required for detecting an SS signal. Some positioning systems (e.g., GPS) calculate correlation values using a serial search method or using a matched filter with analog elements (e.g., the SAW convolver (Misra & Enge, 2001)). In the case of indoor positioning using SS ultrasonic waves (with approximately 1/100000th of the propagation velocity and 1/38000th of the frequency of electrical waves), it is unlikely that one could directly utilize the conventional methods that are applicable to electrical waves. Although methods of correlation calculation for ultrasonic waves traveling through liquids or solids such as metals have been investigated (Teramoto & Yamasaki, 1988), they are still difficult to apply to indoor positioning systems with ultrasonic waves through the air because of the differences of velocity, wavelength, and attenuation rate of the ultrasonic signal. Therefore, we focused on a digital-matched filter to enable the system to both acquire the signal in a relatively short time and apply ultrasound easily.

The method for correlation calculation uses a digital-matched filter; because it requires product-sum operations on received data within a cycle of SS signals in a sampling period, it needs a large amount of calculation. Compared to outdoor positioning, the phase of SS signals shift more noticeably if the object is moving; thus, measured results must be more accurate. The carrier wavelength of ultrasound used was approximately 8.5 [mm]; filtering is required at a higher frequency than the carrier frequency if one is to apply a digital-matched filter in these conditions. Therefore, correlation calculations can be considered a bottleneck to the realization of real-time processing. S/N ratios and the number of channels in the PN sequence increase, as the length of PN sequences increases. In addition, the longer the PN sequences become, the more time is required for processing. Thus, the relation of noise tolerance and CDMA ability to processing time for correlation calculations is a trade-off.

It is difficult to find research on real-time correlation calculation or real-time positioning using SS ultrasonic signals. In general, methods of correlation calculation using product-sum operations (where pipelines use a number of multipliers that corresponds to the length of the PN sequence) have been considered. In addition, there is a method for sequential calculation using one multiplier. In the former case, a result can be obtained in 1 clock cycle, so that real-time calculation can be realized easily. However, it is unrealistic because of the huge number of logic elements required. In contrast, in the latter case, correlation calculations can be realized with a minimum logic size (i.e., they utilize only one multiplier); however, high-speed computation is required to realize real-time positioning. Therefore, certain innovations were required to allow real-time correlation calculation with SS ultrasonic waves for the purpose of self-positioning of humans and robots.

To achieve this, a new algorithm for real-time correlation calculation that uses external memory is suggested. In this chapter, the effectiveness of the proposed algorithm, named SPCM (Stored Partial Correlation Method), is presented in the form of experimental results of correlation calculations using original hardware. We also describe the effectiveness of real-time indoor positioning using SS ultrasonic waves and SPCM hardware based on experimental results of distance measurement.

2. Positioning system using SS ultrasonic waves

2.1 Indoor positioning environment

In our previous study, off-line positioning experiments were conducted that utilized transmitters installed on a positional environment and a receiver mounted on a positioning target (Suzuki et al., 2009). Fig. 1 shows the experimental environment for the indoor positioning system using SS ultrasonic waves. In Fig. 1, there are 4 transmitters Tr_1, Tr_2, Tr_3, and Tr_4 including sensor nodes called SPANs (smart passive / active nodes) (Nonaka et al., 2010), and 1 receiver Rc placed on a robot. This SPAN positioning system can also utilize inverse-GPS based positioning to swap a positional relation between the transmitter and the receiver. There is also a hardware component for controlling the time of transmission and the sampling frequency of the ultrasonic waves, labeled "timing generator" in Fig. 1. We can measure the position of an object with wireless because of the radio transceiver on the timing generator and the receiver unit. In this chapter, real-time correlation calculation is realized at the reception unit in Fig. 1 for real-time positioning.

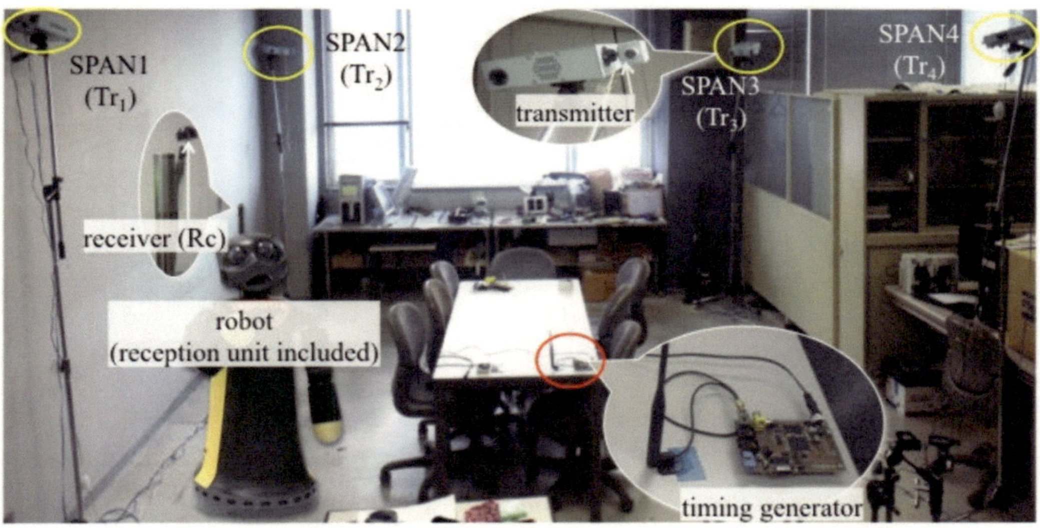

Fig. 1. Experimental environment of the positioning system

2.2 A method for position calculation

Position calculation for the indoor positioning system with SS ultrasonic waves is outlined in Fig. 2. First, spheres are drawn with radii equal to the distance between a receiver Rc and each transmitter (at the center); 2 pairs of the 2 spheres are selected. In Fig. 2, 2 pairs of spheres centered on Tr_1 and Tr_3, Tr_2 and Tr_3 are selected. From these pairs of spheres, $Plane_{13}$ and $Plane_{23}$ in Fig. 2 can be solved as a simultaneous equation. Line of the intersection is also obtained from the 2 planes of $Plane_{13}$ and $Plane_{23}$. Last, points at the intersection of Line with an equation of an arbitrary sphere are solved. Although the 2 intersection points are obtained, in the situation in which transmitters are installed in the four corners of a room, one solution is located outside of the room. Thus, the other solution becomes the position of the receiver Rc.

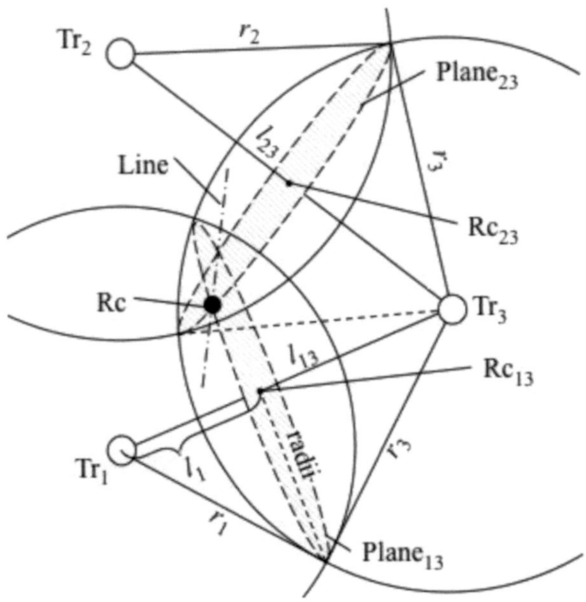

Fig. 2. Diagram for explanation of positioning calculations

Our specific method of calculation is as follows. First, to obtain equations for Plane_{13} and Plane_{23}, the coordinates of points Rc_{13} and Rc_{23} in their respective planes are solved. Here, we focus on Rc_{13}, which is between Tr_1 and Tr_3. In the case in which (x_1, y_1, z_1) and (x_3, y_3, z_3) are defined as the coordinates of Tr_1 and Tr_3, respectively, the distance l_{13} between Tr_1 and Tr_3 is given by Equation 1.

$$l_{13} = \sqrt{(x_1 - x_3)^2 + (y_1 - y_3)^2 + (z_1 - z_3)^2} \tag{1}$$

In the case of defining l_1 as a distance from Tr_1 to the Rc_{13} at the center of Plane_{13}, we can express this as follow from equations for obtaining radii of Plane_{13} using Pythagorean theorem and variables r_1 and r_2.

$$r_1^2 - l_1^2 = r_3^2 - (l_{13} - l_1)^2 \tag{2}$$

Also, Equation 3 can be expressed as follow.

$$l_1 = \frac{r_1^2 - r_3^2 + l_{13}^2}{2l_{13}} \tag{3}$$

Here, we define x_{31}, z_{31}, and z_{31} as $x_{31} = x_3 - x_1, y_{31} = y_3 - y_1$, and $z_{31} = z_3 - z_1$, respectively. Rc_{13} can be defined as in Equation 4.

$$Rc_{13}(x, y, z) = \left(x_1 + x_{31}\frac{l_1}{l_{13}}, y_1 + y_{31}\frac{l_1}{l_{13}}, z_1 + z_{31}\frac{l_1}{l_{13}} \right) \tag{4}$$

As $Plane_{13}$ crosses through Rc_{13} orthogonal to l_{13}, we can apply Equation 5.

$$x_{31}\left\{x-\left(x_1+x_{31}\frac{l_1}{l_{13}}\right)\right\}+y_{31}\left\{y-\left(y_1+y_{31}\frac{l_1}{l_{13}}\right)\right\}$$
$$+z_{31}\left\{z-\left(z_1+z_{31}\frac{l_1}{l_{13}}\right)\right\}=0 \quad (5)$$

$$x_{31}x+y_{31}y+z_{31}z = x_{31}x_1+y_{31}y_1+z_{31}z_1+\frac{l_1}{l_{13}}\left(x_{31}^2+y_{31}^2+z_{31}^2\right)$$
$$= x_{31}x_1+y_{31}y_1+z_{31}z_1+l_1l_{13} \quad (6)$$

The right-hand side of Equation 6 becomes a constant. Let us denote this number by k_{31}.

$$x_{31}x+y_{31}y+z_{31}z = k_{31} \quad (7)$$

Thus, an equation for $Plane_{13}$ is obtained. In the case of the same height for all transmitters, z_{31} =0; then, the equation for $Plane_{13}$ can be expressed as follow.

$$x_{31}x+y_{31}y = k_{31} \quad (8)$$

In addition, $Plane_{23}$ can be obtained from Rc_{23}.

$$x_{32}x+y_{32}y = k_{32} \quad (9)$$

By solving the simultaneous Equations 8 and 9, one can obtain the x-y coordinates of Rc as follows.

$$\mathrm{Rc(x,y)} = \left(\frac{x_{32}k_{31}-x_{31}k_{32}}{x_{32}y_{31}-x_{31}y_{32}}, \frac{y_{32}k_{31}-y_{31}k_{32}}{x_{31}y_{32}-x_{32}y_{31}}\right) \quad (10)$$

The height of receiver Rc(z) can be also obtained using the coordinates of an arbitrary transmitter $Tr_i(x_i, y_i, z_i)$.

$$\mathrm{Rc(z)} = z_i - \sqrt{r_i^2-(x-x_i)^2-(y-y_i)^2} \quad (11)$$

In the situation shown in Fig. 1, 4 results for position are obtained by 4 combinations of transmitters. The measurement position is defined as an average of these results.

2.3 Distance measurement hardware structure in the positioning system

A position is calculated from three or more TOF (time of flight, between the transmitters and the receiver) values. An architecture of measurement TOF for the positioning system is shown in Fig. 3. In a transmission unit, a D/A converter and a FPGA, which are used to generate carrier waves and M-sequences, are included. In a reception unit, an A/D converter and a FPGA, which are used to make correlation calculations, perform peak detection, and take time measurements, are included.

In this system, a SS signal is generated for the multiplication of carrier waves by M-sequences. The SS signal is dynamically generated by the transmission unit shown in Fig. 3 and is outputted by a transducer, after D/A conversion. At the same time as the transmission starts, the time counter is started so as to measure the TOF. Additionally, correlation values are calculated from the sound data by the correlation calculator and the A/D convertor, shown in Fig. 3, which constitutes the on-line, real-time hardware processing. The next peak detector shown in Fig. 3 finds a peak from among the correlation values. The time counter measures the TOF by counting the number of sampling times that elapse from the beginning of a transmission to the arrival of a peak. Thereafter, the 3D position of the receiver can be calculated from three or more distances using the TOF between the transmitters and the receiver. If the correlation calculation part is installed on the hardware, as shown in Fig. 3, real-time positioning is permissible, because other processing such as positioning calculation from distances can be calculated comparatively quickly in software using optimization expressions.

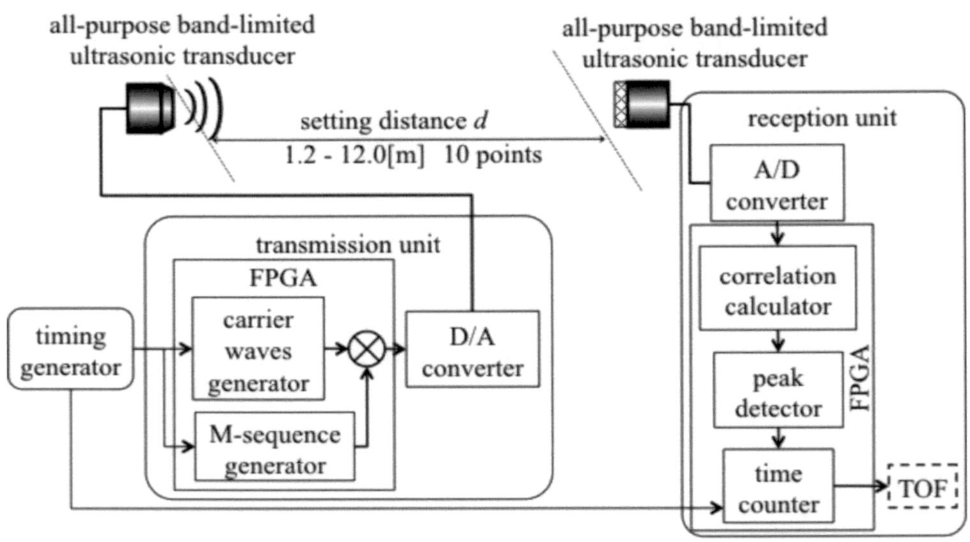

Fig. 3. System architecture for TOF measurement using SS ultrasonic signals

2.4 SS signal

In our indoor positioning system, SS signals are modulated by BPSK (binary phase shift keying) using an M-sequence, a pseudorandom code sequence, as used in the DS (direct sequence) method. Although an M-sequence of '0' or '1' is generated by the shift register, we used the value of '-1' in place of '0' to facilitate signal processing. Fig. 4 shows received SS signals. In Fig. 4, the signals corresponding '1' and '-1' are plotted by solid and dashed lines, respectively. Each dot in Fig. 4 called a 'sampling'; the number of samples including 1 period worth of carrier waves was selected to be 4 [samples]. Here, chip length t_c is defined as the required time to describe 1-chip worth of M-sequence. The chip length can be also defined as $t_c = 4/f$, using carrier frequency $f = 1/t_{cr}$. The length of the SS ultrasonic signals becomes

$2^9 - 1 = 511$ [chip] due to 9-bit shift register used to generate the M-sequence in our system. In this system, the frequency of the carrier waves was set to 40.2 [kHz].

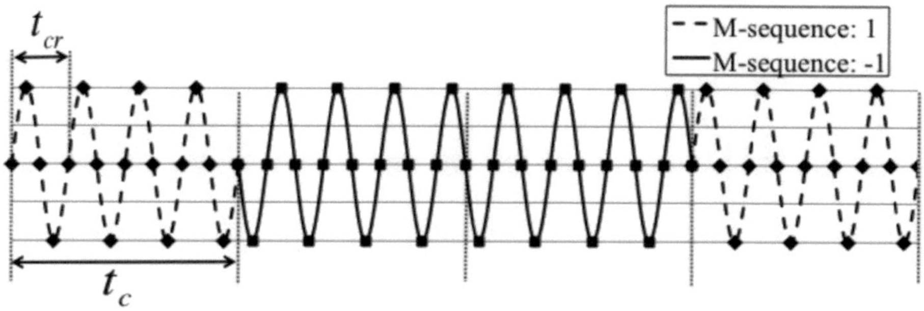

Fig. 4. Received SS signals on the reception unit in Fig. 3

2.5 Real-time peak detection

From the peak detector in Fig. 3, peaks in the correlation values are obtained. Fig. 5 gives the example of a distribution of correlation values obtained from a distance measurement. In this figure, sample values connected by lines are plotted with height as the vertical axis and sampling number as the horizontal axis. A peak value from among the correlation values is obtained for the situation in which the replica signals match the received signals. Here, s_{tc} is defined as the number of samples corresponding to a chip length t_c. From s_{tc} before the peak, the waveform in Fig. 5 becomes am upward sloping line because of the transitional intergradation of the correlation values. The high correlation value of a reflected wave is also shown in Fig. 5, arriving from some indirect pathway, as did the multi pass. The correlation value of the reflected waves can be higher than that of the direct waves; therefore, the sampling with the highest correlation values cannot be decided simply in terms of TOF. We require the detection of measurement time to be defined as the time from the start of a transmission to the first peak, because direct ultrasonic waves have the shortest path.

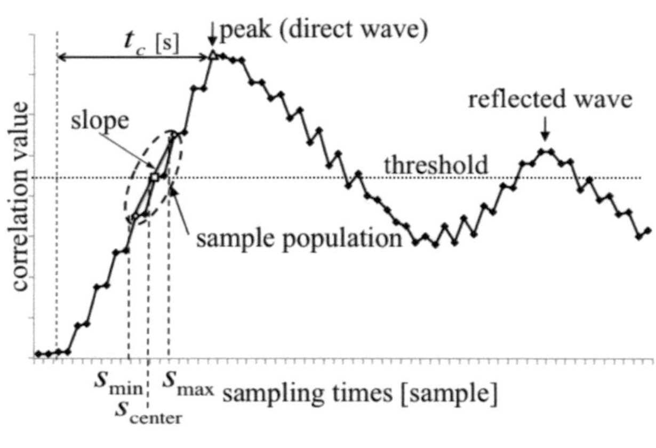

Fig. 5. A method for peak detection

Fig. 5 also describes an algorithm for peak detection. First, a threshold level is selected, based on the noise level without SS signals. Next, after starting the transmission, the first correlation value that is over the threshold is detected. In this procedure, the sampling number of this correlation value is treated as a central sample s_{center}, and a line is traced around a sample population near to s_{center}. The slope of the line is calculated from the sample population using the least-squares method. In the group of samples, s_{max} and s_{min} refer to the minimum and a maximum sampling number, respectively, and are defined as having the same distance from s_{center}. After this step, the slopes are calculated repeatedly, using the sample population shifted forward at increments of one sample. If a slope becomes negative, because we can consider s_{center} as having reached a peak sample, the sampling number of s_{center} is outputted as the peak position.

In the hardware component, a shift register is used for storage of the sample population. The slope a is obtained as follows. First, Let N and $x_i (i = 1, 2, 3...)$ denote the length of the shift register and the sampling distance from s_{center}, respectively. Here, N is restricted to odd numbers. x_1, x_2... are named in ascending order (of sample number) from the minimum onward; x_1 is the sampling distance between s_{min} and s_{center}. x_i is given by Equation 12.

$$-\frac{N-1}{2} \leq x_i \leq \frac{N-1}{2} \tag{12}$$

Next, y_i is defined as the correlation value of x_i. The slope a is calculated using the least-squares method as follows.

$$a = \frac{N \sum_{i=1}^{N} x_i y_i - \sum_{i=1}^{N} x_i \sum_{i=1}^{N} y_i}{N \sum_{i=1}^{N} x_i^2 - \left(\sum_{i=1}^{N} x_i \right)^2} \tag{13}$$

In Equation 13, a summation of $x_i{}^2$ will become a constant number for a fixed bit length N. As K is defined to be a constant and the total of x_i becomes 0, Equation 13 can be given as follows.

$$a = K \sum_{i=1}^{N} x_i y_i \tag{14}$$

We select $N = 5$ for the register length. Peak detection hardware for utilizing this method can be realized comparatively easily using Equation 14.

3. Real-time correlation calculation with SPCM

3.1 Hardware of real-time correlator

To obtain a correlation value, one must perform product-sum operations for all samples within an M-sequence cycle. In the proposed positioning system, 8184 product-sum operations are required within 6.25 [microseconds]. We realize a real-time correlation calculation to construct a hardware-utilizing SPCM. In the SPCM, a preliminary part, calculated from received signals, is processed as a pre-correlation value and saved first. Thereafter, the amount of parallelism for the processing of the correlation calculation is improved by repeatedly using the pre-correlation value results. To produce a large number of pre-correlation values, we utilize a general-purpose, active memory system.

Hardware for the correlation calculation (using SPCM and FPGA) is shown in Fig. 6. The hardware is operated at 50 [MHz]. It is mounted on a FPGA of FLEX10KA (1728 LEs) and has an external memory of PB-SRAM. In this hardware, the transmission of ultrasonic signals triggers the start of correlation calculations. Also, the hardware shown in Fig. 6 consists of a multiplier into which the carrier waves and received signal are inputted, accumulators for 1 chip length worth of M-sequence, a 1-chip data memory to save the result of the previous accumulated result, a pre-correlator, a 4-chip data memory, and a correlator with which a whole correlation value is calculated using the 4-chip data. Although 1-chip memory is installed on the FPGA, 4-chip data is installed on the external memory because of the large amount of pre-correlation data gathered. Block 1, 2, and 3 in Fig. 6 are defined as a part for generating 1-chip data, a part for pre-correlation calculation, and a part for correlation calculation using the obtained pre-correlation values, respectively. Processing of these blocks in parallel is conducted with SPCM.

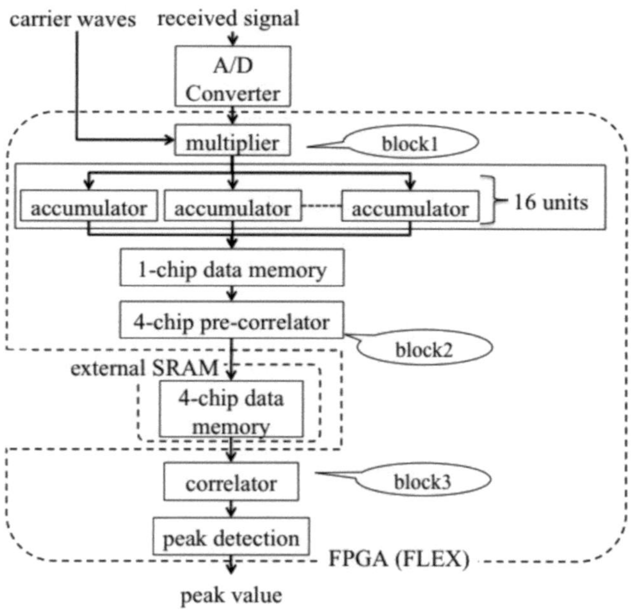

Fig. 6. Hardware layout on the real-time processor

3.2 Generation of 1-chip data

Fig. 7 describes correlation calculation using SPCM including the procedures at (a) block 1, (b) block 2, and (c) block 3. In block 1, received signals are multiplied only by the carrier waves, which are the elements of replica signals. Multiplied results shown in the third row of Fig. 7 (a) are obtained from both the idealized versions of waves of received signals that are shown in the first row of Fig. 7 (a), and the carrier waves generated by a reception unit shown in the second row of Fig. 7 (a). The multiplied data shown in a third row of Fig. 7 (a) are accumulated as 1-chip data (i.e., ocd in Fig. 7). In block 1, segments of M-sequences cannot be detected, because the time for 16 samples is spent on 1 chip. Therefore, 16 accumulators are installed on the hardware to allow the generation of the 1-chip data.

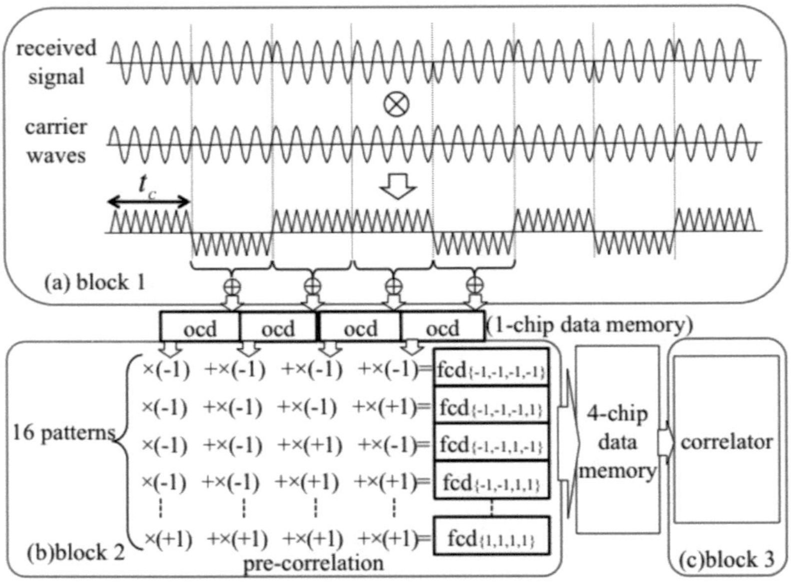

Fig. 7. Work flow of the real-time correlation

3.3 Generation of 4-chip data with pre-correlation calculation

In block 2, the part correlation value for the n-chip time (viz. n-chip data) is generated as a pre-correlation calculation using continuous 1-chip data of n samples for each sampling clock. In this process, the more 'n' increases, the more memory required, and the less time required for calculation. In our hardware, 4-chip data is generated, as $n = 4$. This is chosen in consideration of the operating frequency of the FPGA and the amount of installed memory. 4-chip data are labeled 'fcd' in Fig. 7 (b). In block 2, first, the 4 generated 1-chip data are multiplied by '1' or '-1' following an M-sequence. Next, the sum of the 4 results obtained from this calculation is saved as 'fcd'. In block 2, $2^4 = 16$ patterns of 4-chip data (viz. $\text{fcd}_{\{-1,-1,-1,-1\}}$ to $\text{fcd}_{\{1,1,1,1\}}$) are created based on all combinations of '1' or '-1' for 4 samples of 1-chip data. These 16 results for fcd, shown in Fig. 7, are to be part correlations values for 4 chips corresponding to 4 partial replica signals of {-1,-1,-1,-1} to {1,1,1,1}. All parts of the correlation values for a cycle of one M-sequence are saved in 4-chip data memory. Amount of memory d_{ext} can be expressed as

$$d_{ext} = 2^n s_M l_{pc} \tag{15}$$

In equation 15, s_M is to be the sampling number in the cycle of the M-sequence, 2^n is the number of patterns in the n-chip data generated by pre-correlation calculation, and l_{pc} is the data width of the part correlation value. The amount of memory then becomes 261888 words, because of the 32-bit data width of our hardware.

In block 3, the correlation calculation is conducted using the part correlation values. For example, Fig. 8 and Fig. 9 describe calculation processes in block 2 and block 3, respectively, in the case in which we use a SS signal consisting of 12 chips as 1 cycle of the M-sequence. Fig. 8 explains the procedure inside the 4-chip data memory. 16 data entries regarding fcd, generated in 1 sampling time, are shown in Fig. 8 as a group of 4-chip data (viz. 'fcdg'). A part of the product-sum operation in Fig. 8 is also shown as a pre-correlation calculation in Fig. 8 (b).

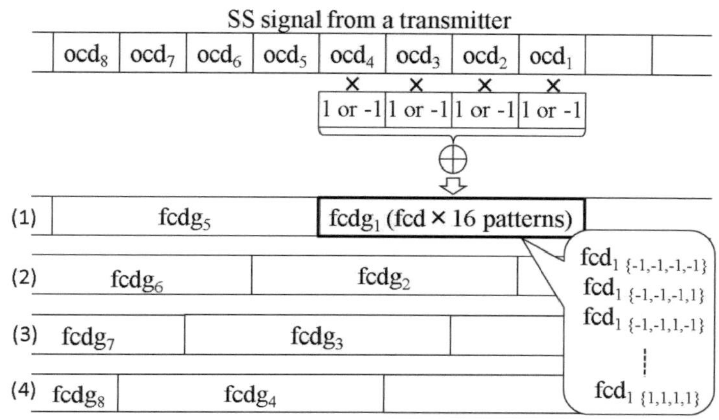

Fig. 8. Production of 4-chip data from 1-chip data

In Fig. 8, 1-chip data is generated in order from right to left, i.e., fcd_1, fcd_2,... When 1-chip data is obtained, a group of 4-chip data is also generated using the newest 4-data of ocd in the following order: $fcdg_1$, $fcdg_2$,... In Fig. 8, $fcdg_5$ (i.e., a group of 4-chip data) is generated after a time corresponding to 4 chips from ocd_4. A part correlation value corresponding to 8 chips can be generated if we lay out each $fcdg_1$ and $fcdg_5$ using different ocd values, as shown in Fig. 8. Also, $fcdg_2$ and $fcdg_6$, $fcdg_3$ and $fcdg_7$, and $fcdg_4$ and $fcdg_8$ can be created from other part correlation values at each 1-chip time. Therefore, Fig. 8 describes 4 groups of (1) $fcdg_1$, $fcdg_5$,... to (4) $fcdg_4$, $fcdg_8$,... Finally, a correlation value for all chips can be calculated using the obtained fcdg in the combination shown in Fig. 8 (1) to (4).

3.4 Correlation calculation using 4-chip data

In block 3, a correlation value is obtained by accumulating 128 continuous, memorized chips of 4-chip data corresponding to replica signals. Fig. 9 depicts the process of correlation calculation on block 3 using 4-chip data. Replica signals of the M-sequence are shown in Fig. 9 (a). The replica signals are divided into 4 sequences from the upper right-most signal in Fig. 9 (a); combinations of 4-sequences {-1,-1,-1,-1} to {1, 1, 1, 1} are named RP_0 to RP_{15}. For example, '1, 1, 1, 1', shown in the top group of replica signals, is called RP_{15}. Fig. 9 (b) describes the arriving scenes of pre-correlation values in chronological order. Here, although pre-correlated signals are generated as $fcdg_1$, $fcdg_2$ in order, Fig. 9 (b) only describes a certain group (1) $fcdg_1$, $fcdg_5$,... in Fig. 8. In Fig. 9, first, because of the correlation calculation with RP_{15} from a replica signal, fcd_1{1,1,1,1} in $fcdg_1$ generated by block 2 is loaded. After 16 sampling times have elapsed, $fcdg_5$ is generated. At this time, $fcdg_1$ must deal with a part correlation value corresponding to RP_3. Therefore, fcd_1 {-1,-1,1,1} in $fcdg_1$ is loaded. In this case, fcd_5 {1,1,1,1} is loaded from $fcdg_5$ from within the latest 4-chip data. Part correlation values for 8 chips are obtained by accumulating fcd_1 {-1,-1,1,1} and fcd_5 {1,1,1,1}. Similar processes to those described above occurred for every generation of a 4-chip data group. Finally, the complete correlation values are obtained by accumulating 4 fcds corresponding to RP_{15}, RP_8, RP_3, and RP_{15} from 4-chip data groups of $fcdg_1$ to $fcdg_{13}$, as shown at the bottom of Fig. 9 (b).

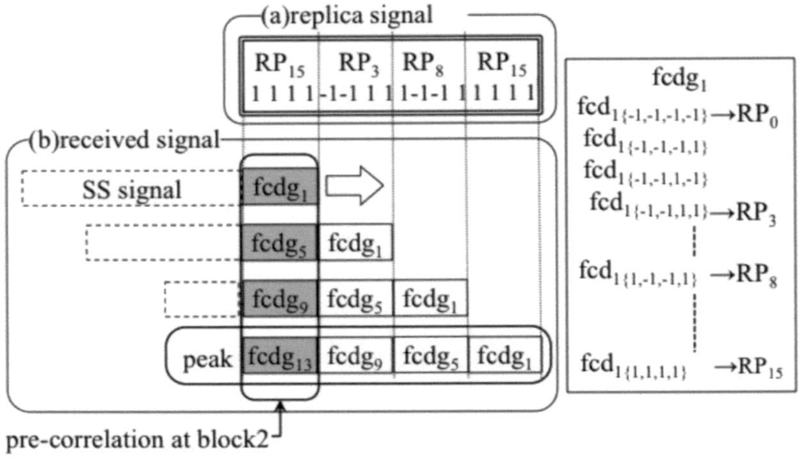

Fig. 9. Selection of 4-chip data using replica signals in block 3

In this method, a part correlation value generated once and saved previously can be reused in the calculation of the total correlation value. The SPCM is only required to calculate the latest 4-chip data as a part correlation value, represented by the gray block in Fig. 9 (b), for obtaining the whole correlation value of the SS signal. SPCM reduces the calculation to 1/16 of its previous value; however, the amount of saved data increases because of the pre-correlation calculation in block 2. Therefore, a real-time correlation calculation can be conducted comparatively easily using SPCM.

In this algorithm, two kinds of carrier waves (sine and cosine) are utilized for the detection of orthogonal components. The whole correlation values are obtained using a root-mean-square value of both correlation values using sine and cosine waves in block 3.

4. Behavior of real-time correlation hardware

4.1 Experiments of correlation calculation using actual received signals

The accuracies and real-time performance of correlation values with SPCM must be verified to allow a discussion of the effectiveness of this method in indoor positioning applications. The proposed real-time correlation mechanism will be mounted on conventional signal reception hardware. Thus, measuring experiments were conducted using the hardware shown in Fig. 3; the received sound data was inputted into the real-time correlation hardware. Additionally, to compare the resultant correlation values gathered by the real-time processing method to those of the conventional off-line correlation calculation that uses a sequential computation, experiments were conducted using the received sound data.

In these experiments, the ultrasonic signals were transmitted by a super tweeter to a condenser microphone to measure with the distance *d* between the receiver and the transmitter installed at 1.2 [m] intervals from 1.2 [m] to 12.0 [m]. Additionally, the TOF (from transmitter to receiver) using the results of real-time correlation calculation and temperatures was measured. Distance errors in the data in this experiment were calculated in real-time.

4.2 Discussion of results of correlation calculation

Calculated correlation values for different sampling times are obtained at the shortest d =1.2 [m] and the longest d =12 [m] distances; they are shown in Fig. 10(a) and Fig. 10(b), respectively. Each figure represents the differences between (1) off-line sequential computation using software, and (2) the real-time method using hardware. In addition, each figure is arranged on the x-axis from low to high sampling time, and on the y-axis from low to high correlation values.

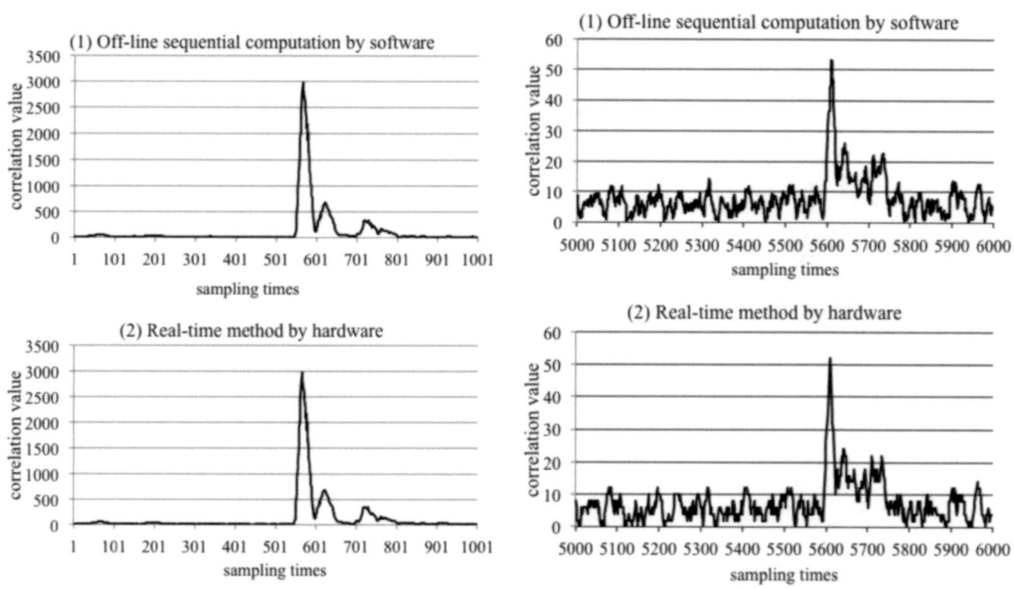

(a) Setting distance d =1.2m (b) Setting distance d =12.0m

Fig. 10. Difference between off-line sequential computation using software and the real-time method using the SPCM hardware

Comparing (1) to (2), the experimental results in Fig. 10(a) and Fig. 10(b), the peak values are plotted at the same time of 566 [samples] and 5609 [samples], respectively. These figures also show that the peak can be detected without the influence of noise from indirect signals, especially at a distance of d = 12 [m] where the lowest S/N ratios were obtained in this experiment; this effect was due to the reduction of signal strength.

4.3 Discussion of real-time performances

Table 1 describes the time required to execute each block in the whole process of the real-time correlation hardware, shown in Fig. 8. For measuring the time required in each block, a counter installed on the FPGA is counted utilizing 50 [MHz] of oscillator. The counted values are outputted to the PC via the PCI bus after real-time processing.

Table 1 shows the time required in each block as the counted values of oscillated frequency. The experimental results show that the required time for the whole process is 5.76 [microseconds], including the required times of 0.44, 1.78, and 5.12 [microseconds] for block 1, 2, and 3, respectively. The required time for the whole process depends on the longest time taken by the block processing, as each block is processed in parallel. Thus, the sum of the required time of block 3 and data transfer time among the blocks yields the overall time required by the whole process, as given in Table 1. Additionally, as shown in Table 1, the time for the whole process is less than the sampling cycle of 6.25 [microseconds]; therefore, the experimental results show that an indoor real-time positioning system using SS ultrasonic signals could be created by the correlation calculation hardware.

	required time [μs]
block 1	0.44
block 2	1.78
block 3	5.12
whole process	5.76

Table 1. Required time for each block

4.4 Measurement errors of the experiments using SS ultrasonic waves

Fig. 11 shows the errors in the distances obtained in real-time at setting distances d between 1.2 [m] and 12.0 [m]. Fig. 11 has an x-axis from $d = 1.2$ [m] to 12.0 [m] and a y-axis from low to high distance error. The distance errors d_{err} in Fig. 11 are approximated using Equation 16 (from sampling time s_d and temperature T).

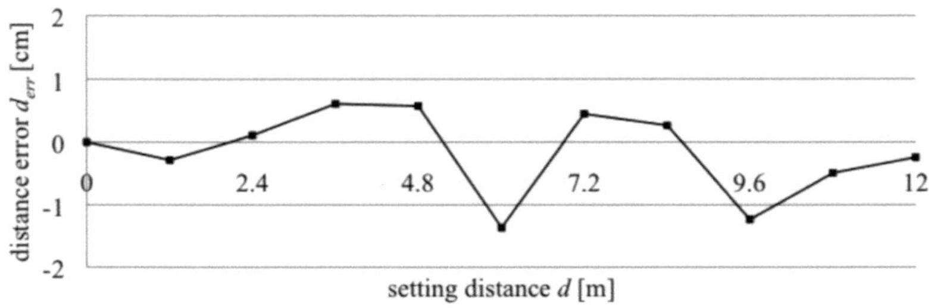

Fig. 11. Experiment errors in the distance measurement using SS ultrasonic waves

$$d_{err} = d - 6.25s_d(331.5 + 0.60714T) \tag{16}$$

Fig. 11 shows that the errors in distance are within ± 1.5 [cm] for all distances measured. We require a 10 [cm] or smaller positioning error for accurate self-location recognition for robots and humans; the obtained experimental measurement error in distance is within this range.

5. Conclusion

This chapter discussed the real-time correlation calculation for SS ultrasonic signals for use in an indoor positioning system. The experimental results were gathered using an original SPCM hardware system with external memory. In our method, positioning processing, signal receiving, correlation calculation, peak detection, and distance measurement are realized by hardware processing on a FPGA mounted on a PCI board.

Because of the real-time correlation calculation, which requires the most time during hardware processing, we proposed and installed a SPCM system with external memory and comparatively small logic elements. SPCM can be divided into 3 blocks: extraction of M-sequence signal by multiplying a received signal by carrier waves and accumulation of the signal in block 1, pre-correlation calculation with 4 chips in block 2, and calculation for whole correlation values in block 3 using pre-correlation values. The amount of correlation calculation can be reduced by 1/64 by the pre-correlation calculation.

A distance measurement experiment was conducted to evaluate both the correlation values and the real-time performances. Experimental results using the original hardware with SPCM are compared to the results of off-line sequential computation using software. Hardware processing time was measured using a counter based on the hardware clock. Additionally, the measurement distance was calculated from TOF data utilizing the SPCM hardware in real-time.

From the experiments, we found that the real-time hardware computed correctly within 5.76 [microseconds], which was less than the sampling time; the distance errors obtained were within ± 1.5 [cm]. Thus, the effectiveness of this hardware has been shown. In the case of 3D indoor positioning with SS ultrasonic signals, more than 3 signals (made by different channels from transmitters using CDMA) must be used. As a correlator is required when calculating signals on 1 channel, downsizing of the logic elements is important for creating useful receivers. At this time, indoor positioning using SPCM is suggested for use in real-time positioning using comparatively small logic elements. This chapter also shows the possibility of real-time position sensing for mobile robots and humans using SS ultrasonic waves.

6. References

Hazas, M. & Hopper, A. (2006). Broadband ultrasonic location systems for improved indoor positioning, *IEEE Transaction on Mobile Computing* Vol. 5(No. 5): 536 – 547.

Lee, C., Chang, Y., Park, G., Ryu, J., Jeong, S.-G., Park, S., Park, J. W., Lee, H. C., shik Hong, K. & Lee, M. H. (2004). Indoor positioning system based on incident angles of infrared emitters, *30th Annual Conference of IEEE Industrial Electronics Society*, Vol. 3, pp. 2218 – 2222.

Misra, P. & Enge, P. (2001). *Global Positioning System Signals, Measurements and Performance*, Ganga-Jamuna Press.

Nonaka, J., Kon, T., Choi, Y. & Watanabe, K. (2010). Implementation of task processing modules to a sensor node of span for offering services in an indoor positioning sensor network, *2010 International Symposium on Smart Sensing and Actuator System*, pp. 67–69.

Petrovsky, I., Ishii, M., Asako, M., Okano, K., Torimoto, H. & Suzuki, K. (2004). Pseudelite application for its, *IEICE Technical Report* Vol. 230(No. 104): 13 –18.

Shih, S., Minami, M., Morikawa, H. & Aoyama, T. (2001). An implementation and evaluation of indoor ultrasonic tracking system, *IEIC Technical Report* Vol. 101(No. 71): 1 – 8.

Suzuki, A., Yamane, A., Iyota, T., Choi, Y., Kubota, Y. & Watanabe, K. (2009). Measurement accuracy on indoor positioning system using spread spectrum ultrasonic waves, *4th International Conference on Autonomous Robots and Agents 2009*, IEEE, pp. 294 – 297.

Teramoto, K. & Yamasaki, H. (1988). Circular holographic sonar utilizing an inverse problem solution, *Transactions of the Society of Instrument and Control Engineers* Vol. 24(No. 7): 655 – 661.

Yamane, A., Iyoda, T., Choi, Y., Kubota, Y. & Watanabe, K. (2004). A study on propagation characteristics of spread spectrum sound waves using a band-limited ultrasonic transducer, *Journal of Robotics and Mechatronics* Vol. 16(No. 3): 333 – 341.

10

Ultrasonic Waves in Mining Application

Ahmet Hakan Onur[1], Safa Bakraç[2] and Doğan Karakuş[1]
[1]*Dokuz Eylul University*
[2]*Turkish General Directorate of Mineral Research and Exploration*
Turkey

1. Introduction

This chapter is aimed to introduce ways of beneficiation from ultrasonic waves in earth science, especially in mining practices. Since rocks are non-homogenous, elasto-plastic material, it has always been difficult to predict the behaviour of rock under any stress loaded environment. Unless removing uncertainties in the rock masses, designers can face to highly surprising and costly operational results in mining practices, so reducing the risk factor becomes vital element of underground constructions. To reduce risks may only be possible by knowing the surroundings where you work in very well. Sometimes it becomes costly to make the mining environment clear, so some practical methods have been trying to develop over years. One of them is acoustic methods based on the theory of elasticity. The elastic properties of substances are characterized by elastic module or constants that specify the relationship between stress and strain. The strains in a body are deformations, which produce restoring forces opposed to the stress. Tensile and compressive stresses give rise to longitudinal and volume strains, which are measured as unit changes in length and volume under pressure. Shear strains are measured by deformation angles. It is usually assumed that the strains are small and reversible, that is, a body resumes its original shape and size when the stresses are relieved. If the stress in an elastic medium is released suddenly, the condition of strain propagates within the medium as an elastic wave.

The principle of the ultrasonic testing method is to create waves at a point and determine the time of arrival at a number of other points for the energy that is travelling within different rock masses. The velocity of ultrasonic pulses travelling in a solid material depends on the density and elastic properties of that material. The quality of some rock masses is sometimes related to their elastic stiffness and rock mass structure, such that the measurement of ultrasonic pulse velocity in these materials can often be used to indicate their quality, as well as to determine their elastic properties. Travelling velocities of ultrasonic pulses are high in homogenous rock masses with high mechanical properties (UCS, tensile strength, cohesion, internal friction angle), which can be used as identification method of the quality of any rock structure. Some methods had been developed to measure rock diggability, stress distribution near a mine opening, bench blasting efficiency due to structural identification of rock masses by comparing the ultrasonic pulse travelling velocities in a reference sample with real in-situ measurements. In laboratory scale, available techniques and measurable rock mass properties are given in Table 1.

Non-destructive Methods	Crack Location	Thickness	Water Content	Hardness	Density	Reinforcement	Delimitation	Other
Acoustic Emission	■							
Electrical		■	■		■	■		
Electromagnetic		■	■			■	■	■
Nuclear	■	■	■		■	■		■
Ultrasonic	■	■				■	■	
Mechanical	■	■		■		■	■	■

Table 1. Non-destructive methods and applications (Moozar, 1993)

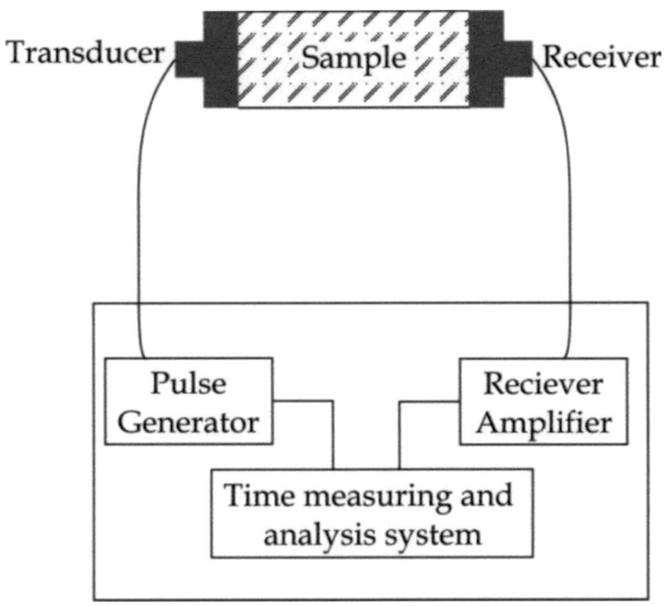

Fig. 1. Pulse velocity test (Malhotra & Carino, 1991)

Discontinuities and their dimensions within any material can be determined by the techniques given in Table 1 (Hasani et al.). There are four main measuring methods, namely; pulse velocity, pulse echo, resonant frequency and sonic tomography. Pulse velocity measures the time of an ultrasonic pulse within a material, hence finding the pulse velocity of the medium. The instruments used with this system include piezoelectric transducers, coupling agents, pulse generator, signal amplifier and an analyses system (Fig. 1.). The Pulse Echo system uses the transmission of low stress pulse energy to its sensor to delineate defects of material. The echoes received from defects are captured on a time domain spectrum and their analysis locates the anomalies within the structure. The resonant frequency test defines dynamic property of a sample and in generally used in laboratory environments. An oscillator outputs vibrations, which is analysed into the materials transverse, longitudinal and torsional frequencies of the material. Sonic tomography analyses seismic P-wave velocities to image sections of a material. The idea behind the previous technique and the ultrasonic pulse velocity method is similar; however sonic tomography can use a large number of transmitters and receiver at the same time. Its sensivity allows this method to analyse between different anomalies within a structure and process a sectional view of a sample (Moozar, 1993).

There has been big development on measurement devices, since their first introduction by Leslie & Cheesman (1949). Then, measurement techniques pervaded in rock mechanic application given rise to ASTM D2845-08 Standard Test Method for Laboratory Determination of Pulse Velocities and Ultrasonic Constants of Rock. This test method describes equipment and procedures for laboratory measurements of the pulse velocities of compression waves and shear waves in rock and the determination of ultrasonic elastic constants of an isotropic rock or one exhibiting slight anisotropy.

This chapter is about to introduce a practical application of ultrasonic waves in marble industry. Miners working in the marble industry have always been interested in identifying structural weaknesses in marble blocks before they are cut in a marble quarry and transported to marble processing plants. To achieve this difficult task, several simple methods have been developed among miners but observation-based methods do not consistently provide satisfactory results. A nondestructive method developed for testing concrete could be used for this purpose. In this chapter, this simple method based on differences in ultrasonic wave propagation in different rock masses will be presented, and the test results performed both in the laboratory and a marble quarry will be discussed.

When ultrasonic testing is applied to marble blocks, its objective is to detect internal flaws that send echoes back in the direction of the incident beam. These echoes are detected by a receiving transducer. The measurement of the time taken for the pulse to travel from the surface to a flaw and back again enables the position of the flaw to be located. This ultrasonic testing technique was originally developed for assessing the quality and condition of concrete. One instrument used for this purpose is known as PUNDIT. The apparatus has been designed especially for field testing, being light, portable and simple to use. Simple correlations between concrete strength, concrete aggregate gradation, water-cement ratio and curing time have been analyzed using PUNDIT.

The possibility of identifying these structural defects using ultrasonic pulses will be discussed and results obtained from these measurements will be introduced in the scope of this chapter. The shape and size of any abnormality in a block can be determined by direct measurements taken from suitably spaced grids. It is important to find the exact position of the surface in marble blocks so that precautions can be taken before the cutting process starts. As stated before, if any discontinuity surface lies in the pulse path, the measured time corresponds to the pulse that follows the shortest path. This is important because any discontinuity causes a time delay compared with the travel time of pulses in homogenous blocks. This study concentrated on the relationship between structural discontinuities and ultrasonic pulse travelling velocities in non-homogeneous marble blocks. Mathematical formulations were developed to find the exact locations of the surfaces that cause a separation during the cutting process. To verify the mathematical model explained above, a cubic homogenous marble block with a certain cut inside was prepared in laboratory. This chapter also covers the results obtained from model marble block in laboratory as well as the in-situ measurements obtained from industrial size marble blocks. Block subjected to testing of mathematical modelling in the marble plant was observed in detail before and after slice cut and results will be discussed.

2. Ultrasonic waves in mining application

There are two main mining applications of ultrasonic waves: one is to define the mechanical properties of intact rock and other is to define geological structures of the rock masses. There are several studies on determination of rock characterisation such as uni-axial compressive strength, static modulus of elasticity via non-destructive techniques especially after development of Portable Non-Destructive Digital Indicating Tester (PUNDIT) (Bray & McBride 1992, Green, 1991, ISRM, 1979, Mix, 1987, Vary, 1991, Chary et all, 2006, D'Andrea et all, 1965, Kahraman, 2007, Vasconles et all, 2008, Bakhorji, 2010, Khandelwala & Ranjithb 1996)

The other area of beneficiation of ultrasonic waves is to achieve rock mass classification based on rock mass structures. Several researchers had done work on indirect in-situ measurements to obtain practical data for the rock mass classification studies, since classical methods are expensive and time taking processes (Lockner, 1993, Galdwin, 1982, Karpuz & Pasamehmetoglu, 1997, Ondera, 1963).

There are some works reported on mineral processing industry about ultrasonic waves such as discharging feeding chutes, vibro-acoustics crushers, increasing shaking table performance, ore washing, milling, screening and optimizing bulb performance in flotation (Ozkan, 2004, Stoev & Martin, 1992, Asai & Sasaki, 1958, Kowalski & Kowalska, 1978, Nicol et all, 1986, Slaczka, 1987, Yerkovic et all, 1993, Djendova and Mehandjski, 1986).

Ultrasonic velocity measurements have previously proven valuable tool in measuring the development of stiffness of cement mixtures, so an engineered mine cemented paste backfill material were tested by ultrasonic wave and it is reported that measurements can be used as a non-destructive test to be correlated with other forms of laboratory testing (Galaa, et al. 2011). Grouted rock bolts are widely used to reinforce excavated ground in mining and civil engineering structures. A research was performed to find opportunities for testing the quality of the grout in grouted rock bolts by using ultrasonic methods instead of destructive, time-consuming and costly pull-out tests and over-coring methods (Madenga, et al., 2009, Zou, et al., 2010, Madenga, et al., 2009). A valuable work was performed by Lee, 2010 to predict ground conditions ahead of the tunnel face. This study investigates the development and application of a high resolution ultrasonic wave imaging system, which captures the multiple reflections of ultrasonic waves at the interface, to detect discontinuities at laboratory scale rock mass model. Another application of ultrasonic waves was introduced by Gladwin, 2011 to determine stress changes induced in a large underground support pillar by mining development at Mt Isa Mine. They introduced an ultrasonic stress monitoring device and compared the results with continuous strain recordings at a nearby site. (Renaud V, et al., 1990) deals with the determination of the excavated damaged zone around a nuclear waste storage cavity using borehole ultrasonic imaging. This analysis is based on a method that is able to sound and image the rock mass velocity field. Another interesting work was published by (Jones, et al., 2010) to monitor and assess the structural health of draglines. They had announced that, by using ultrasonic waves and by studying both the diffraction pattern and the reflected waves, it is possible to detect and size cracking in a typical weld cluster. In the work of Deliormanli at al. 2007, laboratory measurements of P and S wave velocities of marbles from different origins were presented and their anisotropic performances at pressures up to 300 MPa were calculated and compared with the elastic properties.

3. Determination of discontinuities in marble blocks via a non-destructive ultrasonic technique

This ultrasonic testing technique was originally developed for assessing the quality and condition of concrete. One instrument used for this purpose is known as PUNDIT. The apparatus has been designed especially for field testing, being light, portable and simple to use. Simple correlations between concrete strength, concrete aggregate gradation, water-cement ratio and curing time have been analyzed using PUNDIT (Saad & Qudais, 2005). Several articles have been published on the subject of defining the mechanical properties of several different materials apart from rock by nondestructive test methods based on ultrasonic wave propagation (Dereman et al., 1998, Zhang et al, 2006).

There has been a rapid increase in the demand for natural materials to be used in construction engineering, interior decoration, and urban fitting. Over the years, there has been no shortage of quarried blocks, but problems have been encountered in providing sufficient numbers of high quality marble blocks. Blocks of commercial size are directly extracted from the massif. In the case of homogeneous rocks having constant features, structural discontinuities affect the marketability of the blocks. It is important to identify such abnormalities in the marble before the cutting process is performed in order to save money and time. The possibility of finding these structural defects using ultrasonic pulses has been studied, and promising results were obtained. This study concentrated on the relationship between structural discontinuities and ultrasonic pulse travelling velocities in non-homogenous marble blocks. Mathematical formulations were developed to find the exact locations of the surfaces that cause a separation during the cutting process.

3.1 Elastic constants and waves

The principle of the method is to create wave at a point and determine at a number of other points the time of arrival of the energy that is reflected by the discontinuities between different rock surfaces. This then enables the position of the discontinuities to be deduced. The basis of the seismic methods is the theory of elasticity. The elastic properties of substances are characterized by elastic moduli or constants, which specify the relation between the stress and strain. The strains in a body are deformations, which produce restoring forces opposed to the stress. Tensile and compressive stresses give rise to longitudinal and volume strains which are measured as the measured as the change in length per unit length or change in volume per unit volume. Shear strains are measured as angles of deformations. It is usually assumed that the strains are small and reversible, that is, a body resumes its original shape and size when the stresses are relieved. If the stress applied to an elastic medium is released suddenly the condition of strain propagates within the medium as an elastic wave. There are several kinds of elastic waves:

In the longitudinal, compressional of P waves the motion of the medium is in the same direction as the direction of wave propagation. These are ordinary sound waves. Their velocity is given by (New, 1985):

$$V_{\mathrm{p}} = \left(\frac{k4\mu / 3}{\rho} \right)^{1/2} \tag{1}$$

Where ρ is density of the medium and k bulk modulus, μ shear modulus of the medium respectively. In the transverse, shear or S waves the particles of the medium move at right angles to the direction of wave propagation and the velocity is given by (Tomsett, 1976):

$$V_{\mathrm{s}} = \left(\frac{\mu}{\rho} \right)^{1/2} \tag{2}$$

If a medium has a free surface there are also surface waves in addition to the body waves. In the Rayleight waves the particles describe ellipses in the vertical plane that contains the direction of propagation.

Another type of surface waves the Love waves. These are observed when the S wave velocity in the top layer of a medium is less than in the substratum. The particles oscillate transversely to the direction of the wave and in a parallel to the surface. The Love waves are thus essentially shear waves.

The frequency spectrum of body waves in the earth extends from about 15 Hz to about 100 Hz; the surface waves have frequencies lower than about 15 Hz (Parasnis, 1994). For the method described in this study P waves are of importance as exploration tools. For the materials like concrete, marble necessary frequency range changes from 20 - 250 KHz.

3.2 Application of pulse velocity testing

For assessing the quality of marble blocks from ultrasonic pulse velocity measurement, it is necessary to use an apparatus that generates suitable pulses and accurately measures the time of their transmission through the material tested. The instrument indicates the time taken for the earliest part of the pulse the transmitting transducer when these transducers are placed at suitable points on the surface of the material tested. The distance that the pulse travels in the material must be measured to determine the pulse velocity.

$$\text{Pulse velocity} = \frac{\text{Path length}}{\text{Transit time}} \tag{3}$$

Fig. 2. shows how the transducers may be arranged on the surface of the specimen tested. The transmission can either be direct (a), semi-direct (b) or indirect (c).

The direct transmission arrangement is the most satisfactory one since the longitudinal pulses leaving the transmitter are propagated mainly in the direction normal to the transducer face. The indirect arrangement is possible because the ultrasonic beam of energy is scattered by discontinuities within the material tested but the strength of the pulse detected in this case is only about 1 or 2 % of that defected for the same path length when the direct transmission arrangement is used. The purpose of the study was to develop a model in stone quarry so semi-direct and indirect transmissions were taken as the main measurement technique since it is sometimes very difficult to find free faces to place transducers on the production bench.

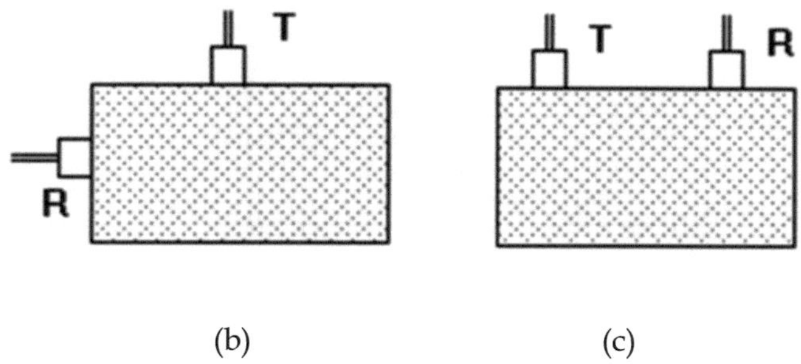

Fig. 2. Methods of propagating ultrasonic pulses

Pulses are not transmitted through large air voids in a material and, if such a void or discontinuity surface lies directly in the pulse path, the instrument will indicate the time taken by the pulse that circumvents the void by quickest route. It is thus possible to detect large voids when a grid of pulse velocity measurements is made over a region in which these voids are located. By using this behaviour, method can be used to test rock strata and provide useful data for geological survey work.

3.3 Laboratory works on simulated model

A concrete model was designed in laboratory to find out the behaviour of ultrasonic pulses travelling through a simulated discontinuity surface inside a concrete block. A wooden surface was settled in the block with the dimension shown in Fig. 3.

Before assessing the effects of simulated discontinuity on pulse velocity, first pulse velocity measurements made nearby simulated surface. This gives the real pulse velocity for prepared concrete block. For this purpose three direct measurements from three free surfaces were obtained. The result is given in Fig. 3.

For later use, a linear equation was set for the line shown in Fig. 4.

$$T = 2.52L - 3.39 \ (\mu s) \tag{4}$$

In equation 4, 2.52 is the slope of the direct line given in Fig. 4. and -3.39 is the value T takes when the length value L is equal to 0.

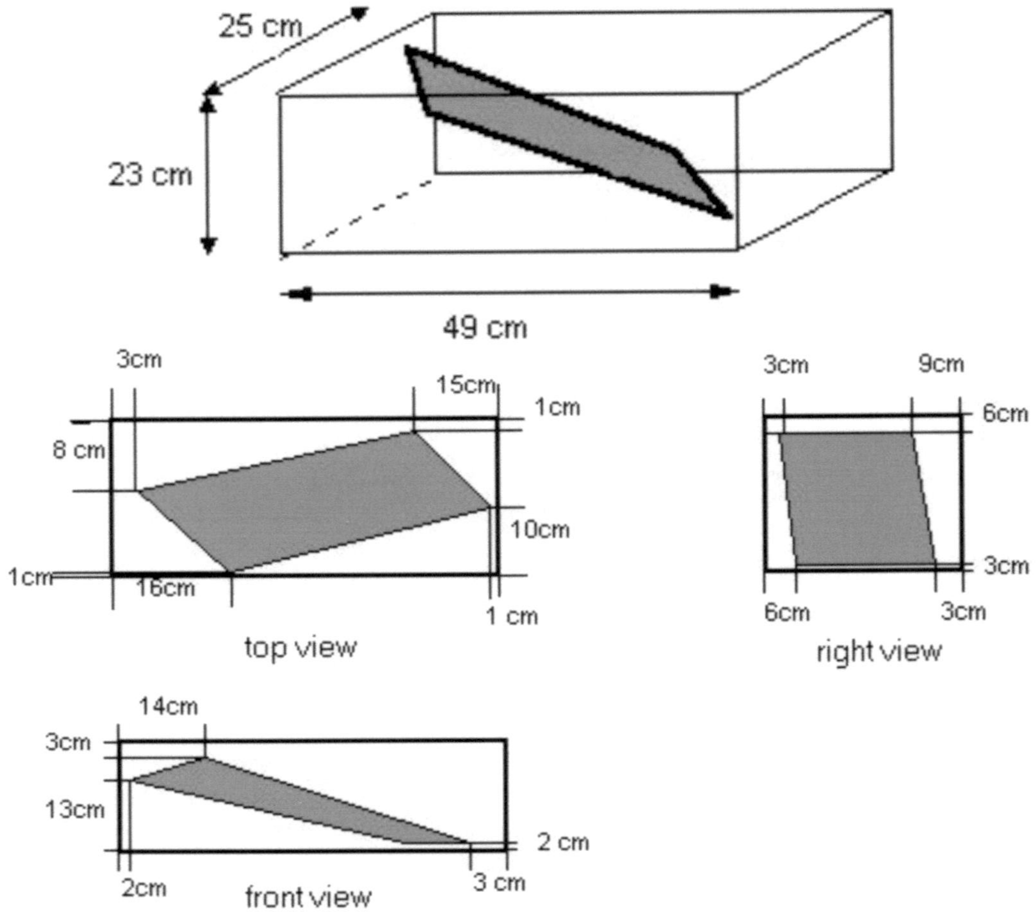

Fig. 3. Prepared concrete block and the dimension of the surface

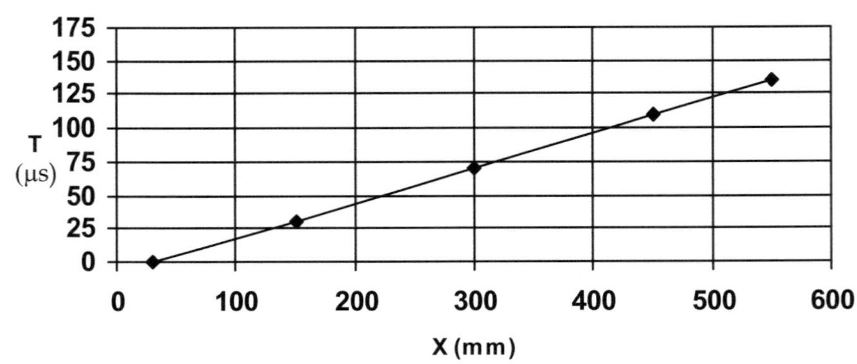

Fig. 4. Pulse velocity determination for homogenous concrete block

1	129.5	130.8	133.4	133.4	130.8	125.5	124.7	125.3	126.3
2	126.8	130.0	130.0	132.5	130.0	126.0	124.4	124.0	125.0
3	125.7	125.7	126.0	129.0	128.5	127.0	124.5	123.5	124.0
4	124.6	124.5	125.5	127.0	129.4	127.0	126.5	126.0	125.3
5	125.3	125.0	126.0	127.2	127.2	127.4	127.0	127.5	128.5
6	125.1	125.2	125.9	126.0	126.3	126.4	126.6	126.7	128.5
7	122.0	124.0	124.7	124.7	125.1	125.3	125.5	126.7	126.5
8	119.5	120.8	121.5	122.3	123.0	122.8	123.0	123.0	123.5
	1	**2**	**3**	**4**	**5**	**6**	**7**	**8**	**9**

Table 2. Transmission times taken from the right face of the concrete block (μs)

1	65.4	65.5	68.5	65.1	66.8	65.0	63.6	62.5	62.3	62.5	62.2	62.4	62.3	62.2	62.3	63.2	63.5	64.4	63.7
2	63.9	64.2	64.8	64.1	65.3	66.0	65.5	63.2	61.8	61.1	61.9	60.4	60.9	60.5	60.4	60.4	60.4	61.9	62.5
3	63.0	62.6	62.3	61.8	62.0	63.7	66.4	66.0	66.0	62.9	62.3	60.7	61.2	60.2	59.7	59.8	59.8	60.4	61.3
4	62.5	61.4	62.0	61.1	60.8	60.5	62.3	63.5	65.0	65.4	65.3	66.0	62.2	60.5	60.7	59.7	59.8	60.2	61.7
5	62.3	61.4	61.3	60.6	60.5	60.5	61.1	60.9	61.3	62.1	62.0	62.4	63.0	62.4	61.6	61.2	61.3	61.6	63.0
6	61.2	61.1	61.2	61.1	60.8	61.3	61.2	61.0	61.0	61.1	60.7	60.7	61.8	60.4	61.3	62.0	63.0	64.0	63.7
7	60.5	60.5	61.0	61.3	61.3	61.7	61.4	61.0	61.0	60.2	60.2	60.4	60.4	60.3	60.4	61.4	62.4	63.0	63.5
8	59.0	59.5	60.0	60.0	59.1	60.3	59.5	60.4	60.0	60.1	60.1	59.6	59.4	60.1	59.4	59.2	59.1	59.2	60.4
	1	**2**	**3**	**4**	**5**	**6**	**7**	**8**	**9**	**10**	**11**	**12**	**13**	**14**	**15**	**16**	**17**	**18**	**19**

Table 3. Transmission times taken from the front face of the concrete block (μs)

1	57.1	56.6	56.4	55.7	55.4	55.9	55.6	56.0	56.1	56.7	56.2	55.8	55.2	54.2	54.7	55.2	55.9	55.9	56.4
2	56.4	56.2	55.3	55.0	55.3	56.1	55.6	56.0	56.0	56.9	57.2	57.1	58.4	54.8	54.9	55.0	55.5	55.9	56.4
3	56.0	56.2	55.8	55.0	55.5	56.3	57.0	58.0	58.0	59.5	59.8	60.4	61.0	58.0	55.4	54.6	54.7	55.9	56.4
4	56.7	56.4	57.0	57.0	58.3	59.8	62.0	62.5	63.2	66.5	66.0	66.0	64.2	60.6	56.0	54.5	54.4	55.2	56.4
5	58.0	59.4	62.4	63.0	66.0	66.0	68.0	68.0	70.0	67.1	73.0	74.0	68.2	63.3	58.4	56.1	56.0	55.4	55.7
6	57.0	56.3	58.0	67.2	73.4	74.4	76.4	75.0	71.0	67.0	67.0	65.5	62.2	61.4	61.2	58.3	57.5	55.8	55.8
7	56.8	55.8	56.1	68.3	82.5	77.0	71.0	66.3	63.0	61.5	60.0	59.5	57.6	56.9	57.4	56.1	55.9	55.7	55.8
8	57.0	55.8	56.0	56.4	71.3	95.0	64.8	60.0	57.0	56.0	55.3	56.0	55.0	54.7	54.9	55.3	55.1	55.1	55.4
9	56.1	56.1	56.1	56.0	58.0	62.0	60.0	56.3	55.0	54.6	54.8	55.5	54.5	54.6	54.8	55.2	55.3	55.2	56.0
	1	**2**	**3**	**4**	**5**	**6**	**7**	**8**	**9**	**10**	**11**	**12**	**13**	**14**	**15**	**16**	**17**	**18**	**19**

Table 4. Transmission times taken from the top of the concrete block (μs)

Measurement of pulse velocities at points that are not affected by the simulated surface provides a reliable method of assessing the pulse velocity behaviour of the homogenous concrete block. It is useful to plot a diagram of pulse velocity contours from the result obtained since this gives a clear picture of the extent of variations. It should be appreciated that the path length can influence the extent of the variations recorded because the pulse velocity measurements correspond to the average quality of the concrete. When an ultrasonic pulse travelling through concrete meets a simulated surface, there is a negligible transmission of energy across this interface so that any air-filled cracks or void directly between the transducers will obstruct the direct beam of ultrasound when the void has a projected area larger than the area of the transducer faces. The first pulse to arrive at the receiving transducer will have been diffracted around the periphery of the defect and the transit time will be longer than in similar concrete with no defect.

In order to detect the simulated surface, pulse velocity measurements were performed over three different directions of concrete block with a grid of 2.5 cm x 2.5 cm and results are given in Table 2, 3, 4.

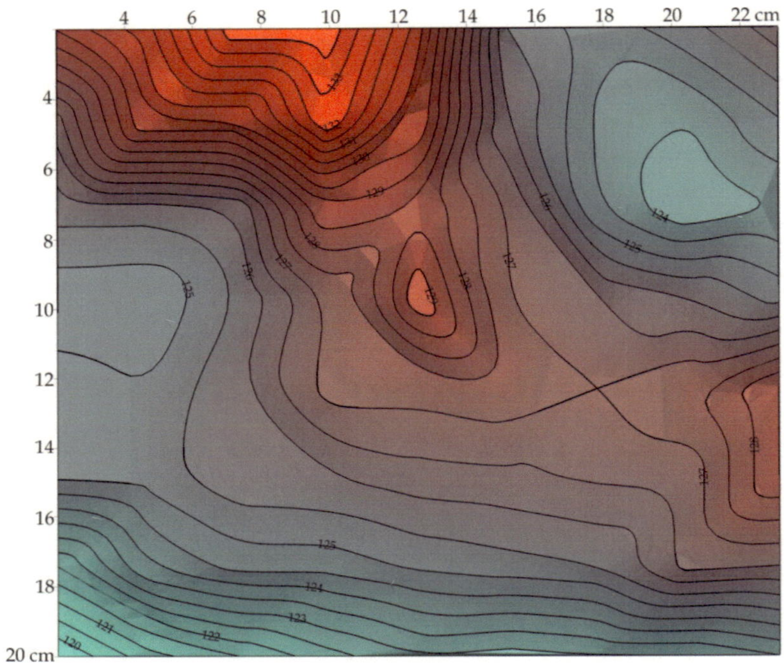

Fig. 5. Contour plotting of transmission times (μs) taken from right face

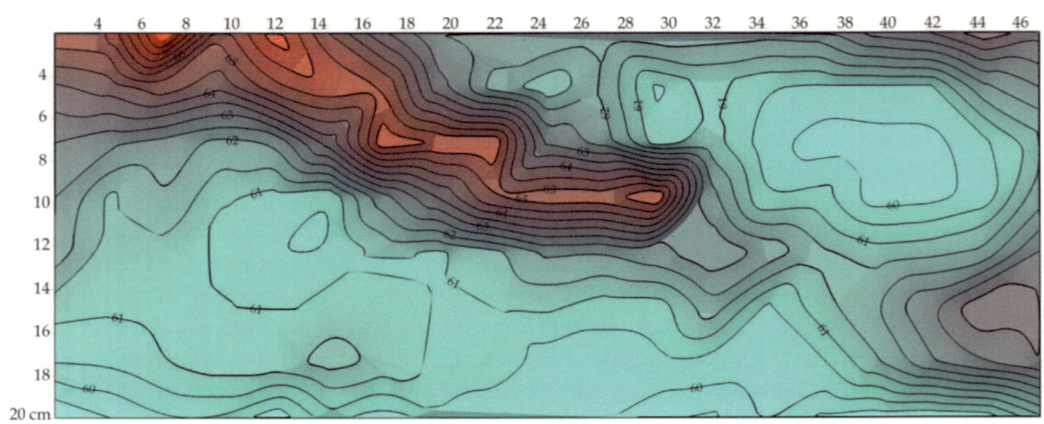

Fig. 6. Contour plotting of transmission times (μs) taken from front face of the model

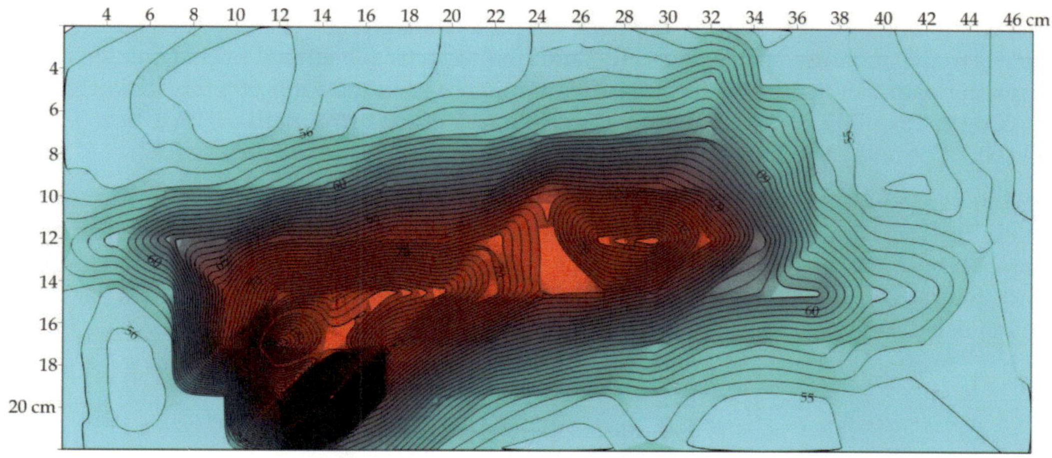

Fig. 7. Contour plotting of transmission times (μs) taken from top of the model

As shown in Fig. 5., Fig. 6. and Fig. 7., it is possible to detect the size and position of the simulated surface. Such estimates are more reliable if the discontinuity surface has a well-defined boundary surrounded by uniformly dense concrete.

3.4 Modeling the boundary of discontinuities

The shape and size of any abnormality in a block can be determined by direct measurements taken from satisfactory grids. It becomes important to find the exact position of the surface in marble blocks so that to take all precautions before cutting process starts. As stated before, if any discontinuity surface lies in the pulse path, the measured time belongs to the pulse that follows the shortest path. This is important because it causes a time delay comparing to travel time of pulses in homogenous blocks. This case is shown in Fig. 8.

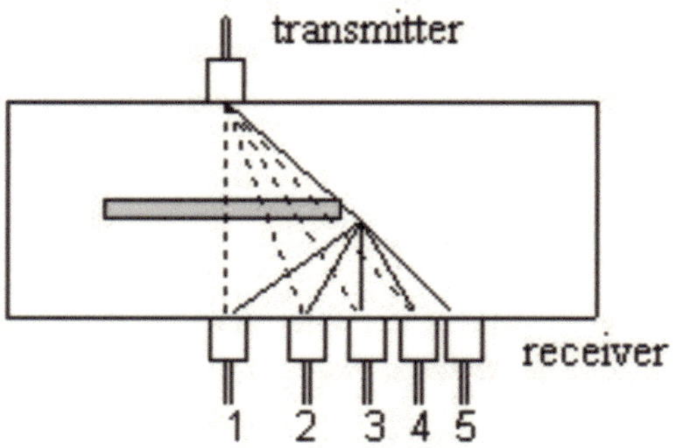

Fig. 8. The cause of time delay of pulses.

Before doing any measurement, pulse velocity behaviour of homogenous material must be obtained as discussed in laboratory work. By doing so, it becomes possible to estimate pulse travel time if no abnormality exists in the block. It is necessary to measure direct distance between transmitter to receiver in order to estimate travel time of pulses. The pulse velocity measuring device gives the minimum travelling time between two points. The pulse velocity is obtained by dividing the path length to transit time. There will be a difference between the measured pulse velocity and the velocity obtaining from equation 3 by applying path length L in the formula. This difference is an indicator of a time delay caused by longer travelling distances due to obstacle in the pulse travel path (dashed lines in Fig. 8.). The time delay will be used later to find the correct position of surface in 3D.

Fig. 9. explains the situation clearly. In this figure, there are two types of curves in the graphics. The linear one represents the direct distance from transmitter to receiver; the parabolic one represents the longer path of pulses following the boundary of discontinuity. Both figures can be obtained from the type of measurement so that receiver moves away from position 1 to position 7 while transmitter stays stable.

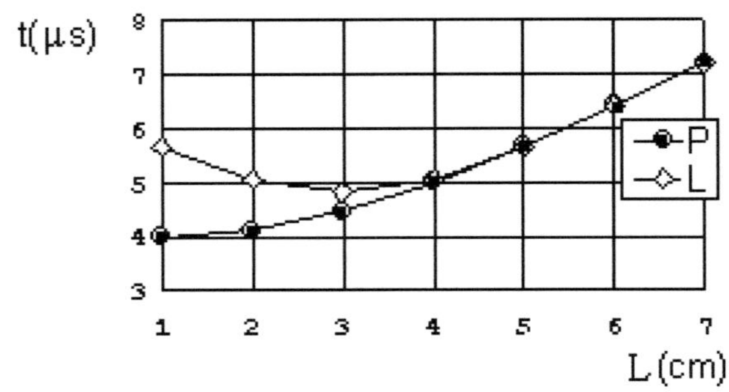

Fig. 9. The difference between direct distance and the pulse travel distance.

The first curve in Fig. 9 is the distance obtained from pulse travel times (shown as L in Fig. 9.) from receiver, the second one is the actual distance (shown as P in Fig. 9.) according to receiver position from 1 to 5. There has been a wide opening between two lines that shows the position of receiver 1 is away from the discontinuity surface. While the receiver moves along a line with equal distance from position 1 to 5 the gap between two lines narrows steadily. This behaviour gives a very important clue about the boundary of the surface. In the second region of the graph started from receiver position 5 to 7, two lines meet and behave in the same way. Because, at receiver position 5 the modelled surface has been passed that indicates there is no surface between transmitter and receiver. This kind of work defines the boundary in longitudinal location (s) but not in vertical one. As shown in Fig. 10. discontinuity can be at any position of (h) that is the vertical distance of the surface from top of the block.

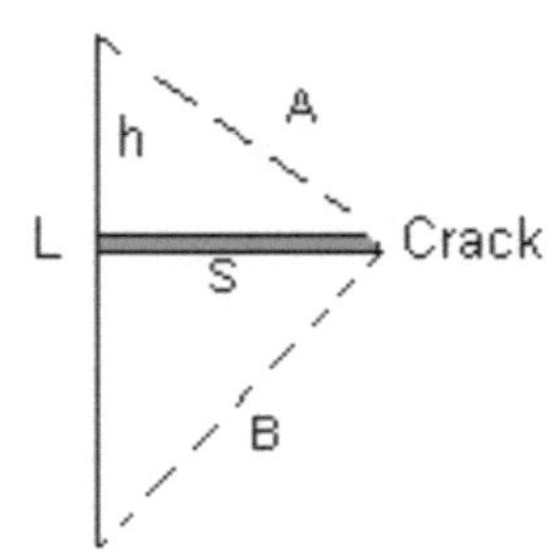

Fig. 10. Pulses moving around the crack

Lets take s as the length of crack, h is the vertical distance from surface, A + B is the path of pulse travelling from transmitter to receiver and L is the shortest distance between two probs. The pulse travels the distance A + B instead of L.

$$Ls = A + B \tag{5}$$

Some simple linear algebra can be used to obtain h and s those are the correct places of the boundary of discontinuity.

$$S^2 = A^2 - h^2 \tag{6}$$

$$S^2 = B^2 - (L-h)^2 \tag{7}$$

$$A^2 - h^2 = B^2 - (L-h)^2 \tag{8}$$

$$A^2 - B^2 = h^2 - (L-h)^2 \tag{9}$$

$$(A+B).(A-B) = h^2 - (L-h)^2 \tag{10}$$

$$A - B = \frac{h^2 - (L-h)^2}{Ls} \tag{11}$$

$$A = \frac{h^2 - (L-h)^2}{2.Ls} + \frac{Ls}{2} \tag{12}$$

$$S = \sqrt{\frac{h^2 - (L-h)^2}{2.Ls} + \frac{Ls}{2} - h^2} \tag{13}$$

Equation 13 is a function of vertical distance h. In this formula both s and h are unknowns. The purpose of all formulations above is to define h and s. In equation 13, if h changes from 0 to L and plotted, Fig. 11. can be obtained. In this process only the measurements that can be obtained are directly measured L, A+B that is estimated from the linear equation set before the homogenous material.

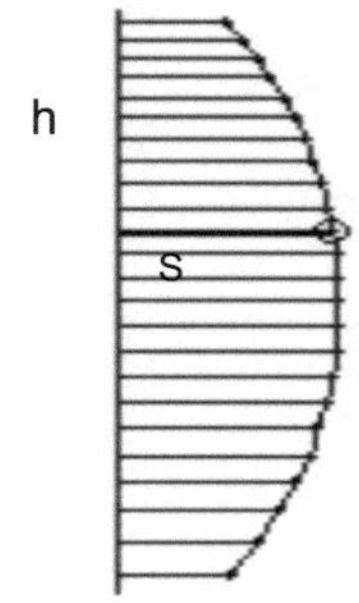

Fig. 11. The curve that shows possible path obtained from a single measurement.

When the receiver moves to position 2 in Fig. 12., the same measurements are made to plot second curve. Both figures are combined to obtain an intersection point that gives the correct position of the boundary, in another saying h and s can be determined (Fig. 12.).

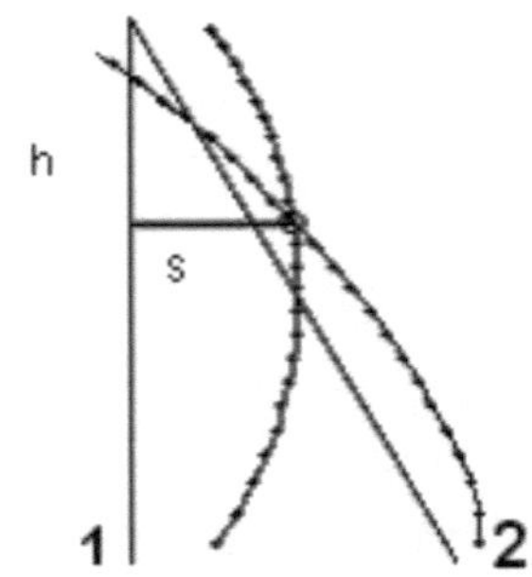

Fig. 12. The combined curves giving h and s

Obtaining h and s is very important because if the receiver moves in four different directions that mean four different h and s can be obtained in different directions. In common, receiver moves in such a way that enables us to find the exact shape of boundary of the discontinuity in 3D.

Finding h and s values are a time taken process so a computer program has been written to analyse and to plot the entire finding. For better understanding, Fig. 13. must be explained first.

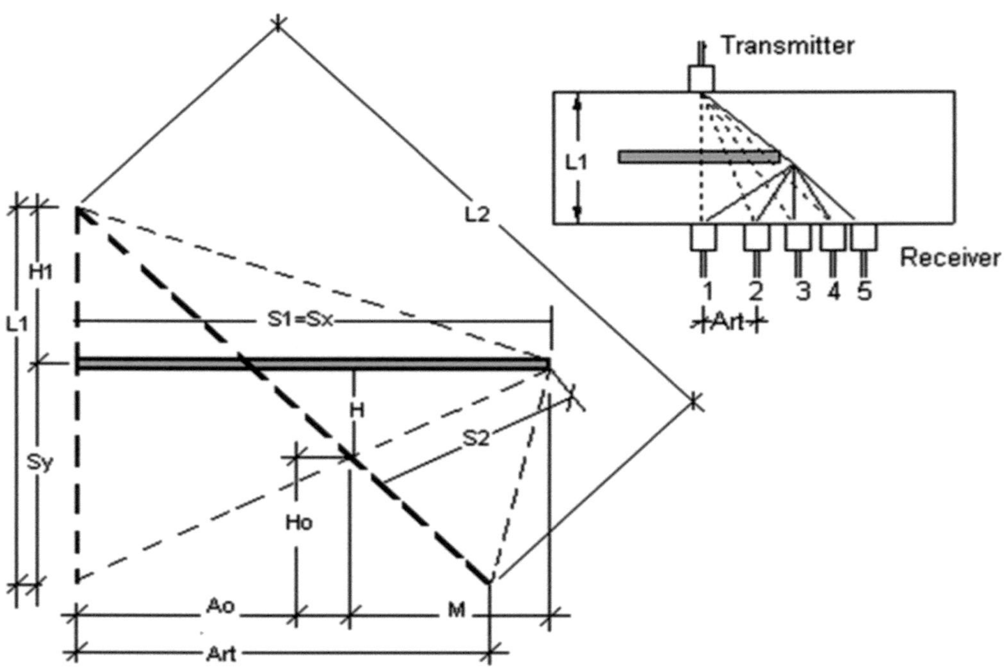

Fig. 13. Pulses travel paths for both receiver positions.

Considering Fig. 13, the values those can be obtained from measurements are:

L1 = The direct distance from transmitter to receiver position 1

L2 = The direct distance from transmitter to receiver position 2

LS1 = The length of pulse travelling path for receiver position 1

LS2 = The length of pulse travelling path for receiver position 2

Art = The distance between receiver 1 and 2

With an iteration of equation 13 for both measurements on the same plane, only one point equals the h and s. The problem is to find out this point that is the boundary of the surface.

3.5 Verification of the model

To verify the model explained in previous section, a cubic homogenous marble block with a certain cut inside was prepared. The dimension of the block and cut is shown in Fig. 14.

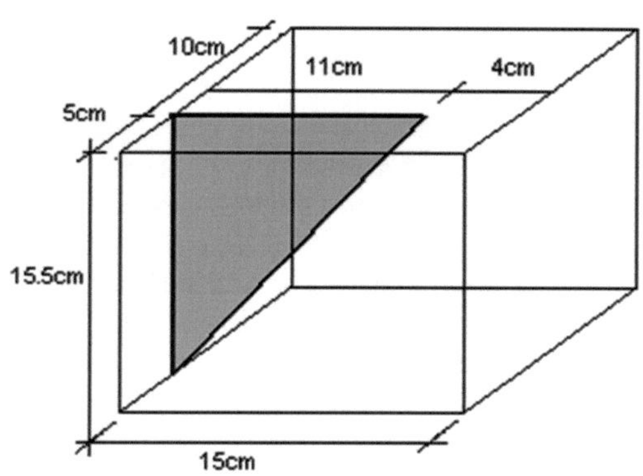

Fig. 14. The block model dimension

The transmitter is placed on the top of the block from the left side and 13 cm away from the front face. The receiver was moved along the front face of the block with a 2.5cm x 2.5cm grid patterns (Fig 15).

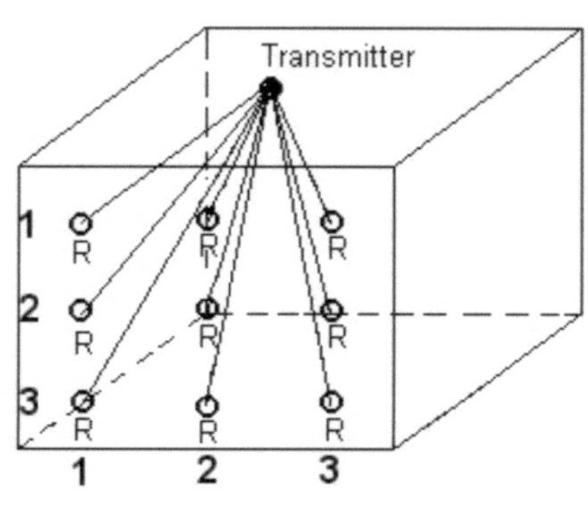

Fig. 15. The position of transmitter and receiver

Before the pulse travelling times are taken, standard linear equation for homogenous marble block was obtained by direct measurements as the following equation:

$$T = 0.877 \times L + 7.287 \text{ (ms)} \tag{14}$$

Table 5 shows the measured times for 9 different receiver position according to Fig. 15.

	Measured travelling time (μs)			Standard travelling time (μs)			Distance between the transmitter and the receiver (cm)			Calculated pulse path length (cm)		
1	27	23	23	20	21	23	15	16	18	18	22	18
2	25	23	24	21	22	24	16	17	19	20	17	18
3	25	24	25	23	24	25	19	21	20	20	19	21
	1	2	3	1	2	3	1	2	3	1	2	3

Table 5. The results of measurements

Table 5 is made up of 4 sections. The first one is the direct measurements taken from the instrument. By using linear behaviour of homogenous block given in formula 12, standard travelling times are calculated for a case if no cut exists in the path of the probes and they are given in the second section of Table 5. The exact distances are measured and given in the third section. The last section is calculated pulse path length obtained from measured travelling time as indicated before. As long as receiver position (1, 1) is concerned, this gives the highest measured travelling time showing (the first section in Table 5) the pulse travels the longest path to reach the receiver. This indicates a crack exists between two probs. When the receiver is moved to position (3, 3), there is no difference between measured travelling time and standard travelling time that indicates no obstacle between two probs. Measured travelling times and distances between probes are given to the computer as data so that all other values can be calculated in order to show the exact place of the cut. The output of the computer located crack position is given in Fig. 16.

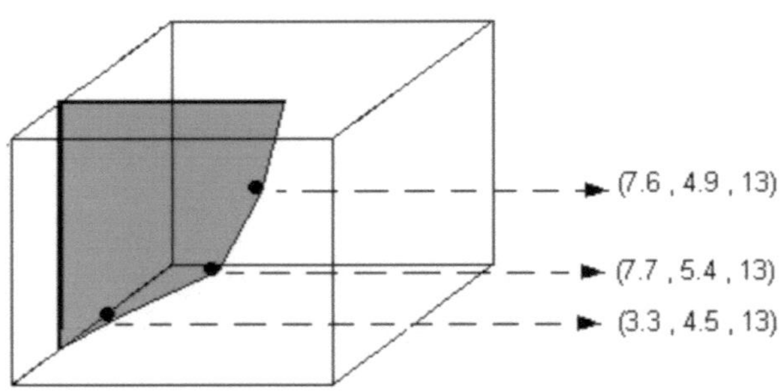

Fig. 16. The computed output of the crack

3.6 In-situ measurements

The same tests were performed on several quarried blocks obtained from a marble factory and a good match was found with the estimated discontinuity surface.

The location of the transmitter and the mesh of receiver are shown in Fig. 17. Working with a block dimension of 155cm x 249cm x 98 cm brought about some difficulties. First, the measured block showed directional anisotropies those affect the liability of the measurements. The second one was that the existence of more than one discontinuity between two probs. The model was developed to search only one discontinuity or cave in the block so better results could be obtained with the moving transmitter. Besides of all these difficulties, a main discontinuity surface was detected and located in the block. From several block measurements only one of them will be given in detail.

Before the measurements were taken, first transmitter was located in such a position that it could indicate some visible cracks from the surface. After locating the transmitter on such an area, the opposite side of the marble block was divided into grids with 20 cm. There are 12 measurement points in longitudinal and 5 measurements in vertical direction. The linear behaviour of homogenous block was measured with indirect method (Equation 15).

$$T = 1.49\,x - 0.56\ (\text{ms}) \tag{15}$$

1	275	279	282	280	280	296	312	334	351	404	390	412
2	248	250	258	296	280	297	311	329	347	363	386	410
3	332	342	300	296	278	287	306	324	347	367	389	411
4	391	384	378	310	280	289	306	328	346	372	393	417
5	416	394	380	286	292	299	316	334	354	375	409	510
	1	2	3	4	5	6	7	8	9	10	11	12

Table 6. Measured travelling times (μs)

1	223	227	235	246	260	277	295	316	337	360	384	409
2	225	229	237	248	262	278	297	317	339	361	385	410
3	231	235	242	253	267	283	301	321	342	365	389	413
4	241	244	251	262	275	291	308	328	349	371	394	418
5	253	257	263	273	286	301	318	337	358	379	402	426
	1	2	3	4	5	6	7	8	9	10	11	12

Table 7. Standard travelling times (μs)

1	150	153	158	166	175	186	198	212	227	242	258	275
2	152	154	159	167	176	187	199	213	228	243	259	275
3	156	158	163	170	179	190	202	216	230	245	261	277
4	162	164	169	176	185	195	207	220	235	249	265	281
5	170	173	177	184	192	202	214	227	240	255	270	286
	1	2	3	4	5	6	7	8	9	10	11	12

Table 8. Distances between transmitter and receiver (cm)

1	184	187	189	188	188	198	209	224	235	271	261	276
2	166	167	173	198	188	199	208	220	233	243	259	275
3	222	229	201	198	186	192	205	217	233	246	261	275
4	262	257	253	208	188	194	205	220	232	249	263	279
5	279	264	255	192	196	200	212	224	237	251	274	342
	1	2	3	4	5	6	7	8	9	10	11	12

Table 9. Calculated pulse path lengths (cm)

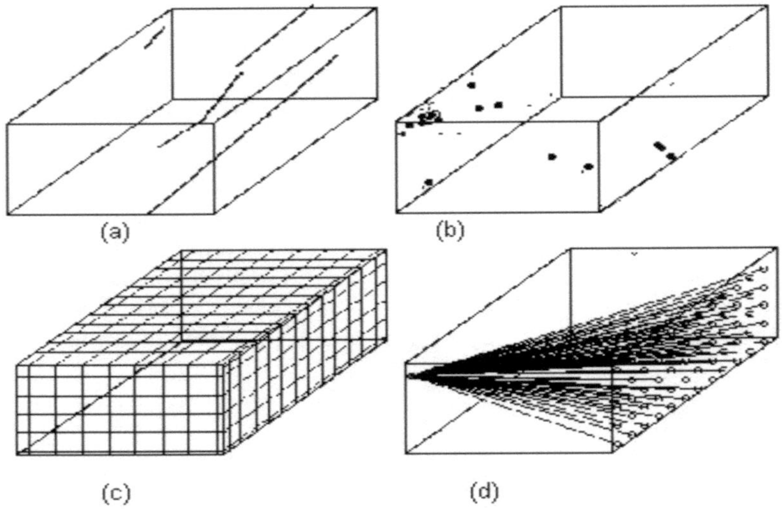

Fig. 17. Dimension and measurement grids of the block.

Table 6 gives measured travelling times taken from the experiment applied on commercial block in the factory before it was sent to sawing machine. Standard expected travelling times representing the block mass behaviour in Table 7 can be obtained by accommodating direct distances between transmitter and receiver (given in Table 8) in to equation 15. Calculated pulse path lengths in Table 9 were obtained from measured travelling times.

The location of transmitter and receiver movements are given in Fig. 17 (d). Because of the necessity of locating the transmitter only one stable position on the block, the experiment was aimed to focus on one discontinuity that could be observed from surface. Several reflection points (Fig. 17 (b)) were observed when formulations given in the text are applied to the model. Fig. 17 (a) gives a diagrammatical illustration of the surface located within the block obtained by connecting the edge points of the discontinuity surface. Better organisation of transmitter position could have been done to obtain better 3D view of the surface within the block by using a single device that have several receivers connecting to same device, but the aim of this study was to show the possibility of exposing hidden surfaces in an enclosed environment.

4. Conclusion

Ultrasonic wave velocity measurements have proven to be a valuable tool in mining industry, since successful applications of this technique have been introducing widely in earth science. Predicting earth conditions before any engineering practices has been one of the most important requirements of mining industry. This is because of difficulties in predicting earth structures before they are reached. In this chapter, some practical techniques have been given related with ultrasonic wave propagation to provide helpful tools to remove discrepancies in mining applications. People dealing with earth science are very familiar with the techniques of seismicity in answering questions such as what it is made up of, how deep it is, what is the position, how big it is. Logic behind the ultrasonic waves is the same as seismic wave propagation in a way that they both have P and S waves in the same form. Measuring the waves reflected from different bodies beneath earth surface and interpreting the data obtained from those measurements give very useful information for engineers.

The importance of determining the marble blocks those are affected by any discontinuity such as void, crack, caves et., in quarry has been presented through this chapter. It is also important to notice that any abnormalities in marble blocks must be pointed out by using a simple method without giving any damage to the main body. The method of testing the quality of concrete is perfectly well adapted to the determination the marble block because of its simplicity. First, experiments performed in the laboratory on simulated block gave promising results that encourage the possibility of using such a technique on marble blocks. But it must be bear in mind that a prepared concrete block differs from the natural stones concerning with homogeneity. Direct measurements give better understanding of the structure of any block measured because it clearly shows the boundary of a surface in the body. Nevertheless, as far as field investigation is concerned, direct measurements become very hard to apply depending on the number of free faces. To develop a measurement technique in field semi-direct and indirect applications have been developed on the blocks obtained from a marble factory. Mathematical model was applied on the blocks but results showed that fixing the transmitter in a stable position does not give a picture of the body if there is a complicated structure as far as discontinuities concern. Although it is possible to find exact positions of the discontinuities, the number of measurements increases in logarithmic scale with the moving transmitter that enables a practical method. However, statistical analysis would better give a high reliability for spotting out this kind of structures, instead of finding the correct location. Authors of this chapter suggests to develop a new measurement technique to allow multiple receivers located on several positions of the block to be analysed and moving transmitter along with the surface to be measured. Much more precise results could be obtained from mobile transmitter unit.

5. References

Asai, K.& Sasaki, N.(1958) Ultrasonic Treatment Of Slurry, *III. International Coal Preparation Congress*, Brussels-Liege, pp. 112-122

Bakhorji, A.M., *Laboratory Measurements of Static and Dynamic Elastic Properties in Carbonate*, PhD Thesis, University of Alberta, Department of Physics, Canada

Bray, D. E. & McBride, D., (1992), *Nondestructive Testing Technique*, John Wiley and Sons Press, ISBN 0-471-52513-8 NewYork, USA

Chary, K.B., Sarma, L.P., Prasanna Lakshmi, K.J., Vijayakumar, N.A., Naga Lakshmi, V & Rao, M.V.M.S. (2006) Evaluation of Engineering Properties of Rock Using Ultrasonic Pulse Velocity and Uniaxial Compressive Strength, *Proc. National Seminar on Non-Destructive Evaluation*, NDE 2006, (December 7-9), Hyderabad, India, pp. 379-385

D'Andrea, D.V., Fischer, R.L., & Fogelson, D.E. (1965) Prediction Of Compressive Strength From Other Rock Properties. *United States Bureau of Mines, Report of Investigation:* United States Bureau of Mines, Boulder, Colorado, USA

Deliormanli, A.H., Burlini, L., Yavuz, A.B. (2007), Anisotropic dynamic elastic properties of Triassic Milas marbles from Mugla region in Turkey, *International Journal of Rock Mechanics & Mining Sciences*, Vol: 44, pp:279 - 288

Dereman, M., Omar, R. Harun, A.G.& Ismail, M.P. (1998), Young's modulus of carbon from self-adhesive carbon grain of oil palm bunches, *J. Mater. Sci. Lett.*, Vol. 17, No. 24 , pp.2059-2060

Djendova, S. Mehandjski, V., 1986, Study of The Effects Of Acoustic Vibration Conditioning of Collector and Frother on Flotation of Sulphide Ores, *International Journal of Mineral Processing*, Vol. 34, No.23, pp. 205-217

Galaa, A.M., Thompson, B.D., Grabinsky, M.W. & Bawden, W.F., 2011, Characterizing stiffness development in hydrating mine backfill using ultrasonic wave measurements, *Canadian Geotechnical Journal*, Published on the web, 10.1139/t11-026

Gladwin M.T., (April 2011), Ultrasonic stress monitoring in underground mining, *Journal of Applied Geophysics*, Volume 73, Issue 4, , Pages 357-367

Gladwin, M.T.(1982) Ultrasonic Stress Monitoring in Underground Mining, *International Journal of Rock Mechanics & Mining Sciences,* Vol. 19, pp. 221 - 228.

Green, R. E., (1991) Introduction to UltrasonicTesting In: *Ultrasonic Testing*, A. S. Birks, R. E. Green, and P. McIntire, (Eds.), American Society for Nondestructive Testing, Metals Park, Ohio, pp. 1–21

Hasani, F.P., Momayez , M., Guevremont, P., Saleh, K. & Tremblay, S. (1996) *Revee De La Literature: Methodes D'inspections Non-Destructives Pour La Detection Fissures Dans Les Ouvrages en Beton*, Hydro-Quebec's Research Institute, IREQ, Report no: IREQ-96-111, Canada

ISRM, (1979) Suggested Methods for DeterminingWater Content, Porosity, Density, Absorption and Related Properties and Swelling and Slake-Durability Index Properties, *Int. J. Rock Mech. Min. Sci. Geomech. Abstr.*, Vol. 16, No. 2, pp. 141–156

Jones, R., Wallbrink,C., Tan, M., Reichl, P. & Dayawansa, D., (March 2010), Health monitoring of draglines using ultrasonic waves, *Ultrasonics Sonochemistry*, Volume 17, Issue 3, Pages 500-508

Kahraman, S. (2007) The Correlations Between the Saturated and Dry P-Wave Velocity of Rocks, *Ultrasonics*, Vol. 46, No. 4, pp. 341-348

Karpuz, C & Pasamehmetoglu, A.G., (1997) ' 'Field Characterization of Weathered Ankara Andesites, *Engineering Geology*, Vol. 46, No. 1, pp. 1-17

Khandelwala, M. & Ranjithb, P.G., (June 1996), Correlating index properties of rocks with P-wave measurements, Ultrasonics, Volume 34, Issues 2-5, Pages 421-423

Kowalski, W., Kowalska, E. (1978) The Ultrasonic Activation of Non-Polar Collectors in The Flotation of Hydrophobic Minerals, *Ultrasonics*,Vol. 16, No. 2, pp. 84- 86

Lee, I. M., Truong Q. H., Kim, D.-H. & Lee, J.S., (October 2010), Discontinuity detection ahead of a tunnel face utilizing ultrasonic reflection: Laboratory scale application, *Journal of Applied Geophysics*, Volume 72, Issue 2, Pages 102-106

Leslie, J.R. & Chessman, W.J. (1949) An Ultrasonic Method of Studying Deterioration and Cracking in Concrete Structures, *Journal of the American Concrete Institute*, Vol.21, No. 1, pp. 17-36.

Lockner, D. (1993) The Role of Acoustic Emission in The Study of Rock Fracture, *Int. J. Rock Mech. Min. Sci. Geomech. Abstr.*, Vol. 30, No. 7, pp. 873-899

Madenga, V., Zou, D.H. & Zhang C., (March 2009), Effects of curing time and frequency on ultrasonic wave velocity in grouted rock bolts, *Tunnelling and Underground Space Technology*, Volume 24, Issue 2, March 2009, Pages 155-163

Madenga, V., Zou, D.H., & Zhang C. (February 2009), Effects of curing time and frequency on ultrasonic wave velocity in grouted rock bolts, *Ultrasonics*, Volume 49, Issue 2, Pages 162-171

Malhotra, V.M. & Carino, N.J. (1991) CRC *Handbook on Nondestructive Testing of Concrete*, CRC Press, ISBN 9780849314858, Florida, USA.

Mix, P. E., (1987) Introduction to Nondestructive Testing:ATraining Guide, JohnWiley and Sons Press, ISBN 0-471-83126-3, NewYork, USA

Moozar, P. L. (2002). *Non –destructive Appraisal of Paste Backfill* , PHd Thesis , Department of Mining and Materials Engineering, McGill University, Montreal, Canada

New, B.M., (1985), Ultrasonic wave propagation in discontinuous rock, *Transport & Road Research Laboratory*, Department of the Environmental, TRRL Laboratory Report No. 720, London

Nicol, S. K., Engel, M. D & Teh, K. C. (1986) Fine Particle Flotation in an Acoustic Field, Technical note, *International Journal of Mineral Processing*, Vol. 17, No. 1-2, pp.143-150

Onodera, T.F. (1963) Dynamic Investigation Of Foundation Rocks, In Situ, *Proceedings of the 5th US Symposium on Rock Mechanics*, Univ of Minn. Pergamon Press, Oxford pp.517 - 533.

Ozkan, Ş. G., Veasey, T. J. (1996) Effect of Simultaneous Ultrasonic Treatment on Colemanite Flotation, *6th International Mineral Processing Symposium*, Kuşadası Aydın-Turkey 24-26 September, pp. 277-281s.

Parasnis, D.S.,(1994), *Principles of Applied Geophysics*, Chapman & Hall, Fourth edition, p.402, London

Renaud V., Balland, C. & Verdel, T., (1990), Numerical simulation and development of data inversion in borehole ultrasonic imaging, *Engineering Fracture Mechanics*, Volume 35, Issues 1-3, Pages 377-384

Saad A, & Qudais, A. (2005) Effect of concrete mixing parameters on propagation of ultrasonic waves, *Construction and Building Materials*, vol. 19, No. 4, pp.257-263

Slaczka, A. (1987) Effects of an Ultrasonic Field on The Flotation Selectivity of Barite from a Barite-Florite-Quartz Ore, *International Journal of Mineral Processing*, Vol. 20, No. 3-4, pp.193-210

Stoev, S. M. & Martin, P. D. (1992) *The Application of Vibration and Sound in Minerals and Metals Industries*, A Techenical Review, Series No:8, MIRO, Lichfield, England

Tomsett, H.N.,(1976), Site testing of concrete, British Journal of Non-Destructive Testing, , vol. 18, pp.82-87.

Vary, A.(1991) Material Property Characterization In: Ultrasonic Testing, A. S. Birks, R. E. Green, and P. McIntire, (Eds)., American Society for Nondestructive Testing, Metals Park, Ohio, pp. 383–431.

Vasconcelos, G., Lourenco, P.B., Alves, C.A.S. & Pamplona, J. (2008) Ultrasonic Evaluation of the Physical and Mechanical Properties of Granites, *Ultrasonics*, Vol. 48, No. 5 pp. 453-466.

Yerkovic, C., Menacho, J., Gaete, L., (1993) Exploring The Ultrasonic Communition of Copper Ores, *Minerals Engineering*, Vol:6, No. 6, pp.607-617

Zhang F., Krishnaswamy S., Fei D., Rebinsky D. & Feng B., (2006), Ultrasonic characterization of mechanical properties of Cr- and W- doped diamond like carbon hard coatings, *Thin Solid Films,* Vol. 503, No. 1-2, pp.250-258

Zou, D.H. S., Cheng J., Yue, R. & Sun X., (October 2010), Grout quality and its impact on guided ultrasonic waves in grouted rock bolts, *International Journal of Rock Mechanics and Mining Sciences & Geomechanics,* Volume 19, Issue 5, Pages 221-228

11

Design and Development of Ultrasonic Process Tomography

Mohd Hafiz Fazalul Rahiman[1], Ruzairi Abdul Rahim[2], Herlina Abdul Rahim[2] and Nor Muzakkir Nor Ayob[2]

[1]*Universiti Malaysia Perlis (UniMAP)*
[2]*Universiti Teknologi Malaysia (UTM)*
[1,2]*Malaysia*

1. Introduction

Process tomography is a process of obtaining the plane-section images of a three-dimensional object. Process tomography techniques produce cross-section images of the distribution of flow components in a pipeline, and it offers great potential for the development, and verification of flow models, and also for process diagnostic.

The measurement of two-component flow such as liquid or oil flow through a pipe is increasingly important in a wide range of applications, for example, pipeline control in oil exploitation, and chemical process monitoring. Knowledge of the flow component distribution is required for the determination of flow parameters such as the void fraction, and the flow regime.

Real-time reconstruction of the flow image is needed in order to estimate the flow regime when it is continuously evolving. This flow image is important in many areas of industry, and scientific research concerning liquid/gas two-phase flow. The operation efficiency of such a process is closely related to accurate measurement, and control of hydrodynamic parameters such as flow regime, and flow rate (Rahiman et al., 2010). Commonly, the monitoring in the process industry is limited to either visual inspection or a single-point product sampling assuming the product is uniformed. This approach for the determination of fluid flow parameters of two-component flow is also known as flow imaging.

1.1 Principle overview

Ultrasonic sensors have been successfully applied in flow measurement, non-destructive testing, and it is widely used in medical imaging. The method involves in using ultrasonic is through transmitting, and receiving sensors that are axially spaced along the flow stream. The sensors do not obstruct the flow. As the suspended solids' concentration fluctuates, the ultrasonic beam is scattered, and the received signal fluctuates in a random manner about a mean value. This type of sensor can be used for measuring the flow velocity. Two pairs of sensors are required in order to obtain the velocity using a cross- correlation method. Ultrasonic sensor propagates acoustic waves within the range of 18 kHz to 20 MHz.

Ultrasonic wave is strongly reflected at the interface between one substance, and another. However, it is difficult to collimate, and problems occur due to reflections within enclosed spaces, such as metal pipes. There are two types of ultrasonic signals that are usually used. They are the continuous signal, and the pulsed signal. Using a continuous signal will provide continuous impact on the crystal whereas by using pulses the interval of the transmission and reception signal can be estimated. Using the ultrasonic method in the air is very inefficient due to the mismatch of the sensors' impedance as compared with air's acoustic impedance. New types of sensors are continually being developed, but the effective ones are expensive. The design of this sensor is critical when it needs to reduce the sensor's ringing.

1.2 The tomography technique

The first step of a tomographic process is to generate the integral measurements using a selected sensor. The second step is to reconstruct the property field (the cross-sectional distribution of the physical properties of the multiphase media) from the measured integral values. This process is called tomographic reconstruction (Warsito et al., 1999).

There are numerous reconstruction algorithms available for tomographic reconstruction. The algorithms based on Fourier's techniques, and the algebraic reconstruction technique (ART) has been widely used in the field of medicine. However, the choice of the reconstruction algorithm is also dependent on the sensor system selected. In engineering related applications, the number of measurement is usually very small to perform a real-time measurement or limited by constraints on the sensor employed. Therefore, the reconstruction results are then further corrected using a mathematical approximation to obtain a better reconstruction.

In contrast to light or other electromagnetic waves, ultrasound needs medium to transmit through, and interrogates the physical properties (i.e. density, compressibility) of the media. Therefore, it is speculated that such a method would be appropriate for application in a medium with relatively homogeneous but high density, which is poorly penetrated by light or other electromagnetic radiation. In addition, in comparison with high-energy electromagnetic radiation, ultrasonic technique consumes much lower energy, low-cost, and simpler to use, and suitable for applications from laboratory scale to industrial plants (Williams & Beck, 1995).

2. Ultrasonic process tomography – An overview

Instrumentation systems employing a variety of ultrasonic techniques have been applied to a wide range of measurements in the chemical, and process industries (Asher, 1983). At least, eight categories of ultrasonic flow meter can be identified with flow meters of time-of-flight type now being employed in single-phase liquid, and gas flow measurement with a great deal of success. It is favoured by most industries due to the benefits as follows (Asher, 1983):

i. Ultrasonic techniques can usually be truly non-invasive.
ii. It has 'no moving parts'.
iii. The radioactive materials are not involved.
iv. The rapid response usually in a fraction of a second.
v. The energy levels required to excite the transducers are very low, and have no detrimental effect on the plant or the materials being interrogated.

vi. A mutually compatible range of techniques can be used to determine a wide range of parameters: these include liquid level, interface position, concentration (or density), temperature, and flow-rate. Hence multiplexed electronics are feasible.

Besides, those benefits have a significant impact, and lead to the development of Ultrasonic tomography. Ultrasonic tomography offers the advantage of imaging two-component flows, and gives the opportunity of providing quantitative and real-time data on chemical media within a full-scale industrial process, such as filtration, without the need of process interruption (Warsito et al., 1999).

The major potential benefits are; it is possible to gain an insight into the actual process; secondly, since ultrasonic tomography is capable of on-line monitoring, it has the opportunity to develop closed-loop control systems, and finally, it can be a non-invasive, and possibly non-intrusive system. The overall anticipated effects are improvements in product yield, and uniformity, minimized input process material, reduced energy consumption, and environmental impact, and the lowering of occupational exposure to plant personnel.

The popular ultrasonic sensing system is the transmission-mode, and the reflection-mode. The transmission-mode technique is based on the measurement of the changed in the properties of the transmitted acoustic wave, which are influenced by the material of the medium in the measuring volume. The change of the physical properties can be the intensity, and/or transmission time (time-of-flight). The reflection-mode technique is based on the measurement of the position, and the change of the physical properties of wave or a particle reflected on an interface. Similar to the reflection-mode technique there are some techniques based on diffraction or refraction of wave at a discrete or continuous interface in the object space.

Utilizing attenuation or time-of-flight of a transmitted energy beam such as light or acoustic waves to produce an image of multiphase flow has been attempted at an early stage. In transparent media, optical methods based on light transmission technique, and the photographical techniques has proven quite effective (Rahim et al., 2011). However, since many real reaction systems are optically opaque, an application of the elegant optical method is severely limited.

Techniques based on reflection or the scattering of optical or acoustical waves were realized by measuring the Doppler shift-frequency of the reflected or scattered signals. An example is the use of laser Doppler anemometry for in-situ measurements of velocity, fluctuating velocity, size, and concentration of particles, bubbles or droplets in multiphase systems. A corresponding example of the Doppler technique utilizing ultrasonic wave is the measurements of bubble velocity in a stirred tank, and a ferment or vessel by Broring et al., (1991).

A combination of the transmission and the reflection modes is found in acoustical imaging techniques, which are widely used in medical, and ocean engineering fields from early stages. An application of ultrasonic imaging velocimetry has been attempted by Kytomaa & Corrington (1994) to investigate a transient liquefaction phenomenon of cohesion-less particulate media. More advanced particle imaging velocimetry techniques were developed by combining the photographic technique and image processing technique or using radioactive particle tracking techniques. Examples are using particle image velocimetry

(PIV), particle streak velocimetry (PSV), and particle tracking velocimetry (PTV) for visualizing the flow pattern of multiphase flows (Rashidi, 1997).

2.1 The attenuation model

The attenuation process may be modelled by Lambert's exponential law of absorption, where the ultrasonic energy intensity of transmitter and receiver are related as in Figure 1 and Equation (1), where L represents the total path length.

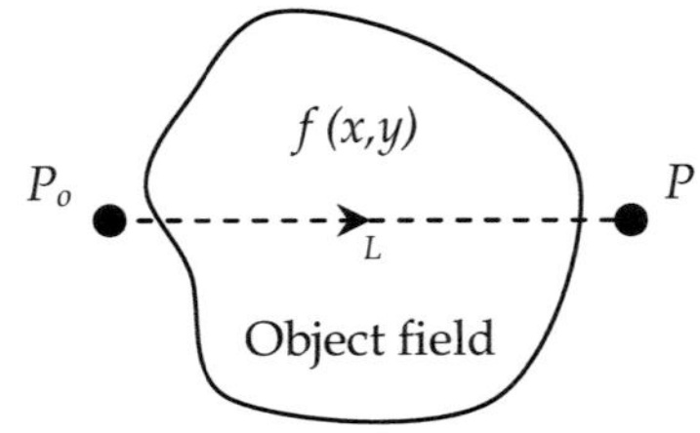

Fig. 1. The ultrasonic attenuation model

$$P = P_o \exp\left(-\int_L f(x,y)dP\right) \tag{1}$$

where P = the measured sound pressure (dB), P_o = the initial sound pressure (dB), L = path length in the object field (m), and $f(x,y)$ = the attenuation function of the object field (dB/m). Because the pressure is proportional to the voltage measured by the transducer, Equation (1) can be written as Equation (2).

$$v_{Rx} = v_{Tx} e^{-\alpha L} \tag{2}$$

where v_{Rx} = the ultrasonic receiver voltage (V), v_{Tx} = the ultrasonic transmitted voltage (V), and α = the attenuation coefficient of the object field (Np/m).

As introduced above, the attenuation will critically depend upon the material through which the ultrasonic wave travels.

2.2 Scattering issues

Acoustic impedance is a ratio of acoustic pressure to acoustic volume flow, and is frequency dependent. The greater the difference in acoustic impedance at the interface, the greater will be the amount of energy reflected. At a water, and gas interface, about 99.94% of the ultrasonic energy will be reflected (Rahiman et al., 2008). However, in some cases, scattering occurs on small gas hold-ups. Small was defined as a sphere with a radius of a where the circumference of the sphere, $2\pi a$, divided by the wavelength λ is much less than 1 (i.e. ka = $2\pi a/\lambda << 1$) where k is the wave number = $2\pi f/c$.

If ultrasonic waves propagate in a bubbly air/water with a wavelength much shorter than the gas radius a, i.e. ka >> 1, the diffraction can be ignored, and these hold-ups will act as many acoustics opacities. This is because when ka >> 1, the surface of the sphere appears as a flat surface with respect to the wavelength, and the scattering becomes the same as a reflection from a flat surface.

The relationship of the simplified ultrasonic transmission model is depicted in Equation (3), and shown graphically in Figure 2.

$$V_G = V_C - V_R \tag{3}$$

where V_G is the sensor loss voltage due to the gas opacity, V_C is the calibration voltage, and V_R is the receiving voltage. For measuring the concentration profile, the parameter V_G is to be resolved.

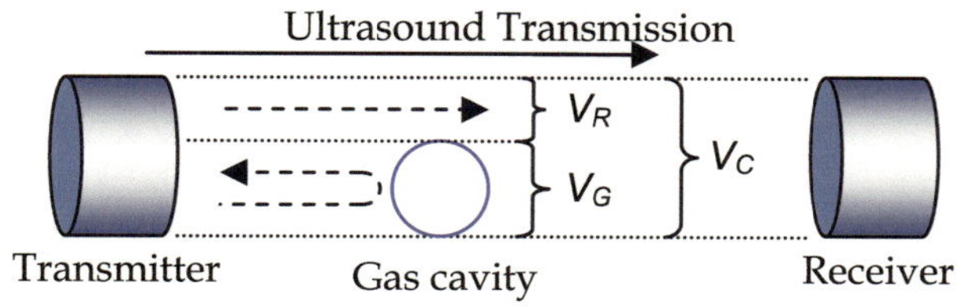

Fig. 2. The simplified ultrasound transmission model

3. The measurement system

One of the most important parts in an ultrasonic tomography system is the front-end, which is the transducer array, and associated electronic hardware. This is important for acquiring the data needed to produce a meaningful image. This is fundamental to the success or failure of an acoustic imaging system. Therefore, given the object to be imaged, and the specifications to be achieved, the design of the front-end of an acoustic imaging system should be regarded as a first priority.

Ultrasonic transducer is a device capable of converting electrical energy into high-frequency sound waves, and also converting sound waves back into electrical energy. Ultrasonic transducer contains piezoelectric crystal materials that have the ability to transform mechanical energy into electrical energy, and vice versa. In reality, when a crystal element is pulsed with a voltage profile, a wave starts travelling from each face of the crystal element. The vibrational mode of the crystal can therefore, only be considered from a transient wave propagation viewpoint. Resolution and penetrating power of an ultrasonic wave depends on the resulting wavelength of excitation inside the material in question. Greater wavelengths or lower frequencies generally penetrate much further into a material (Kannath & Dewhurst, 2004). Higher frequency ultrasonic excitations with smaller wavelengths generally decay more rapidly inside a material, but resolution capability is improved.

For the presented systems, the active element for the transducers is the wide-angle beam ceramic piezoelectric with resonance frequency of 333 kHz. Transducer elements employed in ultrasonic imaging arrays may be designed to produce either a narrow focused beam or a divergent beam. The beam pattern used for two-dimensional imaging is the cylindrically diverging or 'fan-shaped' beam pattern. This is to ensure a maximum number of transducers located around the pipe circumference, which receive the directly transmitted wave.

3.1 Fan-shaped beam sensor array

An example of a fan-shaped beam sensor array is shown in Figure 3, which consists of 32-ultrasonic transceivers that is mounted on an experimental column. The transceivers enable the transmission, and reception on the same sensor. The transmission and reception of ultrasonic waves are, however, controlled by the electronic analogue-switches which direct the signal to the corresponding channels.

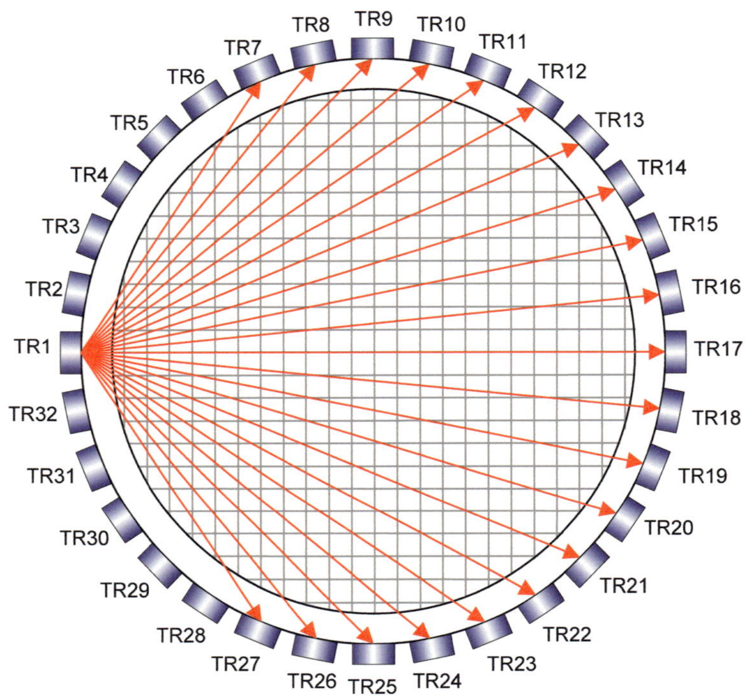

Fig. 3. Ultrasonic transceiver sensor array

One of the significant advantages in employing ultrasonic techniques is it enables measurement to be made without breaking into the process vessel, and therefore, measurements can be made where for a reason of safety hygiene, continuity of supply or cost it is not possible to break into the process vessel. However, the invasive transducers actually contact the flow inside the pipe, for obvious reasons it is not favoured by most industries (Sanderson & Yeung, 2002).

In Figure 3, the transmitter is modelled as a point source, which propagates within angle a in the image plane, and the receiver is modelled as a circular arc with radius of curvature r. The wavefronts are taken to be circular arcs of uniform ultrasonic energy. When ultrasound is propagating in the flow medium, areas occupied by the discontinuous component block the transmitted ultrasound. As a result, an effect analogous to the shadowing of visible light by an opaque object can be seen in Figure 4. An example of transmitting, and receiving ultrasonic signals is shown in Figure 5.

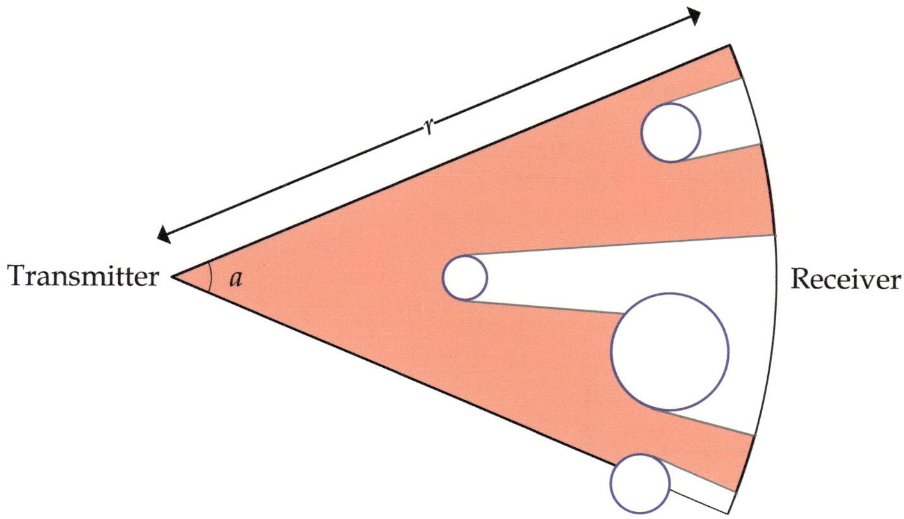

Fig. 4. The discontinuous component viewed by the ultrasonic receiver

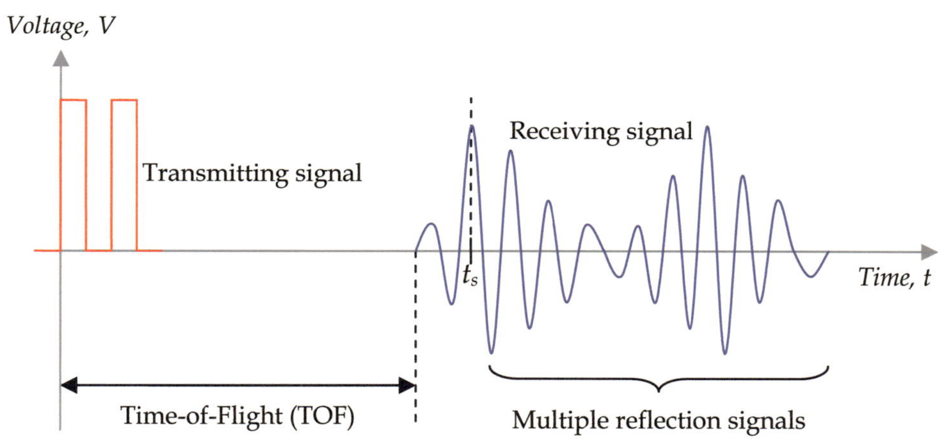

Fig. 5. An example of transmitting, and receiving an ultrasonic signal

In our work, a transmission-mode method emphasizing the receiver amplitude, and the arrival time analysis has been used. Arrival time analysis is based on the simple fact that it takes some finite time for an ultrasonic disturbance to move from one position to another inside the experimental pipe. In Figure 5, the *observation time* denoted by t_s was the first peak after the time-of-flight corresponding to a straight path. By sampling amplitude of this observation time for every receiving sensor due to projection of transmitters, the information via transmission-mode method can be obtained. As the distance between the transmitting sensor, and the receiving sensor increases, the ultrasound will consume longer time-of flight to reach to the point of interest, and therefore, set out a longer observation time. This time-of-flight may then be assumed to be proportional to the distance that they had travelled (Moore *et al.* 2000).

3.2 The system block diagram

The system block diagram is shown in Figure 6. The microcontroller will generate a burst tone of a two-cycle ultrasonic pulses at 333kHz with duty-cycle of 50% at each cycle. The delay between each burst tone was 6.667ms which is for the reverberation effect delays of the receiver before the next transmitting transceiver excited. An illustration of the pulses is shown in Figure 7. The reverberation effect delays are needed to avoid overlapping echoes at the receiving transceiver due to two separate ultrasound excitation. To select the corresponding transmitting, and receiving transceivers, several analogue switches have been utilized.

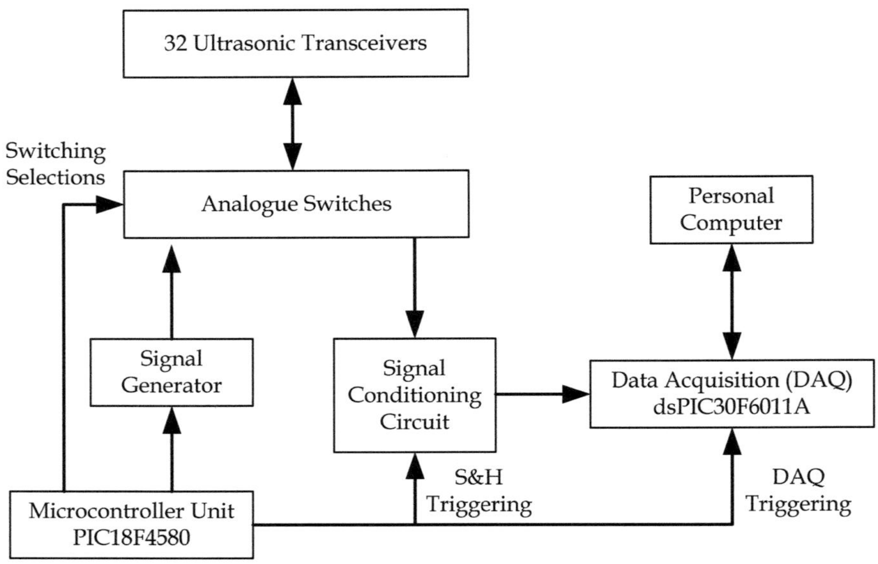

Fig. 6. The ultrasonic system block diagram

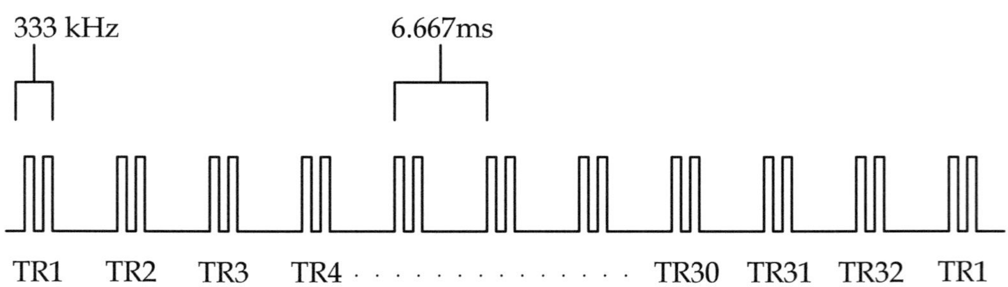

Fig. 7. The burst tones generated by microcontroller

The signal generator was designed using a low-noise high-speed op-amp, TLE2141 that acts as comparator. The comparator will generate a 20Vp-p burst tone of 333 kHz. The burst tone excitation is designed so that it is long enough for transient effects but short enough for the burst to be received without multiple reflections as describe previously. The circuit is shown in Figure 8.

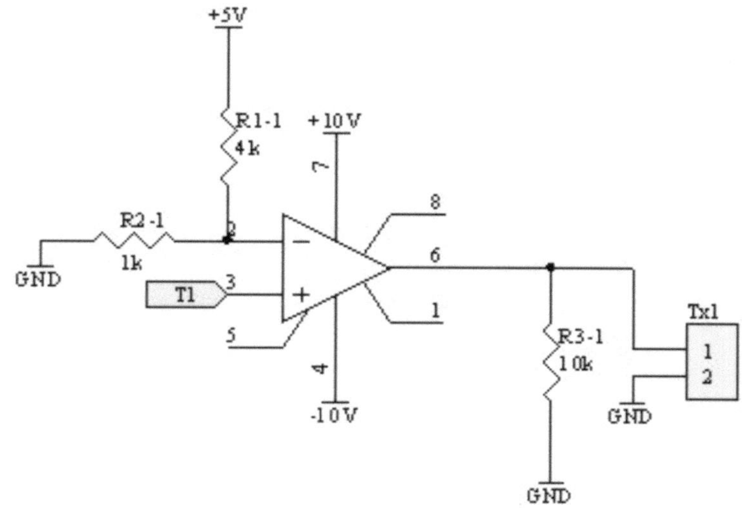

Fig. 8. The signal generator circuit

The signal conditioning circuit consists of two components where the first component is the amplifier, and the second component is the signal processing circuit that is the sample and hold circuit. The amplifier was built using a dual-wide gain bandwidth op-amp to avoid signal distortion during amplification. The amplifier uses two stages inverting op-amp design. Both the first stage and the second stage amplifier gain were set with -150. The signal conditioning circuit is shown in Figure 9.

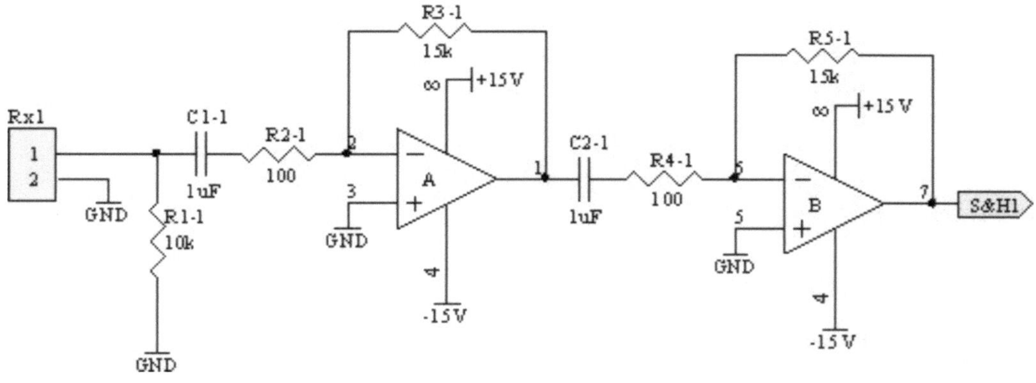

Fig. 9. The signal conditioning circuit

When the components to be imaged are gases, there may be no directly transmitted signals from the transmitter to the receiver because of the obstacles. By reflecting at the gas component surfaces, the receiver may still detect some signals but at a later time though because direct transmission takes the shortest path, and hence the shortest time. Thus, if the *observation time* is monitored, it is possible to test whether there are any objects between the transmitter, and the receiver. By sampling the signals at this observation time for every receiving transceiver, the spatial information in the measurement area can be obtained. To discriminate the exact information (information by the observation time) the sample and hold method is used. A sample and hold circuit, also called a track-and-hold circuit is a circuit that captures, and holds an analogue voltage in a specific point in time under control of an external circuit (microcontroller). The operation of sample and hold is shown in Figure 10, and the circuit diagram for sample and hold is shown in Figure 11.

The holding signal is also known as the *sensor value* voltage will be sampled by the DAQ using the dsPIC30F6011A IC which served as the analogue-to-digital converter as well. The information captured by the DAQ will be processed, and send to the PC for generating the corresponding tomography images.

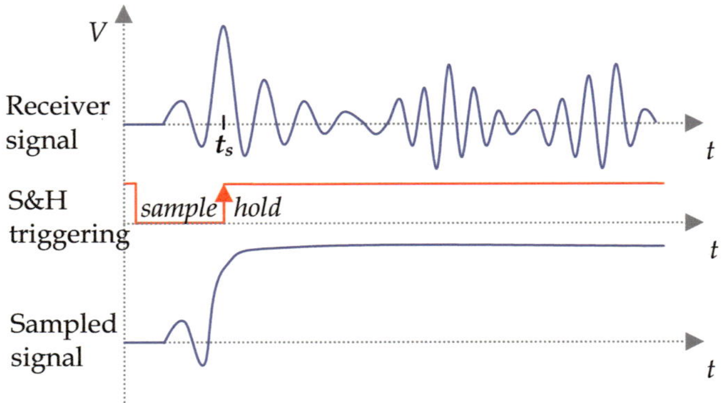

Fig. 10. The sample and hold operation

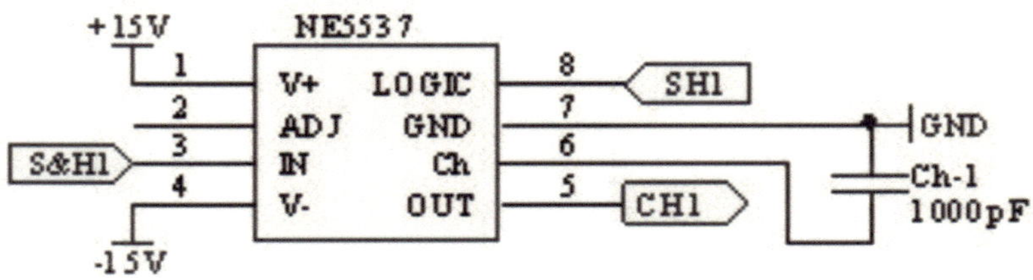

Fig. 11. The sample and hold circuit

4. Image reconstruction

In this work, the tomographic images are derived by using a back projection algorithm. In order to derive this algorithm, which results in the solution to the inverse problem, the forward problem must be solved first.

4.1 The forward problem

The forward problem determines the theoretical output of each of the sensors when the sensing area is considered to be two-dimensional. The cross-section of the pipe is mapped onto a 64 x 64 rectangular array consisting of 4096 pixels. The forward problem can be solved by using the analytical solution of sensitivity maps, which produces the sensitivity matrices. Each transmitting sensor is virtually excited, and the affected pixels are taken into account (as shown in Figure 12).

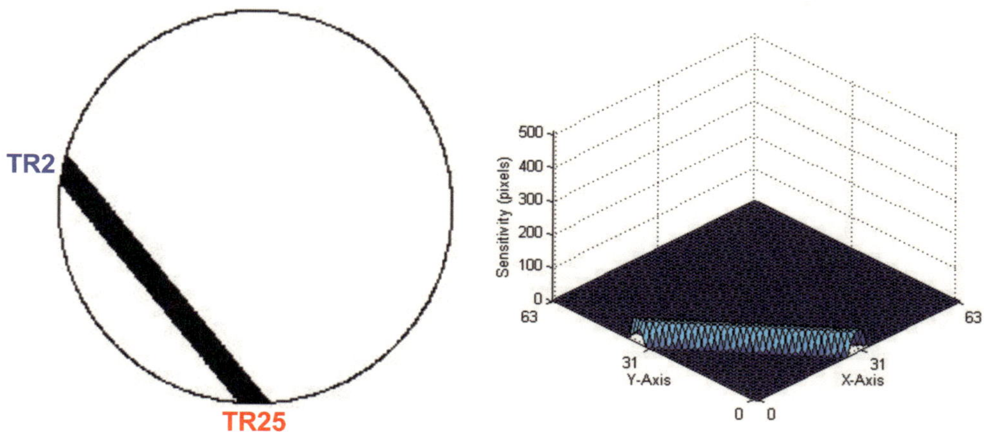

Fig. 12. The sensitivity map for projection TR25 to TR2

4.2 The inverse problem

The inverse problem is to determine from the system response matrix (sensitivity matrices), a complex transformation matrix for converting the measured sensor values into pixel values. It is known as the tomogram. To reconstruct the tomogram requires an image reconstruction algorithm. The details for tomogram reconstruction are presented in the following section.

4.3 Image reconstruction algorithm

To reconstruct the cross-section of an image plane from the projection data, the back-projection algorithm has been employed. Most of the work in process tomography has focused on the back-projection technique. It is originally developed for the X-ray tomography, and it also has the advantages of low computation cost (Garcia-Stewart et al., 2003). The measurements obtained at each projected data are the attenuated sensor values due to object space in the image plane. These sensor values are then back projected by multiply with the corresponding normalized sensitivity maps. The back projected data values are smeared back across the unknown density function (image), and overlapped to each other to increase the projection data density. The process of back-projection is shown in Figure 13, and Figure 14.

The density of each point in the reconstructed image is obtained by summing up the densities of all rays which pass through that point. This process may be described by Equation 4. Equation 4 is the back-projection algorithm where the spoke pattern represents blurring of the object in space.

$$f_b(x,y) = \sum_{j=1}^{m} g_j(x\cos\theta_j + y\sin\theta_j)\Delta\theta \tag{4}$$

where $f_b(x, y)$ = the function of reconstructed image from the back-projection algorithm, θ_j = the j-th projection angle, and $\Delta\theta$ = the angular distance between projection, and the summation extends over all the *m-projection.*

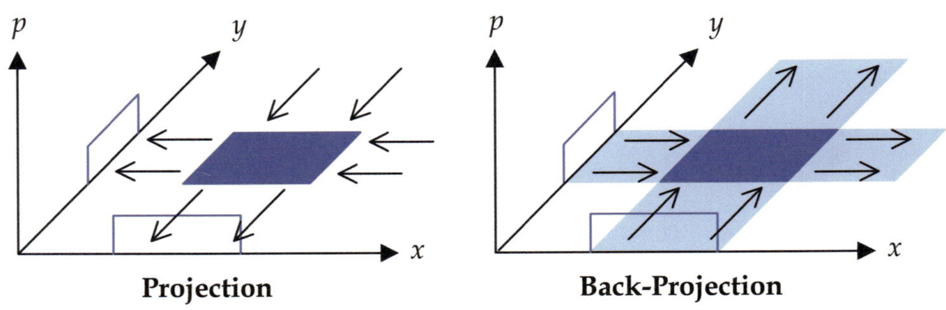

Fig. 13. The back-projection method

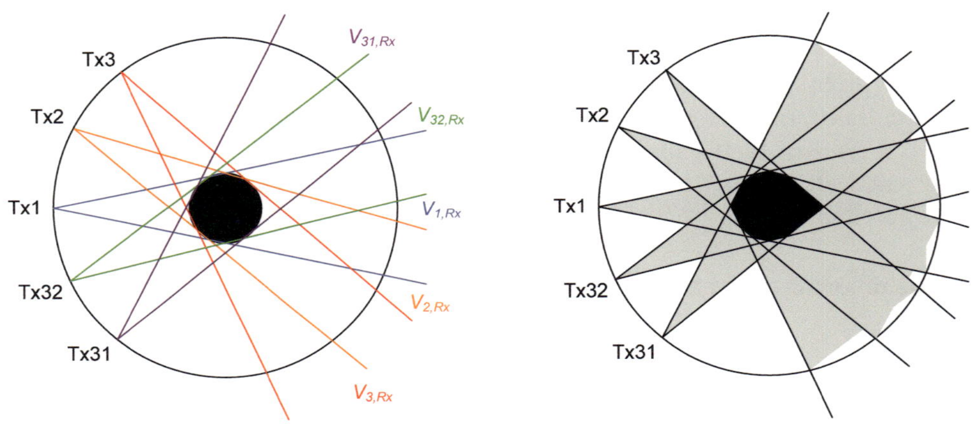

Fig. 14. The fan-shaped beam back-projection

4.3.1 Linear back projection algorithm

The Linear Back Projection algorithm (LBP) is computationally straight forward to implement, and is a popular method for image reconstruction. Sensitivity maps, which were derived for the individual sensors are used by the LBP algorithm to calculate concentration profiles from measured sensor values. The process of obtaining concentration profile using LBP can be expressed mathematically as in Equation 5.

$$V_{LBP}(x,y) = \sum_{Tx=0}^{m} \sum_{Rx=0}^{n} S_{Tx,Rx} \times \overline{M}_{Tx,Rx}(x,y) \tag{5}$$

where $V_{LBP}(x,y)$ is the voltage distribution on the concentration profile matrix, $S_{Tx,Rx}$ is the sensor loss value, and $\overline{M}_{Tx,Rx}(x,y)$ is the normalized sensitivity matrices.

4.4 Image reconstruction results

The tomogram reconstruction on three phantoms of a small gas hold-up, a large gas hold-up, and a dual gas hold-up are shown in Figure 15, 16, and 17 respectively. The results showed several tomograms reconstructed using; (a) LBP, and (b) LBP with a threshold ratio of 0.55.

(a) LBP (b) LBP (with threshold ratio 0.55)

Fig. 15. Small gas hold-up

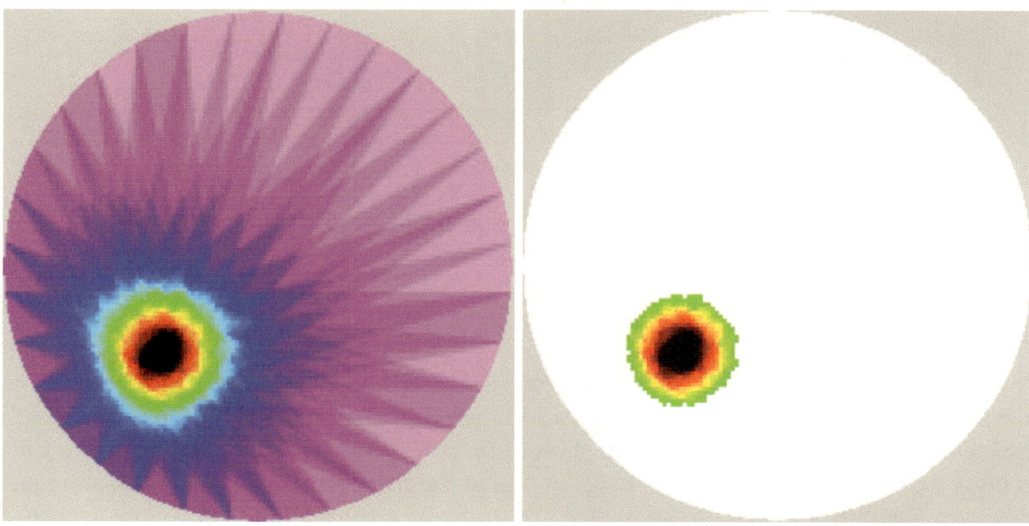

(a) LBP (b) LBP (with threshold ratio 0.55)

Fig. 16. Large gas hold-up

The tomogram in (a) showed that the LBP smears out, and introduces false images elsewhere. As seen in Figure 15(a), 16(a), and 17(a) the reconstructed images clearly contain qualitative information about the gas hold-up, but it is hard to distinguish the gas hold-up boundaries. However, the tomograms in Figure 15(a), 16(a), and 17(a) showed that the area of high gas concentration is clearly visible, and could be distinguished from the background image, and the shapes of the reconstructed images are reasonably accurate.

A thresholding technique has been used. To obtain the optimal threshold ratio, a further analysis has to be conducted by measuring the lowest error corresponds to the threshold ratio. By thresholding the image with a ratio of 0.55, the reconstructed images have tremendously improved. The smearing effects by back projection technique, which had caused non-uniformity of background image has been eliminated, and this is shown in Figure 15(b), 16(b), and 17(b). As a result, the information of liquid, and gas such as position, and shape could be easily obtained.

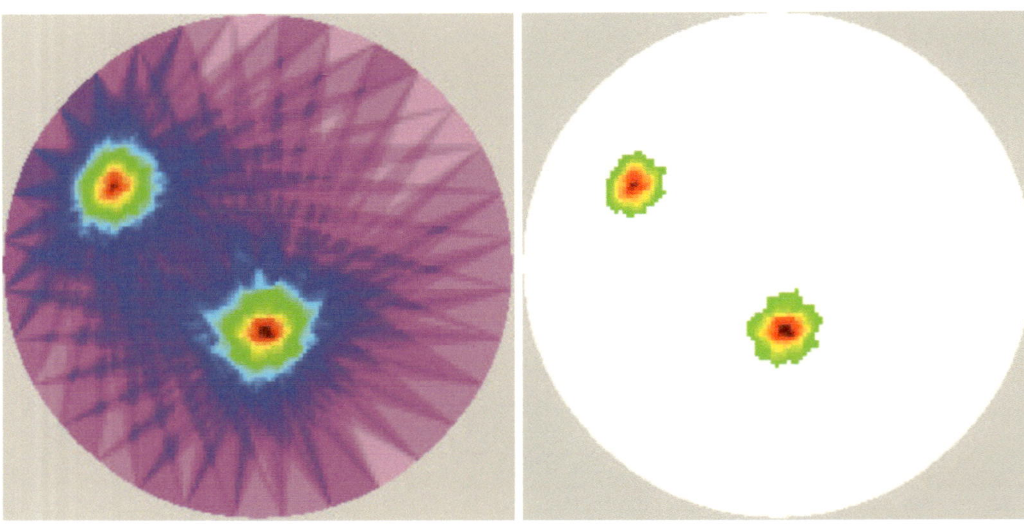

(a) LBP (b) LBP (with threshold ratio 0.55)

Fig. 17. Dual gas hold-up

5. Conclusions

The ultrasonic process tomography has been designed, and developed. The results show that the system could be used to identify, and locate the size, and position of gas bubbles in the measurement column. Experiments showed that, the image reconstructed by LBP results in blurring the image. This blurry image is due to the nature of back projection technique. However, the blurry image can be reduced by applying a thresholding technique. Hence, a clearly visible gas hold-up can be identified. This information is useful, and could be used for measuring the liquid/gas concentrations.

6. Acknowledgment

Authors are grateful to the financial support by Ministry of Higher Education Malaysia, Research University Grant from Universiti Teknologi Malaysia (Grant No. Q.J130000.7123.00J04), and Universiti Malaysia Perlis.

7. References

Asher, R. C. (1983). Ultrasonic Sensors in the Chemical, and Process Industries. *Journal Science Instrument Physics*. Vol. 16, Pp. 959-963

Broring, S., Fischer, J., Korte, T., Sollinger, S., & Lubbert, A. (1991). Flow Structure of The Dispersed Gas-phase in Real Multiphase Chemical Reactor-Investigated by A New Ultrasound-Doppler Technique. *Canadian Journal of Chemical Engineering*. Vol. 69, Pp. 1247-1256

Garcia-Stewart, C. A., Polydorides, N., Ozanyan, K. B. & McCann, H. (2003). Image Reconstruction Algorithms for High-Speed Chemical Species Tomography. *Proceedings 3rd World Congress on Industrial Process Tomography*. Banff, Canada. 80-85

Kannath, A. & Dewhurst, R. J. (2004). Real-Time Measurement of Acoustic Field Displacements Using Ultrasonic Interferometry. *Measurement Science Technology*. Vol. 15, Pp. 59–66

Kytomaa, H. K. & Corrington, S. W. (1994). Ultrasonic Imaging Velocimetry of Transient Liquefaction of Cohesionless Particulaled Media. *International Journal of Multiphase Flow*. Vol. 20, Pp. 915-926

Moore, P. I., Brown, G. J. & Stimpson, B. P. (2000). Ultrasonic Transit-Time Flowmeters Modelled With Theoretical Velocity Profiles: Methodology. *Measurement Science Technology*. Vol. 11, Pp. 1802–1811

Rahim, R. A., Rahiman, M. H. F., Zain, R. M., & Rahim, H. A. (2011). Image Fusion of Dual-Modal Tomography (Electrical Capacitance, and Optical) for Solid/Gas Flow. *International Journal of Innovative Computing, Information, and Control*. Vol. 7, No. 9, Pp. 5119- 5132

Rahiman, M. H. F., Rahim, R. A., & Ayob, N. M. N. (2010). The Front-End Hardware Design Issue in Ultrasonic Tomography. *Sensors Journal*. Vol. 10, Issue 7, Pp. 1276-1281

Rahiman, M. H. F., Rahim, R. A., & Zakaria, Z. (2008). Design, and Modelling of Ultrasonic Tomography for Two Component High Acoustic Impedance Mixture. *Sensors & Actuators: A. Physical*. Vol. 147, Issue 2, Pp. 409-414

Rashidi, M. (1997). Fluorescence Imaging Techniques: Application to Measuring Flow, and Transport in Refractive Index-Matched Porous Media. *Chemical Engineering Technology*. Vol. 21, Pp. 7-18

Sanderson, M.L. & Yeung, H. (2002). Guidelines for the Use of Ultrasonic Non-invasive Metering Technique. *Flow Measurement, and Instrumentation*. Vol. 13, Pp. 125-142

Warsito, W., Ohkawa, M., Kawata, N., & Uchida, S. (1999). Cross-Sectional Distributions of Gas, and Solid Holdups in Slurry Bubble Column Investigated by Ultrasonic Computed Tomography. *Chemical Engineering Science*. Vol. 54, Pp. 4711-4728

Williams, R. A., & Beck, M. S. (1995). Process Tomography-Principles, Techniques, and Applications. Oxford, UK: Butterworth-Heinemann

12

Suppression of Corrosion Growth of Stainless Steel by Ultrasound

Rongguang Wang
Department of Mechanical Systems Engineering, Faculty of Engineering,
Hiroshima Institute of Technology,
Japan

1. Introduction

Metals are smelted from natural minerals and they tend to return back to their original or stable states including oxides and hydroxides. One of the later processes is corrosion, which occurs from metal surfaces with surrounded air or electrolytes. With the propagation of corrosion, the metal becomes thin as well known as the uniform corrosion of carbon steels, holes appear from the surface such as pitting corrosion of stainless steels, or cracks initiate and propagate from or in the metal like stress corrosion cracking, hydrogen embrittlement or corrosion fatigue. In many cases, cracks initiate from pitting corrosion. Either of the above phenomena leads certainly to the strength loss of metals and accordingly largely shorten the structures' life. Before carrying out a maintenance of metallic structures, the ultrasound is frequently used to investigate the degradation level of the structures by comparing the emitted and the reflected waves. Many researches and applications have been carried out on this topic.

On the other hand, in comparison with the prediction of the residual life and the maintenance of structures, it is much more important to stop or retard the initiation of corrosion; at least the propagation of corrosion should be effectively suppressed to thoroughly prolong the structures' life. This is generally solved by increasing the thickness of the applied steels, using high grade steels containing expensive or rare elements of chromium, nickel and molybdeum, or changing the surroundings of air or electrolytes. In this topic, a new method of using ultrasound to suppress the corrosion of stainless steel was introduced.

2. Corrosion process of stainless steel with formation of corrosion products

The high corrosion resistance of stainless steels originates from the formation of a passive film on the surface. In general, much oxide and hydroxide of chromium are contained in the passive film. However, the pitting corrosion or the crevice corrosion often occurs when the passive film is locally broken down by the chloride ions in solutions. The researches to retard the pitting initiation and growth have elapsed for near one century since the birth of stainless steel. In general, during the corrosion reaction, the metal surface will be separated into cathode zones and anode zones. On the cathode zones, the reduction reactions occur generally with the transformation of hydrogen ions to hydrogen gas or the transformation of oxygen gas to hydroxide ions, where the loss of metal does not occur. On the anode zones, the oxidization reactions occur with the transformation of metals to metallic ions into the solution, which means the loss of metal. The total reaction on the anode zones is equal to that on the cathode ones. When either of the anode or the cathode reaction was suppressed, the corrosion propagation should be slowed down. In the pitting corrosion, the area of cathode zones is much larger than the anode ones, which results in that even small cathode reaction rate can induce deep pit formation.

The main characteristic of pitting corrosion is that the thickness of the steel almost does not verify, while in some places holes form from the surface and sometimes penetrate the whole steel. The formation of the holes (hereinafter called as pits) includes the initiation stage and the growth stage. It is very important to suppress both the initiation and the growth of stainless steels to prolong the life of machines and structures. The stainless steel generally contains chromium and nickel with the balance of iron. The widely used typical austenite stainless steel is the type 18-8 steel (also being called as SUS304 steel or Type 304 steel), where 18mass% Cr and 8% Ni were contained with a little of carbon less than 0.08%. In the initiation of pitting corrosion of stainless steel, the Cl^- ions generally concentrate to weak sites of the passive film, under which always impurities such as MnS, MnO, Al_2O_3, TiO_2 and others exist (Ryan et al., 2002; Shimizu, 2010a, 2010b; Yashiro & Shimizu, 2010). In comparison to other zones with good state of passive film, the potential of such weak sites is somewhat lower, which promotes to form micro cells where the weak sites act as the anodes and other wide zones become the cathodes. This induces the acceleration of anode reaction at such weak sites, where pits initiate. The pit growth is always accompanied by the corrosion product of oxides and hydroxides of metals covering the pit, which accelerates the pit growth by promoting the accumulation of hydrogen and chloride ions into the pits, i.e., excessive hydrogen ions are produced in the pit with the hydrolysis reaction and the chloride ions migrate from outside to neutralize the excessive positive charges (Hisamatsu, 1981).

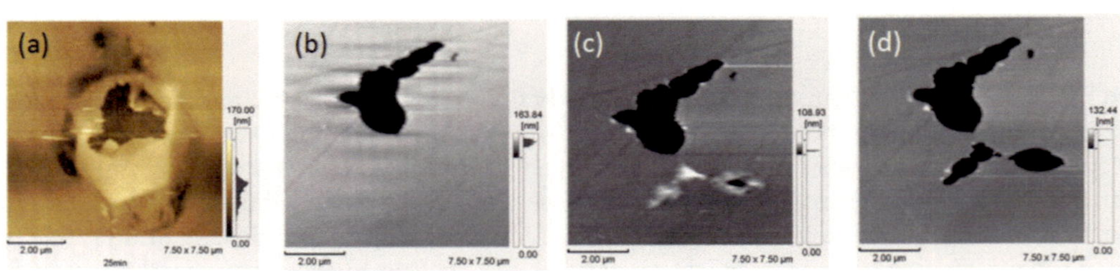

Fig. 1. Corrosion products on Type 304 stainless steel in 3.5% NaCl solution, in-situ observed by atomic force microscopy. Copyright 2005 Elsevier

The corrosion product on Type 304 steel in 3.5 mass% NaCl solution during the pitting corrosion was in-situ observed at room temperature by atomic force microscopy (AFM), as shown in Fig.1 (Zhang et al., 2005). A corrosion product crust covered perfectly a small pit at its initiation stage. Part of the corrosion product was then removed by the scanning probe of AFM (Fig.1 (a)). After the break of the corrosion product crust, the pit did not grow more. Fig.1 (b-d) shows several irregular pits formed near chromium carbides. With the increase in the corrosion time, a corrosion product and a small elliptical pit with corrosion product around it were observed (b, c). Almost all the corrosion products were removed by the probe of AFM and two small pits clearly appeared there (d). After then, their shapes did not change anymore. The retardation of pit growth is explained by the decrease in the concentration of chloride and hydrogen ions in the pit due to the destruction of the crust and the stirring of the solution in the pit by the probe. As a result, re-passivation on the inner surface of the pit easily occurred and the corrosion almost stopped.

3. Promotion or suppression of corrosion of steels by ultrasound (Wang & Kido, 2009)

If there is a simple method to remove the corrosion product covering pits, the growth of pits can be suppressed even for the cheap and low grade stainless steels. Here comes the idea of using ultrasound (US) to achieve this aim. The ultrasound in liquids always induces acoustic cavitation (Chouonpa Binran Henshu Iinkai(CBHI), 1999). In the acoustic cavitation, bubbles generate, grow and collapse due to the extremely increased internal tensile and compressive stress in the liquid. The cavitation power relates to the frequency and the amplitude of ultrasound as well as the type of liquid. It is widely used to clean solid surfaces, disperse powders and accelerate chemical reactions in liquids. The acceleration of chemical reactions is mainly caused by the high internal pressure and high temperature in the cavitation.

It is frequently reported that the erosion or corrosion rate on metal surfaces in specific liquids can be promoted by the acoustic cavitation. Alkire et al studied the passivity of iron in 2 N H_2SO_4 solution and found that iron became active when a focused ultrasound was applied in the solution (Alkire & Perusich, 1983). Al-Hashem et al investigated the acoustic cavitation corrosion behaviour of cast Ni-Al-Cu alloy in seawater. Both the cathodic and anodic current of the alloy increased by an order of magnitude and the rate of mass loss increased near 186 times under the application of a 20 kHz ultrasound (Al-Hashem et al., 1995). Whillock et al measured the corrosion behaviour of 304L stainless steel in 2 N HNO_3 solution containing a small amount of Cl^- at 323 K. The corrosion rate increased in the active state and the breakdown of passive film was promoted when a 55 kHz and 380 kW/m^2 ultrasound was applied with a vibrator-to-specimen distance of 1.1 mm (Whillock & Harvey, 1996). Kwok et al studied the cavitation erosion and corrosion characteristics on various engineering alloys including the grey cast iron, mild steels and stainless steels in 3.5% NaCl solution at 300 K when a 20 kHz ultrasound was applied. They found that corrosion mainly occurred on mild steel and grey cast iron but was negligible on stainless steel. The stainless steel only suffered pure mechanical erosion in the presence of cavitation (Kwok et al., 2000). Whillock et al also investigated the corrosion behaviour of 304L stainless steel in an ultrasonic field with different frequencies, acoustic powers and vibrator-to-specimen distances (Whillock & B.F. Harvey, 1997). At 20 kHz, the corrosion rate increased continuously with the increase in the power, while at the frequency of 40 to 60 kHz, the corrosion rate increased to the maximum and thereafter decreased with the increase in the power. The corrosion rate increased with the decrease in the vibrator-to-specimen distance, high corrosion rate in excess of 800 mm/year were obtained when the distance was 0.1 mm.

On the other hand, several papers reported that the corrosion on stainless steel can also be suppressed by the application of ultrasound in chloride containing solutions (Nakayama & Sasa, 1976; Whillock & Harvey, 1996; Wang & Nakasa, 2007; Wang & Kido, 2008; Wang, 2008). Nakayama and Sasa measured the polarization curves of a 304 type stainless steel in 0.1 N NaCl solution when applying a 200 kHz and 38-46 kW/m^2 ultrasound with a vibrator-to-specimen distance of 60 mm (Nakayama & Sasa, 1976). They found that the critical pitting potential became noble in the applied ultrasound field. Whillock et al found that ultrasound can encourage the passivation of 304L stainless steel in 2 N HNO_3 containing small amounts of Cl^- (Whillock & Harvey, 1996). Wang et al investigated the detail of the suppression effect of ultrasound on stainless steel in chloride solution (Wang & Nakasa, 2007; Wang & Kido, 2008; Wang, 2008; Wang & Kido, 2009; Wang, 2011).

The above conflicted influences of ultrasound on the corrosion behaviour of metals should be caused by (1) the type of metal, (2) the type of solution, (3) the acoustic power (frequency and amplitude) of ultrasound and (4) the ultrasound vibrator-metal distance. Especially, for stainless steels pitting corrosion and crevice corrosion usually occur when the passive film is locally broken down in chloride containing solutions. When the acoustic cavitation is strong enough, the passive film can be damaged and thus corrosion is activated. However, when the acoustic cavitation is not strong enough to damage the passive film the corrosion will not be accelerated. The suppression effect of acoustic cavitation on the pitting corrosion of stainless steel should be similar with that of the scanning of AFM probe in (a) the removal of the corrosion products (including the removal of the metallic cover under relatively larger power of ultrasound) and (b) the stirring of the solution in the pits. Of course, both of the removal and the stirring effects finally depend on a suitable power of the acoustic cavitation.

4. Suppression of pitting corrosion on stainless steel by ultrasound (Wang, 2008)

A trial was carried out to use ultrasound to remove the corrosion products covering pits, and the growth of pitting corrosion on Type 304 stainless steel was investigated in 3.5% NaCl solutions at 308 K. Pitting corrosion tests were carried out by using a corrosion cell attached with potentiostat apparatus (Hokuto Denko. Co., HAB-151) and an ultrasound cleaner (Yamato Co., Branson 2510J-MTH, 100 W, 42 kHz) as shown in Fig.2 (a). The counter electrode was platinum and the reference electrode was saturated calomel electrode (S.C.E.). The polarization was started from the cathode side ($E_{S.C.E.}$= –600 mV) to the anode side at a constant potential increasing rate of 50 mV/min under the control of the potentiostat. With the linear increase in potential, the current changes from the cathode and passive zone to the pitting zone where the current density largely increases with a small increase in potential (Fig.2 (b)). When the anodic current density reached a value of i_h = 20 A/m^2 in the pitting zone, the potential was immediately held constant for 600 s. In this way, the deviation of current density caused by different pitting potentials on different surfaces will be small. The ultrasound (nominate intensity: 3 kW/m^2) was applied in 3 types of conditions, i.e., (i) without ultrasound (hereinafter called as condition A); (ii) applying ultrasound simultaneously with the holding of potential (condition B); (iii) applying ultrasound from the beginning to the end of the polarization (condition C). The potential corresponding to the above constant current density was kept for 600 s in all conditions.

For the polarization curves in conditions A, B and C, the current density changes from the cathode to the anode near the potential E = -150 mV irrespective of the application of ultrasound. No significant difference was found in the corrosion potential and in the current density in the cathode zone and the anodic passive zone. This indicates that the ultrasound is not strong enough to break the passive film down, or the partially broken passive film can be easily self-repaired.

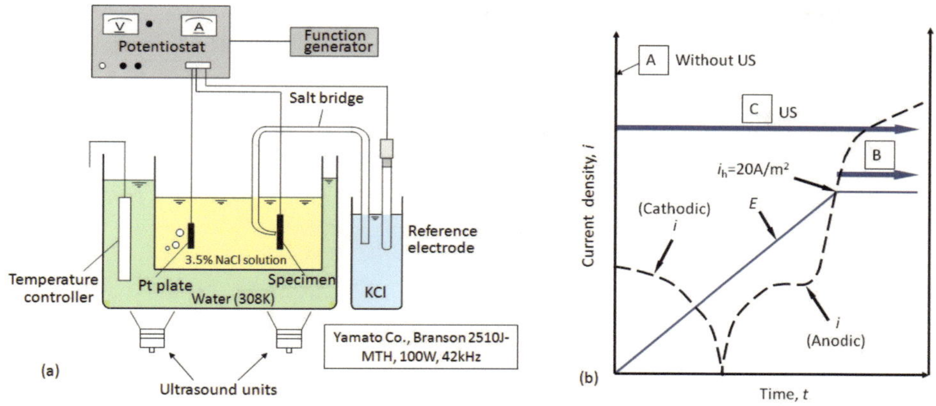

Fig. 2. (a) Polarization apparatus and (b) timing of ultrasound. Copyright 2008 Elsevier

The influence of ultrasound on the corrosion behaviour was observed mainly in the pitting zone. Fig.3 (a) shows parts of the polarization curves from $i_h = 20$ A/m², where the vertical axis is shown in a normal decimal scale. It is clear that the current density in condition A is much higher than those in conditions B and C. This means that the pitting growth can be largely suppressed by applying ultrasound in solution. On the other hand, no large difference can be found in conditions B and C, which means that ultrasound is only effective after corrosion has occurred. Fig.3 (b) shows the typical surface morphologies of specimens after polarization and the electric charge calculated by the integration of the current density during the 600 s holding of potential after the current density reached $i_h = 20$ A/m². The area of pits was measured from enlarged photos. The electric charge means the dissolution amount of metal ions into solution. Fewer pits were found on the surface with ultrasound, corresponding to the smaller electric charge.

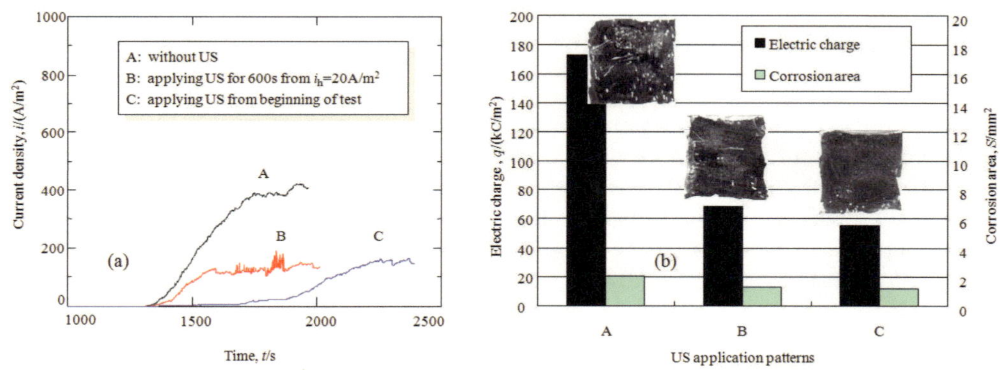

Fig. 3. (a) Change of anodic current density from $i_h = 20$ A/m², and (b) electric charges for the 600 s holding from $i_h = 20$ A/m². Copyright 2008 Elsevier

The mechanism is schematically shown in Fig.4. When the ultrasound is not applied, pits grow on the specimen surface (a) and corrosion product covers the pit (b) (Wranglen, 1985; Zhang et al., 2005). The growth of pit is accelerated due to the hydrogen ions produced by hydrolysis reaction and the chloride ions attracted from solution (c) (Wranglen, 1985). When the ultrasound is applied to the specimen surface, the corrosion product is removed by the cavitation of ultrasound (b′). The concentration of corrosive hydrogen ions and chloride ions decreases due to the stirring effect of cavitation, and the growth rate of pits decreases (c′).

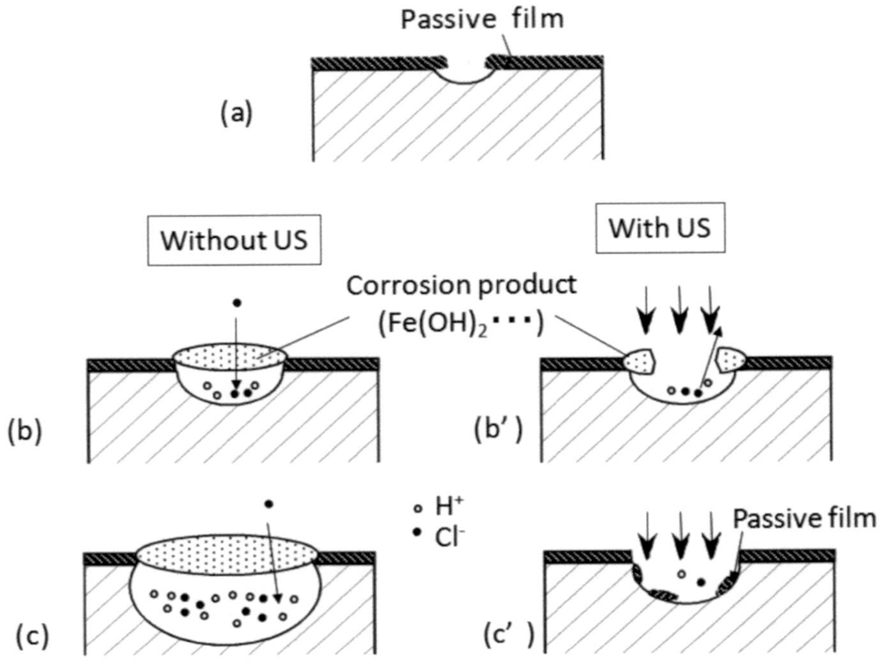

Fig. 4. Decrease in growth of corrosion pits by ultrasound. Copyright 2007 The Japan Institute of Metals

5. Suppression of crevice corrosion on stainless steel by ultrasound (Wang & Kido, 2008)

The mechanism of crevice corrosion is similar with pitting corrosion. In this section, the influence of ultrasound on the crevice corrosion will be introduced. Type 304 specimens were polished with the # 600 emery paper on both sides and then assembled by the following JIS G0592 standard (Japanese Industrial Standards Committee (JISC), 2002) to produce a crevice between the twice polished surfaces. The crevice was wetted by the 3.5% NaCl solution and tightened hardly by titanium nuts and washers to form the artificial crevices. Of course, there were also two small crevices between steel / washer. The corrosion test was carried out in a corrosion cell connected to a potentiostat (Hokuto Denko Co.; HAB-151) and an ultrasound cleaner (Yamato Co., Branson 2510J-MTH; 100 W, 42 kHz), as is shown in Fig.5. Hereinafter, i_h and i_u mean the current density when the potential was held constantly and the current density when the ultrasound was triggered, respectively. Δt_h and Δt_u mean the period for holding the potential constantly and the period for applying ultrasound, respectively.

Fig. 6 shows the change of the current density when the current density reached i_h = 10 A/m^2 and the potential at that moment was held for Δt_h = 0.9~1.2 ks. The arrow (⇒) shows the period for the application of ultrasound. The solid line (a) shows the result without ultrasound, the broken line (b) shows the result when ultrasound was stopped after $\Delta t_u = 0.6$ ks but the potential was continuously kept constantly for further $\Delta t_h = 0.3$ ks, the dotted line (c) shows the result with ultrasound during the overall potential holding period ($\Delta t_h = 1.2$ ks). In the case of (a), the current density largely increased near $t = 0.9\sim1.1$ ks ($E = 470\sim530$ mV) after the passive zone. After the polarization, no pitting corrosion can be observed and only crevice corrosion appeared (see Fig.7), i.e., the increase in the current density means the happening of crevice corrosion. The potential was about $E_h = 488\sim570$ mV at the moment of the crevice corrosion current density $i_h = 10$ A/m^2. During this period, the current density continued to rise slowly and finally kept almost stable at a value of about 200 A/m^2. In the case of (b), the current density almost kept stable at the level of about 40 A/m^2. This low and stable current density should be attributed to the dilution of the enriched Cl^- and H^+ in the crevice caused by the stirring effect of ultrasound. When the ultrasound was stopped after 0.6 ks, the current density increased. It will be due to the re-enrichment of Cl^- and H^+ because the disappearance of ultrasound. However, in the case of (c), the current density kept stable at a low level.

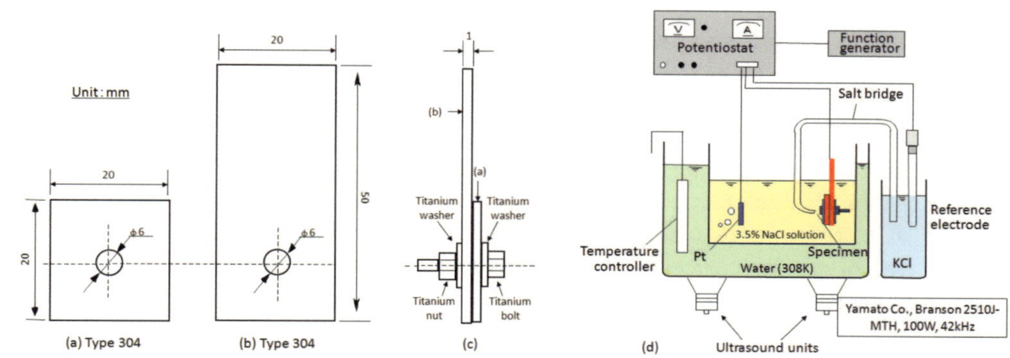

Fig. 5. (a-c) Specimen with crevice and (d) apparatus for polarization test. Copyright 2008 The Japan Institute of Metals

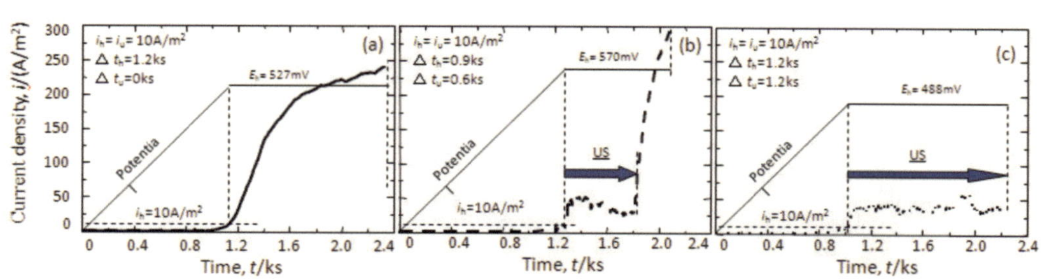

Fig. 6. Polarization curves without ultrasound (a), with ultrasound (b) $i_h = i_u = 10$ A/m^2, $\triangle t_h = 0.9$ ks, $\triangle t_u = 0.6$ ks, and with ultrasound (c) $i_h = i_u = 10$ A/m^2, $\triangle t_h = 1.2$ ks, $\triangle t_u = 1.2$ ks. Copyright 2008 The Japan Institute of Metals

Fig.7 shows the steel surface between crevices of (i) steel / steel and (ii) steel / washer (titanium) after the polarization. Crevice corrosion clearly appeared in any case. The difference between (a) and (b) is not remarkable probably because the applied period of ultrasound was short in (b), however, the corrosion area in (c) is clearly small.

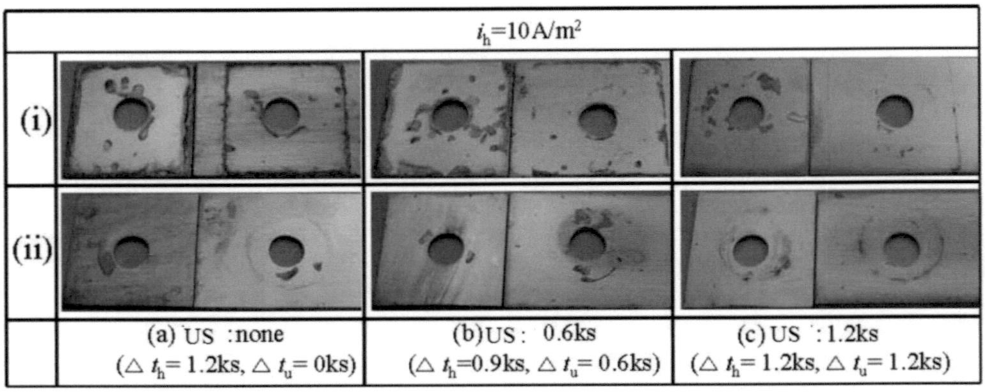

Fig. 7. Specimen surfaces after polarization without and with ultrasound ($i_h = i_u$= 10 A/m²). Copyright 2008 The Japan Institute of Metals

Fig.8 shows the electric charge q, which is integrated with time during the period of potential holding for Δt_h = 0.6 ks, 0.9 ks or 1.2 ks. In the case of Δt_h = 0.6 ks, q without ultrasound was about 65 kC/m², however, it decreased to about 15 kC/m² when the ultrasound was applied for 0.6 ks. In the case of Δt_h = 1.2 ks, q without ultrasound was about 190 kC/m², it decreased to about 45 kC/m² when the ultrasound was applied for 1.2 ks. In addition, in the case of Δt_h = 0.9 ks and Δt_u = 0.6 ks, q was about 165 kC/m². In any cases, about 77% of the electric charge decreased due to the application of ultrasound. Of course, the effect of ultrasound would be small if the ultrasound was midway stopped. The corrosion area of crevice corrosion is also shown in the right side of Fig.8 with hatching sticks. The corrosion area changed almost at the same ratio with the electric charge, although the depth of the corrosion zone would be different.

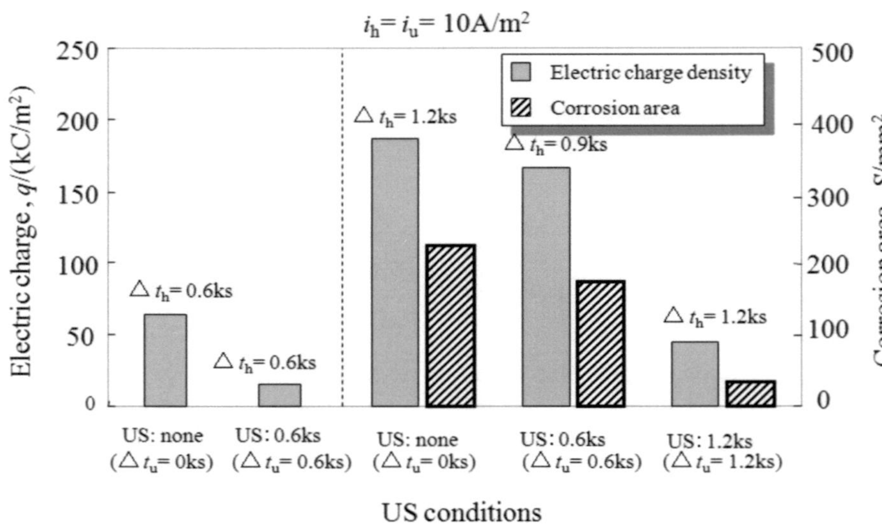

Fig. 8. Electric charge during the holding of potential without and with ultrasound ($i_h = i_u$ = 10 A/m²). Copyright 2008 The Japan Institute of Metals

Fig. 9 (a) shows the change of current density for another type of crevice spcimen, where only one side of the specimen was polished when the potential was held constantly at i_h = 10 A/m². On this specimen, both pitting corrosion (on the not polished side) and crevice corrosion occurred. After the current density reached i_u = 120 A/m² the ultrasound was applied for 60 s and then stopped for 60 s. Such ultrasound application / stop repeated for 4 cycles. In the first application of ultrasound, the current density largely decreased to about 50 A/m². It gradually recovered during the 60 s stop state of ultrasound to the previous level before the application of ultrasound. Such change almost synchronized with the cyclic application and stop of ultrasound. Fig.9 (b) shows the change of current density, after the current density reached i_h = 50 A/m² and the potential at that moment was kept constant. The ultrasound was applied for 60 s after the current density reached i_u = 250 A/m², and then stopped for 270 s. Such ultrasound application was twice cycled. The current density can be largely decreased to about 40 A/m² during the 60 s application of ultrasound. During the stop of ultrasound for 270 s, the current density gradually increased to the previous level. Almost the same change occurred in the second ultrasound cycle. According to the change of electric charge q calculated from Fig.9, either of the happening of pitting corrosion or crevice corrosion, the decrease in the corrosion amount by ultrasound was about 44~55% in comparison with those without ultrasound. Therefore, even if the ultrasound was not continuously applied, an intermittently application for 60 s can bring about large suppression of corrosion.

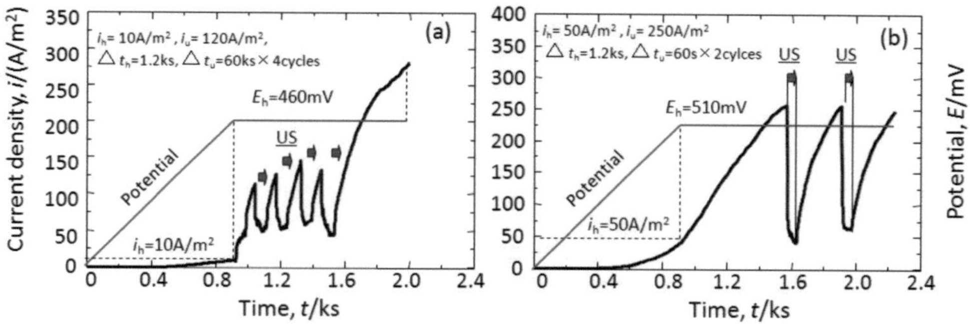

Fig. 9. Polarization curves of both pitting corrosion and crevice corrosion with ultrasound. (a): i_h = 10 A/m², i_u = 120 A/m², $\triangle t_h$ = 1.2 ks, $\triangle t_u$ = 60 s × 4 cycles; (b): i_h = 50 A/m², i_u = 250 A/m², $\triangle t_h$ = 1.2 ks, $\triangle t_u$ = 60 s x 2 cylces. Copyright 2008 The Japan Institute of Metals

The schematic drawing of decrease in growth of crevice corrosion by application of ultrasound is shown in Fig.10. In the case of without ultrasound, the metallic ions (M^{n+}) in the crevice will be accumulated by the flow of the passive current before crevice corrosion is triggered, and hydrogen ions increase due to the hydrolysis reaction of M^{n+}. Furthermore, Cl^- ions were attracted from solution out of the crevice to neutralize the excessive plus charge. Thus, the crevice corrosion is induced ((a)). That is, H^+ and Cl^- ions enriched in the crevice with the anodic crevice corrosion and the corrosion is accelerated ((b), (c)). On the one hand, when the ultrasound was applied, the concentration of M^{n+}, Cl^- and H^+ inside the crevice can be diluted by the stirring effect of ultrasound, which results in the slowing down of the corrosion ((b'), (c')). In some cases, re-passivation of the crevice surface might occur (c). Of course, the promotion of the diffusion of oxygen into the crevice will also decrease the corrosion rate

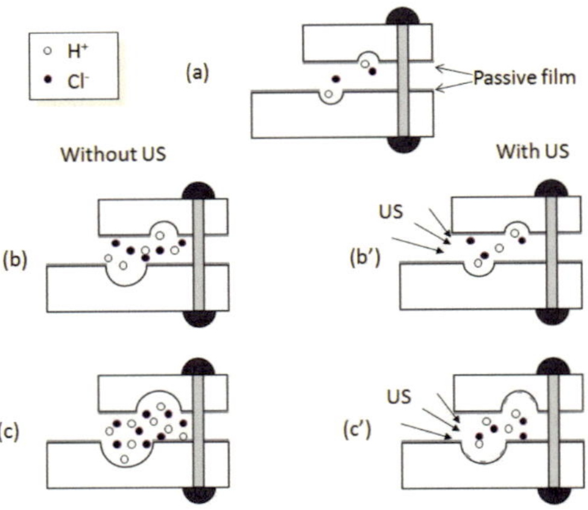

Fig. 10. Schematic drawing of decrease in growth of crevice corrosion by application of ultrasound. Copyright 2008 The Japan Institute of Metals

6. Influence of power and distance of ultrasound on pitting corrosion of Type 304 steel (Wang & Kido, 2009)

In this section, the influence of the transmitted acoustic power of ultrasound on the specimen and the vibrator-to-specimen distance on the pitting corrosion will be introduced. The corrosion tests were carried out in a corrosion cell connected to a potentiostat (Hokuto Denko. Co., HAB-151) and an ultrasound vibrator (Kaijo Co., 4292C; 19.5kHz; 130 mm x 150 mm) (Fig.11). The input power to the ultrasound vibrator can be adjusted by a controller (Kaijo Co., TA-4021) from I = 0 to I = 10, where the full input power of I = 10 to vibrator is 200 W corresponding to a mean input intensity of 10 kW/m^2 from the vibrator. The specimen was immersed in 3.5% NaCl aqueous solution in the corrosion cell facing to the ultrasound vibrator and the temperature was tried to be kept stable at 305±2 K by an auto-heater (thermostat without cooling function), but the increased temperature in the solution was not further adjusted when ultrasound was applied.

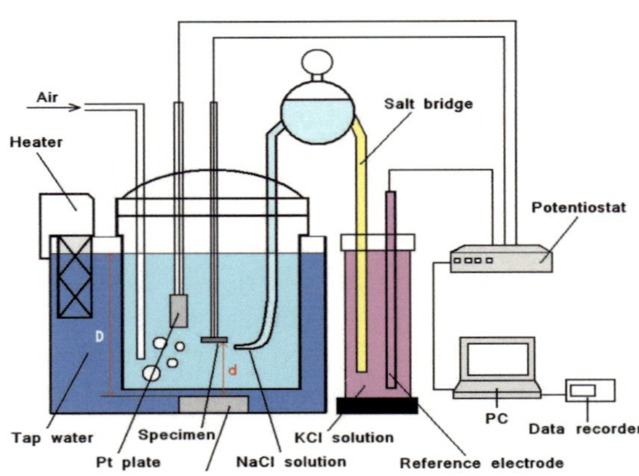

Fig. 11. Apparatus for polarization and applying ultrasound. Copyright 2009 Elsevier

The ultrasound was applied in simultaneously with the holding of potential. The distance from the ultrasound vibrator to the specimen surface (d) and the input power to vibrator (I) were changed as d = 76 mm (I = 0, 2, 4, 6 and 8) and I = 8 (d = 19, 39, 76 and 95 mm). Note that the wavelength (λ) of the ultrasound in the frequency of 19 kHz is about 76 mm, which was calculated from the speed of sound of 1480 m/s. After the polarization tests, specimen surfaces were observed using an optical microscope and the corrosion areas were accordingly measured.

Fig.12 (a) shows the typical change of the current density in the pitting corrosion zone from i_h = 20 A/m^2 during the period of simultaneously holding potential and applying ultrasound. The input power to ultrasound vibrator changed from I = 0 to I = 8 at a constant distance of d = 76 mm. The distance is just equal to the wavelength of ultrasound in the solution. In Fig.12 (a), the anodic current density increased gradually with time during the period of holding potential without the application of ultrasound (I = 0), however, the value largely decreased when the ultrasound was applied under each input power to vibrator. No large difference of the current density under I = 1, 2 and 4 can be seen, but the current density sharply decreased under I = 6. The largest decrease of the current density was obtained when applying ultrasound under I = 8 in this work. The smallest value of current density was near 1×10^{-3} A/m^2, meaning the passivation of pits. Fig.12 (b) shows the electric charge obtained from the integrity of the current density during the potential holding. Each value of the electric charge is averaged from at least 3 tests. The value of electric charge was about 22 kC/m^2 without ultrasound (I = 0), which is smaller than the value obtained in the previous report (Wang, 2008) perhaps because the chemical composition of the specimens in this work is different from those used in the previous work. The electric charge decreased to about 6 ~ 9 kC/m^2 when applying ultrasound under I = 1, 2 or 4. Large decrease in the electric charge was obtained under I = 6 and I = 8, especially the electric charge under I = 8 was the smallest one in this work (1 kC/m^2). Such results gave the detail of the corrosion rate and meaning that the pitting corrosion of Type 304 stainless steel can surely be suppressed by the application of ultrasound in the solution and the suppression effect become remarkable with the increase in the input power to ultrasound vibrator.

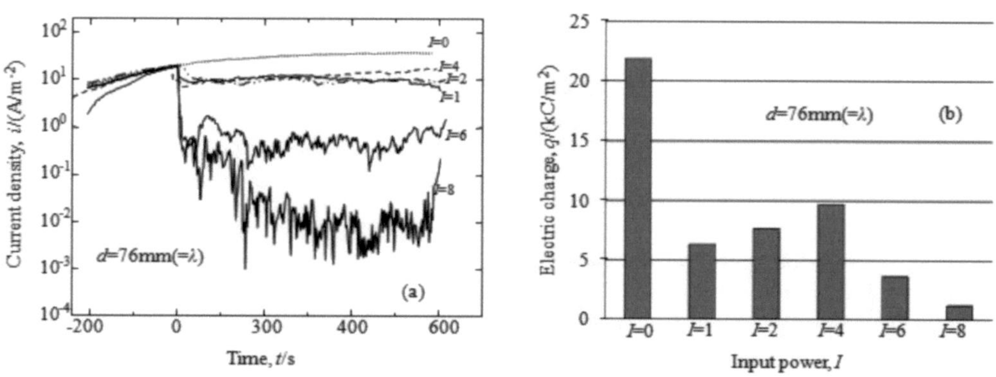

Fig. 12. Pitting current density (a) and accumulated electric charge (b) during the period of simultaneously holding potential and applying ultrasound with different input powers at a constant distance of 76 mm. Copyright 2009 Elsevier

Fig.13 (a) shows the surface morphology of specimens after the polarization tests in Fig.12. The area ratio and the mean depth of pits are shown in Fig.13 (b) and (c). The depth of pits was obtained by focusing on the bottom of pits and the flat specimen surface by moving an optical lens. Pits appeared on each specimen surface after the polarization, but the sum, size and depth gradually decreased with the increase in the input power to ultrasound vibrator. Although the difference of current density under $I = 1$, 2 and 4 was indistinct, in Fig.13 it is known that the pitting corrosion can be suppressed more with the increase in the input power to ultrasound vibrator. The least and smallest pits were found when applying ultrasound under $I = 8$. It means that the initiation and growth of pits were suppressed both in the width and the depth by ultrasound.

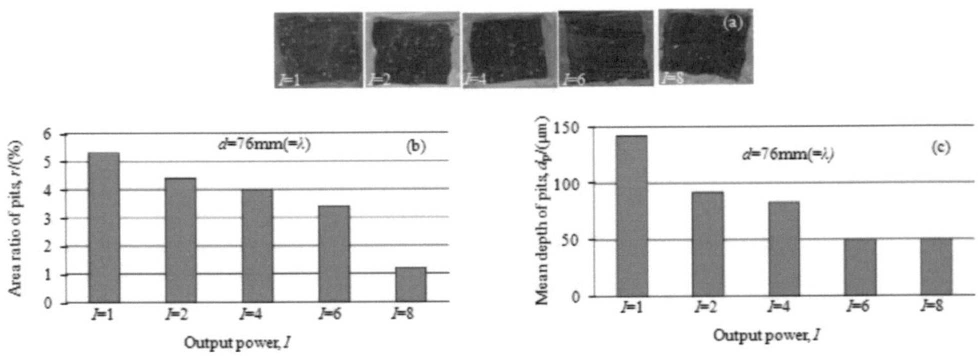

Fig. 13. Surface morphology (a), area ratio of pits (b) and mean depth of pits (c) on specimen surface after simultaneously holding potential and applying ultrasound with different input powers at a constant distance of 76 mm (= λ). Copyright 2009 Elsevier

Fig. 14 (a) shows the pitting current density during the period of simultaneously holding potential and applying ultrasound at different distances under a constant input power of $I = 8$ to vibrator. For each case with the application of ultrasound, the current density decreased by comparing to that without ultrasound. No large difference of the current density can be seen at the distance of $d = 19$ and 38 mm, but the current density at $d = 76$ mm was the smallest one while the current at $d = 95$ mm was the largest one. Fig.14 (b) shows the electric charge during the potential holding under $I = 2$ and 8 at different distances. According to the left part of Fig.14 (b) (averaged results under $I = 2$), by comparing to that at $d = 19$ mm (= $\lambda/4$) the current density increased at $d = 38$ mm (= $\lambda/2$). However, it largely decreased at $d = 76$ mm (= λ). The current is the largest at the distance of $d = 95$ mm (= $5\lambda/4$). In the right side of Fig.14 (b) (averaged results under $I = 8$), almost the same tendency was obtained with that under $I = 2$. Of course, the absolute value under $I = 8$ is much smaller than that under $I = 2$.

Fig. 15 (a) shows the specimen surface after the corrosion test with changing the distance under $I = 8$. Fig.15 (b) and (c) show the area ratio and the mean depth of pits in these cases. It is clear that both the area ratio and the mean depth decreased with the increase of distance from $d = 19$ mm to 76 mm, except the largest value appeared at the distance of $d = 95$ mm. It suggests that the suppression effect of acoustic cavitation on the pitting corrosion depends not only on the distance but also on the transmission phase of the ultrasound wave. Larger suppression effect appears at the vibrator-to-specimen distance equals to the transmission wavelength of ultrasound.

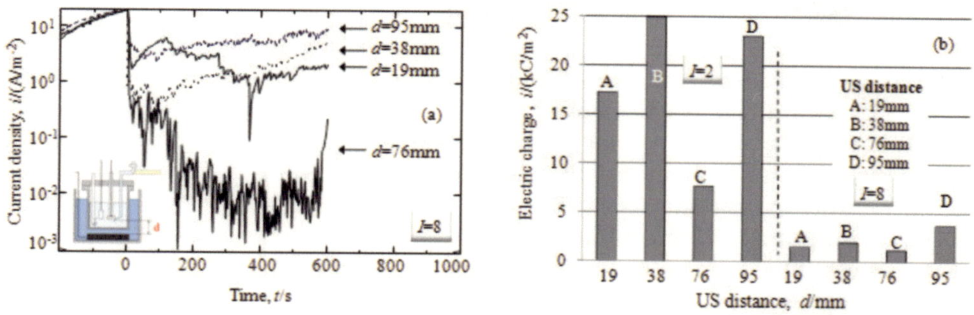

Fig. 14. Pitting current density (a) and accumulated electric charge (b) during the period of simultaneously holding potential and applying ultrasound with different distances at a constant input power of I = 8 or I = 2. Copyright 2009 Elsevier

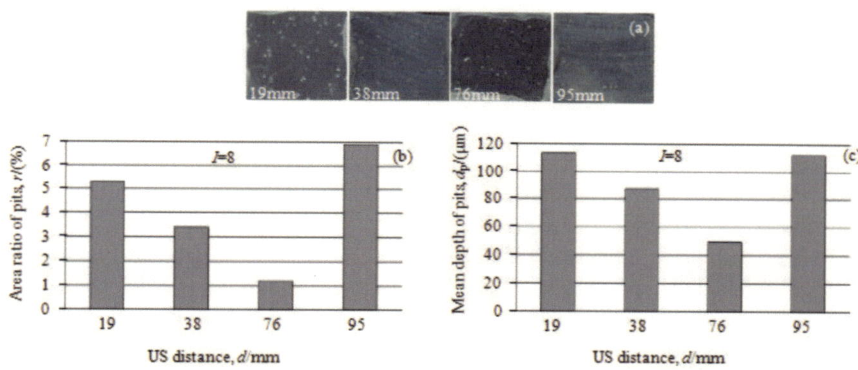

Fig. 15. Surface morphology (a), area ratio of pits (b) and mean depth of pits (c) on specimen surface after simultaneously holding potential and applying ultrasound with different distances at a constant input power of I = 8. Copyright 2009 Elsevier

Although the input power to ultrasound vibrator can be exactly set from the ultrasound controller (see Fig.11), the transmitted power dissipated near the specimen surface was not known. Here, the transmitted acoustic power near the specimen surface was measured by a calorimetry method as follows (CBHI, 1999; Whillock & Harvey, 1996). The temperature of 5 mL pure water contained in a small glass tube (inner diameter φ = 12 mm) was inserted into the water bathe and measured at 10 s intervals by a digital thermometer during the sonication for a period of 120 s (before the application of ultrasound, the power of the auto-heater was cut). The absorbed acoustic power density (p) was calculated using:

$$p = mC_{\mathrm{p}}(\mathrm{d}T / \mathrm{d}t) / A \tag{1}$$

where m is the mass of water (unit: kg), C_p is the heat capacity of water (4180 J/(K • kg)) and $\mathrm{d}T/\mathrm{d}t$ is the temperature rise per second during the initial 20 s (K/s), A is the cross-sectional area of the glass tube (about 113 mm^2). The result is shown in Fig.16. In Fig.16 (a), at the constant distance of d = 76 mm, almost no temperature increase can be detected under the input power of I = 1 and I = 2 to vibrator, meaning the transmitted power is weak. In the case of I = 4 the temperature slowly increased 0.4 K after 120 s, while the increased temperatures are respectively 0.8 K and 1.6 K in the case of I = 6 and I = 8. The above result shows the transmitted power of ultrasound to the specimen surface increases with the increase in the input power to vibrator. In each case of I = 4, 6 and 8, the temperature increased sharply at the initial 20 s but the increase became slow after then. This should be due to the happening of heat transfer in the solution during a relatively long period. Fig.16 (b) shows the temperature change under the constant input power of I = 8 to vibrator with changing the distance from d = 19 to 95 mm. In the case of d = 19 mm and d = 95 mm, the increase in temperature is almost the same after 120 s. This suggests that the transmitted power depends on the transmission phase of ultrasound wave. In the case of d = 38 mm (= $\lambda/2$), the temperature almost did not change at the initial 40 s but after then largely increased more than 2.8 K. In the case of d = 76 mm, the largest increase of temperature obtained (more than 3.7 K). Note that two different curves of temperature are shown in Fig.16 (a) and (b) at the same condition of I = 8 and d = 76 mm.

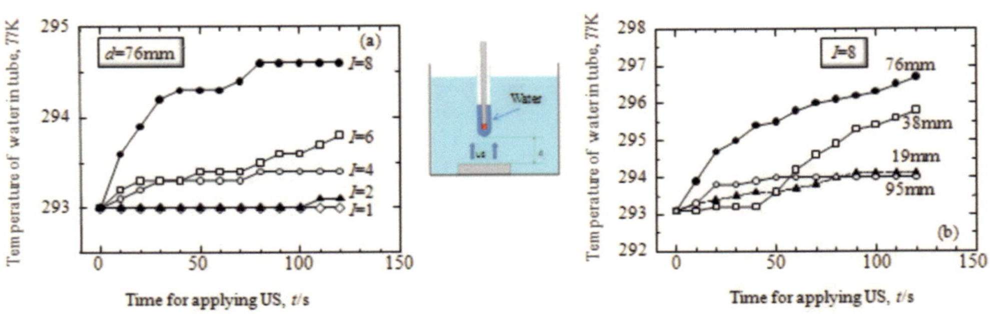

Fig. 16. Temperature change of a 5 mL pure water during the application of ultrasound in different cases. Copyright 2009 Elsevier

Table 1 shows the transmitted power of ultrasound near the specimen surface using the increase in temperature during the initial 20 s by equation 1). Comparing to the results in Fig.12~15, it is clear that the suppression effect on the pitting corrosion increased with the increase in the transmitted power of ultrasound. During a long period measurement, the increased temperature in the solution will also largely influence the corrosion rate. However, this would not change the conclusion of the suppression effect of ultrasound in the input power to the vibrator and the distance. It is clear that the transmitted power depends on both the input power to the vibrator and the ultrasound phase.

Fig.17 shows the change of current density when applying ultrasound in solutions with and without the addition of 0.5% ethanol at the constant distance of d = 76 mm under the input power of I = 2 and I = 8 to vibrator. In the case of I = 2, the current density in the ethanol-added solution decreased comparing to that in the ethanol-not-added solution. This means that the enhanced cavitation in this solution enhanced the suppression effect on the pitting corrosion under a weak ultrasound. On the other hand, in the case of I = 8, the current in ethanol-added solution became larger and unstable comparing to that in the ethanol-not-added solution. (two curves in the ethanol-not-added solution were shown in Fig.17 (b), including the lowest and highest current in all measurements.) This means that the suppression effect of corrosion with the addition of ethanol decreased under higher input power to ultrasound vibrator.

Condition	d=76 (mm)					I=8			
	I=1	I=2	I=4	I=6	I=8	d=19 (mm)	d=38 (mm)	d=76 (mm)	d=95 (mm)
Increased temperature in 20 s (K)	-	-	0.2	0.3	0.9	0.4	0.2	1.7	0.8
Acoustic power density (kJ/m^2)	-	-	1.85	2.78	8.33	3.70	1.85	15.73	7.40

Table 1. Increased temperature and adsorbed acoustic power density in a 5 mL water when ultrasound is applied. Copyright 2009 Elsevier

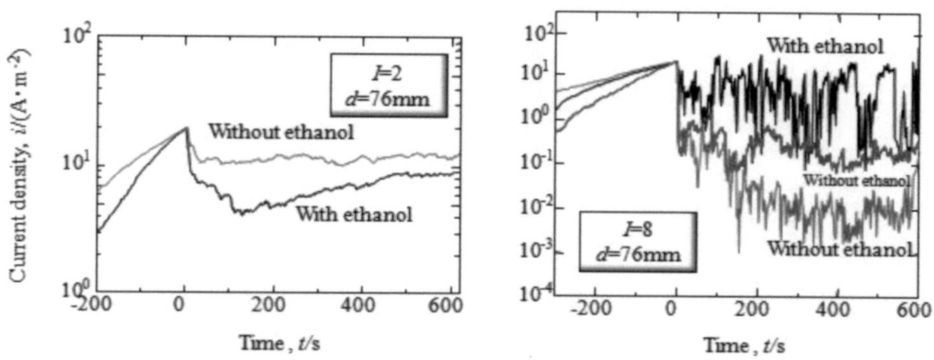

Fig. 17. Pitting current density during the period of simultaneously holding potential and applying ultrasound with different input powers at distance of d = 76 mm and input power of I = 8 in 3.5% NaCl solution containing 0.5% ethanol . Copyright 2009 Elsevier

When the acoustic cavitation is not strong enough to damage the passive film, the suppression effect of acoustic cavitation on pitting corrosion will increase with the increase in the stirring effect of solution in pits after removing the corrosion products (or the metallic cover). The suppression effect should be related to (1) the bubbles' size decided by the tensile stress and (2) the collapse power (shock wave power or cavitation power) decided by the compressive stress in the ultrasound field. Both of the stresses are determined by (i) the amplitude and (ii) the phase of the ultrasound wave. The collapse of larger bubbles brings about larger collapse power.

The removal of corrosion products or metallic covers can be promoted by larger collapse power of the cavitation under larger input power to vibrator and the solution in pits can be completely stirred. This is the reason that the suppression effect on corrosion can be enhanced when increasing the input power to vibrator from I = 1 to I = 8. However, the stirring of solution in pits after removing corrosion products or metallic covers should depend on both the bubbles' size as well as the collapse power. The schematic drawing is shown in Fig.18. Note that not all the pits are covered by metallic covers (a). Near the specimen, micro-jets to the specimen surface will appear from each collapsing bubble. When the input power to vibrator is small the bubbles' size and the collapse power are small. Part of corrosion products will be cleaned out but no damage occurs on the metallic cover. However, the collapse of bubbles smaller than pits perhaps gives relatively effective stirring effect of solution in pits (b). With the increase in the input power to vibrator (c), the removal of corrosion products increases but the metallic covers still remains there because of the strong strength connecting with the substrate. This results in that the stirring effect does not remarkably increase and thus the suppression effect on pitting corrosion did not largely increase when the input power to vibrator increased from I = 1 to I = 2 and I = 4. With further increase in the input power to ultrasound vibrator (d), the extremely enhanced cavitation power on the removal of corrosion products and metallic covers, and the stirring effect will be significantly enlarged (with the neglect of the weakness in the enlarged size of bubbles). That is the reason that the corrosion was greatly suppressed under I = 6 and I = 8. Especially the current density decrease to a level of 1×10^{-3} A/m^2 under I = 8, meaning the growth of the pit almost stopped.

On the other hand, the transmitted power to the specimen generally decreases with the increase in the vibrator-to-specimen distance because of the amplitude attenuation of ultrasound wave. It resulted in the decrease in the suppression effect when increasing the distance from d = 19 mm to d = 38 mm and d = 95 mm. On the other hand, the phase change of the wave should be also considered in the explanation of the result. In this section, the largest suppression effect of pitting corrosion was obtained at a distance of d = 76 mm, which is just equal to wavelength of ultrasound in the solution. In another word, the phase of the wave there is the same with that on the vibrator surface. This should be related to the formation of a "*standing wave field*" of ultrasound in the solution with overlapping the forward wave from the vibrator and the backward (reflected) wave from the liquid / air surface (CBHI, 1999; Mitome, 2008). In the standing wave field, anti-nodes with strong cavitation generally appears with an interval of $\lambda/2$ along the transmission direction. However, in this work only at a distance of λ (76 mm) the suppression effect of pitting corrosion is large while the suppression effect at a distance of $\lambda/2$ is much small (with good correspondence to the temperature measurement). The reason has not been clearly know, perhaps due to the unfixed distance from the vibrator to the solution surface in the above measurement.

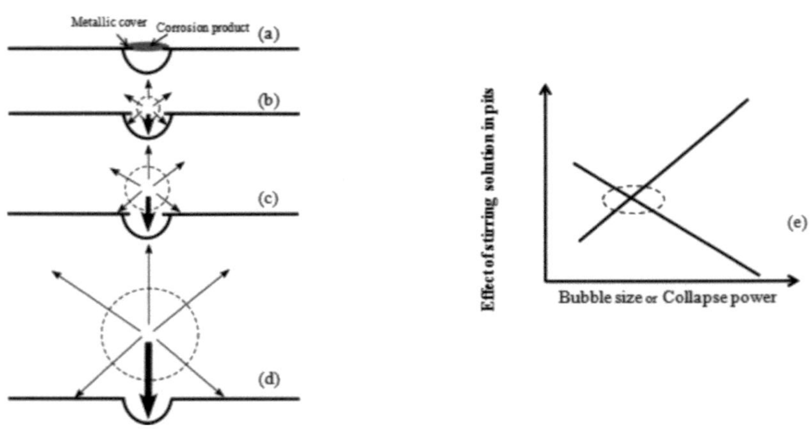

Fig. 18. Schematic drawing of relation of effect of stirring solution in pits with change of bubble size and collapse power. (a) Pit is covered by metallic cover and corrosion product; (b) corrosion product is removed by smaller collapsing bubble; (c) (d) corrosion product and metallic cover is removed by larger collapsing bubble; (e) stirring effect and bubble size or collapse power. Copyright 2009 Elsevier

Except the amplitude of ultrasound, the power of cavitation is also influenced by the evaporability of solution (CBHI, 1999). Ethanol is evaporable specie to improve the evaporability of the solution to result in the increase in stronger cavitation. Accordingly, the corrosion behaviour of stainless steel in the ethanol-added solution changed much with the application of ultrasound. In the case of $I = 2$, the current density is smaller in the ethanol-added solution than that in the ethanol-not-added solution. This means that the improved cavitation in ethanol-added solution gave fully stirring of the solution in the pits and suppressed the growth of pits. On the other hand, in the case of $I = 8$, the suppression of corrosion became weak after adding ethanol in the solution. This should be due to the activation of passive films on the surface by the excessively enhanced cavitation, which bring about promotion of pitting corrosion. This also means that strong acoustic cavitation can also promote corrosion, which corresponds well with other reports described before.

7. Influence of frequency of ultrasound on the pitting corrosion of Type 304 steel (Wang, 2011)

The influence of the ultrasound frequency as well as the distance from vibrator to specimen on the growth of pitting corrosion will be introduced in this section. The vibrators were used with different frequencies of f_1 = 19.5 kHz (Vibrator: Kaijo Co., 4292C), f_2 = 50 kHz (4492H) or f_3 = 420 kHz (4711C). The input power to the ultrasound vibrator was set at $I = 8$ by a controller (Kaijo Co., TA-4021). The distance from the vibrator to the specimen (d) was varied with the wavelength of each type of ultrasound (λ_1 = 76 mm , λ_2 = 29.6 mm , λ_3 = 3.5 mm). In addition, the distance (D) from vibrator to the solution surface was set as several integral times of the half-wavelength of each ultrasound, i.e., 1.5 λ_1 (D_1 =114 mm), 3 λ_2 (D_2 = 89 mm) or 27.5 λ_3 (D_3 = 96 mm). Note that the distance of D from vibrator to the solution surface in the above sections was not precisely fixed; only in this section the *standing wave* from the solution surface can be discussed. When the anodic current density reached the value of i_h = 20 A/m^2 in the pitting growth zone, the potential was immediately held constant for 600 s. The ultrasound was applied simultaneously with the holding of potential for 600 s.

Fig.19 (a) shows the accumulated electric charge during the period of simultaneously holding potential and applying ultrasound at $I = 8$ with different frequencies (a: 19.5 kHz; b: 50 kHz; c: 420 kHz) and different vibrator-to-specimen distances. Although the input powers to the vibrators are the same, the suppression effects on the pitting corrosion showed large difference when changing either the frequency or the vibrator-to-specimen distance. At each frequency, there is an optimum distance where the largest suppression effect was obtained. The optimum vibrator-to-specimen distance is respectively 57 mm (= $3\lambda_1/4$) in the case of f_1 = 19.5 kHz, 29.6 mm (= λ_2) in the case of f_2 = 50 kHz, and 17.6 mm (= 5 λ_3) in the case of f_3 = 420 kHz. This difference should be attributed to the balance of the formation of standing wave and the energy attenuation of ultrasound in the solution. In addition, the suppression effect in the case of 19.5 kHz is much larger than (100 times of) those in the cases of 50 kHz and 420 kHz. The latter two effects are almost the same.

From the observation of specimen surfaces after the polarization, the area ratio and the mean depth of pits are obtained (Fig.19 (b, c)). Note that the values were obtained from the surfaces after further ultrasonically cleaning with multi-vibrators (Fig.2). Pits appeared on each surface after the polarization and the pitting corrosion was surely suppressed by each type of ultrasound. However, the suppression of corrosion does not directly decrease the sum of pits. In several different distances of 19.5 kHz and 420 kHz the sum did not decrease, while in the case of 50 kHz the sum increased reversely in comparison with that without ultrasound. On the other hand, the total area of pits (Fig.19 (b)) changed, corresponding well with that of the accumulated electric charge for all ultrasound conditions. Large difference cannot be found in the depth of pits under different ultrasound conditions (Fig.19 (c)). The pits in the case of 50 kHz looked like a little shallower than others. This should be attributed to the low precision of the dial gauge (minimum value: 0.01 mm) and attribute to the residual metallic covers on pits (Laycock et al., 1998) (see description later in Fig.21 and 22). In general, the depth of pits with ultrasound is much shallower than those without ultrasound.

Fig.19 (d) shows the cavitation power during applying ultrasound with different frequencies and distances from the vibrator. According to the former report (Wang & Kido, 2009) , at the frequency of 19.5 kHz and at the distance of 76 mm, the cavitation power kept increasing with the increase in the input power from I = 0 to 8. In this work, at the frequency of 19.5 kHz and the input power of I = 8, the cavitation power kept increasing with the increase in the distance from 19 mm and reached the maximum value of about 9.5 kJ/(m^2 • s) at the distance of d = 57 mm. However, the power decreased with further increasing the distance and obtained the value of 2.2 kJ/(m^2 • s) at d = 95 mm. The largest cavitation power at d = 57 mm corresponds to the largest suppression effect on the growth of pitting at this frequency. On the other hand, the cavitation power at the frequency of 50 kHz did not show large change from the distance of 14.8 mm to 59.2 mm, which also corresponds well to the not largely changed suppression effect on pitting corrosion in this case. In the case of 420 kHz, the cavitation power at 17.6 mm and 35.2 mm was almost the same and that at 70.4 mm became a little small. This corresponds to the suppression effect on corrosion, where the largest effect was obtained at 17.6 mm in this case. On the other hand, the power at 50 kHz was the largest, with about 7~35 times of that of 19.5 kHz and 420 kHz. The power at 420 kHz was the smallest.

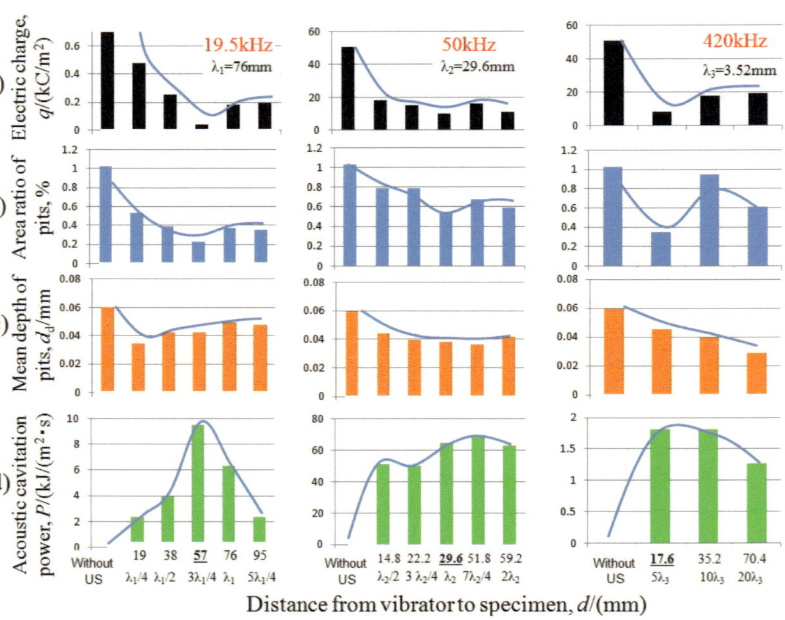

Fig. 19. Electric charge during potential holding and ultrasound applying (a), pit area ratio (b), pit depth (c) after polarization and cavitation power (d) during ultrasound applying. Copyright 2011 Japan Society of Corrosion Engineering

It is known as above that the suppression effect on the pitting corrosion increased with the increase in the cavitation power when the power is less than 10 kJ/(m^2 • s) in the case of 19.5 kHz and 420 kHz. The optimum condition to suppress corrosion appeared at 19.5 kHz at the distance of 57 mm. However, when the cavitation power is larger than 50 kJ/(m^2 • s) in the case of 50 kHz, the much large cavitation power is not helpful to increase the suppression effect. This result should be attributed to the simultaneous damage of the passive film on the specimen surface. Of course, during a long period measurement, the increased temperature in the solution will also influence the corrosion rate to some extent. That is, the transmission power depends on both the input power to vibrator and the ultrasound wave phase.

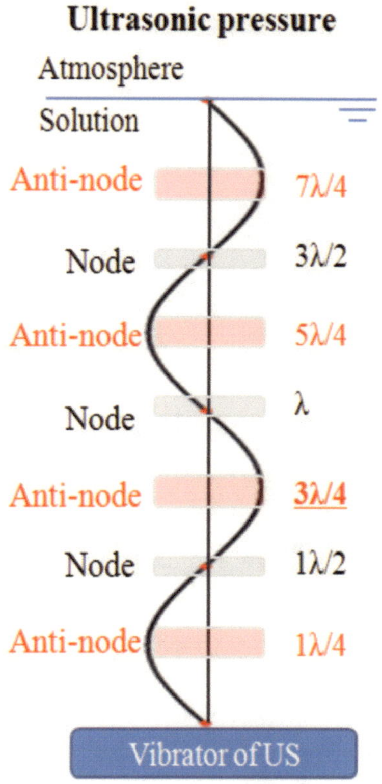

Fig. 20. Standing wave of ultrasonic pressure ultrasound from the vibrator to the solution surface. The large ultrasonic pressure generally appears at anti-nodes.

Fig.20 shows the simulated standing wave of ultrasonic pressure in the solution from the vibrator to the solution surface. Since the distance between the vibrator and the solution surface is set as several times of the half-wavelength of ultrasound, the large ultrasonic pressure generally appears at positions from either the solution surface or the vibrator surface with the interval of half-wavelength. Such positions of $\lambda/4$, $3\lambda/4$, $5\lambda/4$ and $7\lambda/4$ from the vibrator are called as anti-nodes, where large cavitation theoretically occurs. This can be used to well explain the largest cavitation power and the largest suppression effect of corrosion at the $3\lambda_1/4$ in the case of 19.5 kHz. However, large cavitation power and suppression effect did not appear at other anti-nodes, which should also be influenced by the energy attenuation of ultrasound in the solution. In the case of 50 kHz and 420 kHz, large cavitation power and suppression did not obtain at either of anti-nodes. In the case of 420 kHz, the wavelength is too small to precisely discuss the influence of position.

In the case of 19.5 kHz, the suppression effect of corrosion increased with the increase in the cavitation power when the power is less than 10 kJ/(m^2 • s). The effect should be attributed to the increased stirring effect of solution in pits after removing the corrosion products and the metallic covers. The cavitation power is generally related to the bubbles' size decided by the tensile stress and the sum of the collapsed bubbles. In general, the bubble size becomes smaller with the increase in the frequency of ultrasound and thus the happening of cavitation becomes difficult. This should be the reason to get the small cavitation power in the case of 420 kHz, which brought about weak suppression effect of corrosion. On the other hand, the much large cavitation power in the case of 50 kHz should be attributed to the increased sum of collapse of bubbles rather than the balance of the decreased bubbles' size.

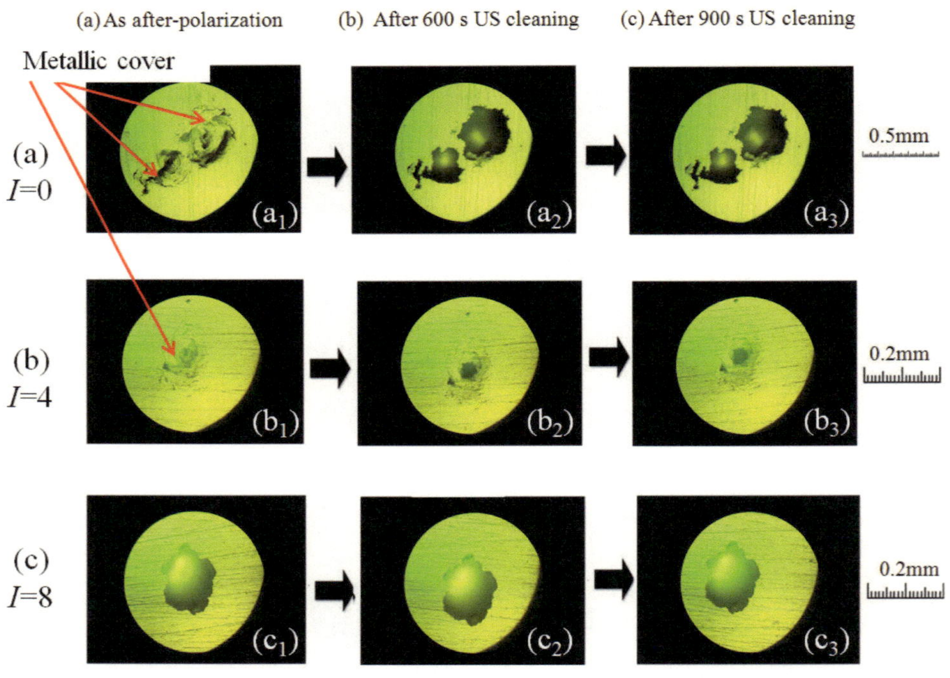

Fig. 21. Surface morphology of specimens after simultaneously holding potential and applying ultrasound at $I = 0$ (a), $I = 4$ (b) and $I = 8$ (c) at frequency of 19.5 kHz. The surfaces a_1, b_1 and c_1 are cleaned only by running distilled water; a_2, b_2, c_2 and a_3, b_3, c_3 are further cleaned in distilled water by an ultrasound cleaner for a total time of 600 s and 900 s.

Fig. 21 shows the specimen surfaces after the polarization with ultrasound at $I = 0$, 4 and 8 at the frequency of 19.5 kHz. The surfaces (a_1, b_1, c_1) are rinsed only by running distilled water, while the surfaces (a_2, b_2, c_2 and a_3, b_3, c_3) are further cleaned in distilled water in an ultrasound cleaner with multi-vibrators (Fig.1) for 600 s or 900 s. Corrosion products were not found on the specimen without ultrasound (a), indicating that not all corrosion products reside on pits during the corrosion process. On the other hand, the metallic cover was left on the pits when ultrasound was applied at I = 0 and 4 (a, b), while it disappeared at I = 8. This means that the metallic cover forms during the corrosion and they can be removed when the applied ultrasound power is strong enough. Such removal of metallic cover and corrosion products is the reason of the suppression of the pitting corrosion. The depth of pit after further ultrasound cleaning was measured again and the results were shown in Fig.22. Of course the mean depth became larger than that before the ultrasound clean, which surely decreased with the increase in the input power from I = 0 to 8 but didn't verify much with the change of vibrator-to-specimen distance.

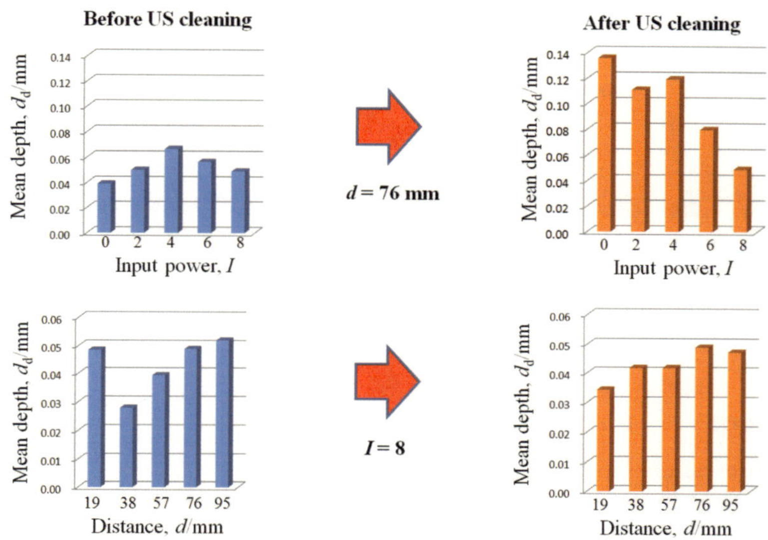

Fig. 22. Mean depth of pits before and after ultrasound cleaning on specimens, which were polarized by simultaneously holding potential and applying ultrasound with frequency of f_1 = 19.5 kHz.

Fig. 23 shows the re-passivation behaviour of the specimen when the potential was swift back from 500 mV at 19.5 kHz and d = 57 mm. In the case of without ultrasound (a), the re-passive potential appeared at -93 mV with a necessary accumulated charge of 218 kC/m^2, while the potential became 316 mV with only a much smaller charge of 1 kC/m^2. This promoted re-passivation behaviour by ultrasound is attributed to the remove of corrosion products and the metallic cover with the disturbing of the solution in pits.

8. Future works

According to the above results, it is clear that the pitting corrosion can be suppressed by either type of ultrasound. However, the influence on the initiation of pits has not been known, which will be investigated in the near future. Except the above physical effect of ultrasound application in solution, a chemical effect should also be considered. When the ultrasound is applied to an aqueous solution, water can be decomposed to H • and OH • radicals, and the solution becomes weak acidic (Jana & Chatterjee, 1995). According to the report of Jana, about 3×10^{21} /m^3 radicals are produced by the ultrasound (frequency: 20 kHz, intensity: 190 k W/m^2) and four OH • transfer one Fe^{2+} ion to one Fe^{3+}. The ultrasound intensity in this research was about 3~10 kW/m2, and the application time is short. Therefore, the amount of radicals produced will be not large. However, it is possible that there is a contrary influence of radicals on the pitting corrosion behaviour considering the breakdown of passive film by lowered pH solution by H • radical and the re-passivation of the broken passive film by the OH • radical with strong oxidation ability.

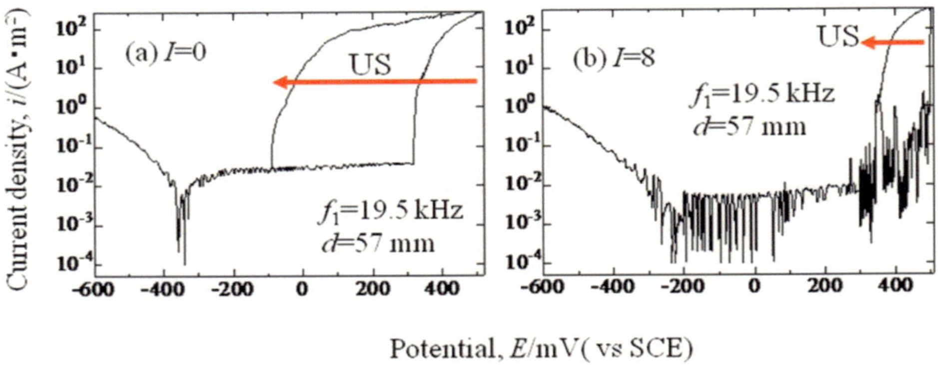

Fig. 23. Polarization curves with reversely sweeping of the potential at 500 mV, simultaneously without or with ultrasound with f_1 = 19.5 kHz at d = 57 mm and I = 8.

9. Conclusions

The influence of ultrasound in solution on the corrosion behaviour of stainless steel was introduced basing on our results and other literatures. It has known that when the acoustic cavitation caused by ultrasound is strong enough, the passive film can be damaged and thus corrosion is activated. However, when the acoustic cavitation caused by ultrasound is not strong enough to damage the passive film the corrosion will not be accelerated. As for the suppression effect of ultrasound on the stainless steel of Type 304 stainless steel, the following conclusions were obtained.

1. In case of pitting corrosion of Type 304 steel, the corrosion product was in-situ confirmed on the growing pits at the early stage. When the corrosion product was removed by the probe of AFM, the growth rate of pits largely decreased, which was explained by the decrease in the concentration of chloride and hydrogen ions in pits.
2. The cathode current, passive current and corrosion potential in the polarization curve were not almost changed by the application of ultrasound. However, the growth of pitting corrosion and crevice corrosion of Type 304 stainless steel can be suppressed by ultrasound with 19.5 kHz ultrasound. The change of the current density almost synchronized with the cyclic application and stop of ultrasound. In the case of 19.5 kHz at a constant vibrator-to-specimen distance of d = 76 mm, the suppression effect on pitting corrosion increased with the input power to vibrator.
3. The suppression effect of the growth of pitting corrosion was different when changing either the frequency or the distance from ultrasound vibrator to the specimen with either of frequencies of 19.5, 50 and 420 kHz. The largest suppression effect in this work was obtained at 19.5 kHz at the vibrator-to-specimen distance of 57 mm at the input power of I = 8 to vibrator.

4. The suppression effect of the growth of pitting corrosion became large with the increase in the cavitation power when the power is less than 10 kJ/(m^2·s). However, in the case of 50 kHz, the cavitation was strong enough to damage the passive film, which weakens the suppression effect on the corrosion. In the case of 420 kHz, the effect on the suppression of corrosion was weak due to small cavitation power.
5. The ultrasound promotes the re-passivation of pits by not only removing corrosion products but also removing the metallic cover.

10. Acknowledgment

This work is a summary of our recent researches on *suppression of corrosion of stainless steel by ultrasound*, which were carried out in Hiroshima University (2004.4~2005.3) and in Hiroshima Institute of Technology (2005.4~2011.7). Professor K.Nakasa gave important advices and discussion in this work. Dr. Q.Zhang helped the in-situ observation of pitting corrosion by atomic force microscope. The author is very grateful to Mr. T.Tamai, Mr. A.Yamamoto, Mr. T.Kangai, Mr. H.Morishita, Mr. S.Nagai, Mr.Y.Odagami, Mr. K.Etsuki, Mr. M.Kimura, Mr. H.Doi, Mr. G.Nishida, Mr. Y.Shibatani and Mr. S.Fujii for their assistance of experiments. Part of this work was supported by MEXT.HAITEKU, 2004~.

11. References

Ryan, M.P.; Williams, D.E.; Chater, R.J.; Hutton, B.M. & Mcphail, D.S. (2002). Why stainless steel corrodes. *Nature*, Vol.415, No.6873, (February 2002), pp.770-774, ISSN 0028-0836

Shimizu, K. (2010). New Role of a Low-Voltage, Ultra-High Resolution FE-SEM for Corrosion Studies (2) – Application examples -. *Zairyo-to-Kankyo / Corrosion Engineering of Japan*, Vol.59, No.7, (July 2010), pp.245-250, ISSN 0917-0480

Shimizu, K. (2010). New Role of a Low-Voltage, Ultra-High Resolution FE-SEM for Corrosion Studies (1) – Sample Surface Preparation for Ultra-High Resolution FE-SEM -. *Zairyo-to-Kankyo / Corrosion Engineering of Japan*, Vol.59, No.10, (October 2010), pp.360-365, ISSN 0917-0480

Yashiro, H. & Shimizu, K. (2010). *Poc. 57th Japan Conf. Materials and Environments, JSCE*, pp.175-178

Hisamatsu, Y. (1981). Locallized Corrosion of Iron-Nickel-Chromium Alloys – Pitting, Crevice Corrosion, and Stress Corrosion Cracking -. *Bulletin of the Japan Institute of Metals*, Vol.20, No.1, (1981), pp.3-11, ISSN 1340-2625

Zhang, Q.; Wang, R.; Kato, M. & Nakasa, K. (2005). Observation by atomic force microscope of corrosion product during pitting corrosion on SUS304 stainless steel. *Scripta Materialla*, Vol.52, No.3, (February 2005), pp.227-230, ISSN 1359-6462

Wang, R. & Kido, M. (2009). Influence of input power to vibrator and vibrator-to-specimen distance of ultrasound on pitting corrosion of SUS304 stainless steel in 3.5% chloride sodium aqueous solution. *Corrosion Science*, Vol.51, No.8, (August 2009), pp.1604-1610, ISSN 0010-938X

Chouonpa Binran Henshu Iinkai (1999). *Hand Book of Ultrasonic Wave*, Maruzen, ISBN 4-621-04663-0 C 3055

Alkire, R.C. & Perusich, S. (1983). The effect of focused ultrasound on the electrochemicalpassivity of iron in sulfuric acid. *Corrosion Science,* Vol.23, No.10, (October 1983), pp.1121-1132, ISSN 0010-938X

Al-Hashem, A.; Caceres, P.G.; Riad, W.T. & Shalaby, H.M. (1995). Cavitation corrosionbehavior of cast nickel–aluminum bronze in seawater. *Corrosion,* Vol.51, No.5, (May 1995), pp.331-342, ISSN 0010-9312

Whillock, G.O.H. & Harvey, B.F. (1996). Preliminary investigation of the ultrasonicallyenhanced corrosion of stainless steel in the nitric/chloride system. *Ultrasonics Sonochemistry,* Vol.3, No.2, (1996), pp. S111-S118, ISSN 1350-4177

Kwok, C.T.; Cheng, F.T. & Man, H.C. (2000). Synergistic effect of cavitation erosion andcorrosion of various engineering alloys in 3.5% NaCl solution. *Materials Science Engineering* A, Vol.290, No.1-2, (October 2000), pp.145-154, ISSN 0921-5093

Whillock, G.O.H. & Harvey, B.F. (1997). Ultrasonically enhanced corrosion of 304Lstainless steel II: the effect of frequency, acoustic power and vibrator-to specimen distance. *Ultrasonics Sonochemistry,* Vol.4, No.1, (January 1997), pp.33-38, ISSN 1350-4177

Nakayama, T. & Sasa, K. (1976). Effect of ultrasonic waves on the pitting potential of 18–8 stainless steel in sodium chloride solution. *Corrosion,* Vol.32, No.7, (July 1976), pp.283-285, ISSN 0010-9312

Wang, R. & Nakasa, K. (2007). Effect of ultrasonic wave on the growth of corrosion pits on SUS304 stainless steel. *Materials Transactions,* Vol.48, No.5, (May 2007), pp.1017-1022, ISSN 1345-9678

Wang, R. & Kido, M. (2008). Influence of application of ultrasound on corrosion behavior of SUS304 stainless steel with crevice. *Materials Transactions,* Vol.49, No.8, (August 2008), pp.1806-1811, ISSN 1345-9678

Wang, R. (2008). Influence of ultrasound on pitting corrosion and crevice corrosion of SUS304 stainless steel in chloride sodium aqueous solution. *Corrosion Science,* Vol.50, No.2, (February 2008), pp.325-328, ISSN 0010-938X

Wang, R. (2011). Growth behavior of pitting corrosion of SUS304 stainless steel in NaCl aqueous solution when applying ultrasound with different frequencies. *Zairyo-to-Kankyo / Corrosion Engineering of Japan*, Vol.60, No.2, (February 2011), pp.66-68, ISSN 0917-0480

Wranglen, G. (1985). *An Introduction to Corrosion and Protection of Metals*. Chapman and Hall, ISBN 0412260409

Japanese Industrial Standards Committee (2002). *JIS G0592:2002,* Method of determining the repassivation potential for crevice corrosion of stainless steels

Mitome, H. (2008). Generation of acoustic cavitation and its application. *Journal of the Japan Society of Mechanical Engineering,* Vol.111, No.1074, (May 2008), pp.32-35, ISSN 0021-4728

Laycock, N.J.; White, S.P.; Noh, J.S.; Wilson, P.T. & Newman, R.C. (1998). Perforated covers for propagating pits. *Journal of the electrochemical society,* Vol.145, No.4, (April 1998), pp.1101-1108, ISSN 0013-4651

Jana, A.K. & Chatterjee, S.N. (1995). *Ultrasonics Sonochemistry,* Vol.2, No.2, (1995), pp.s87–s91, ISSN 1350-4177

13

New Trends in Materials Nondestructive Characterization Using Surface Acoustic Wave Methodologies

T. E. Matikas and D. G. Aggelis
Materials Science & Engineering Department,
University of Ioannina, Ioannina,
Greece

1. Introduction

The surface of the materials is usually the most sensitive part due to exposure to environmental influence, as well as higher bending and torsional loads than the interior. Therefore, degradation is bound to initiate from the surface in most engineering components. Surface wave propagation in heterogeneous media is a topic concentrating many efforts in the engineering community. The main aim is quality characterization via nondestructive evaluation (NDE) methodologies by correlation of propagation characteristics with material properties. In the present chapter surface waves are examined in structural materials of outmost significance such as aerospace composites and concrete.

Concrete structures are exposed to deterioration factors like weathering, corrosive agents, thermal expansion and contraction or even freezing and thawing. Additionally, they support operation loads, own weight and possibly dynamic overloading by earthquakes. Most of the above factors affect primarily the surface of structures, which is directly exposed to the atmospheric conditions and sustain maximum flexural loads. Deterioration therefore, is bound to start from the surface in most cases. This deterioration may be manifested in the form of large surface breaking cracks and/or distributed micro-cracking in the surface layer of the material. Inspection techniques based on the propagation of elastic waves have been long used for the estimation of the quality and general condition of the material [1,2] either in through the thickness or in surface mode. Surface wave propagation is complicated in that different kinds of waves co-exist. Normally the Rayleigh waves occupy most of the energy, while the longitudinal are the fastest [3,4]. Therefore, measuring the transit time of the first detectable disturbance of the waveform, leads to the calculation of the longitudinal wave velocity. This is referred to as pulse velocity [1-3] and it is widely used for rough correlations with quality. Other forms of wave speed are the phase velocity, which is calculated either by some characteristic point in the middle of a tone-burst of a specific frequency [5,6], or by spectral analysis of a broadband pulse [7]. Additionally, group velocity is calculated by the maximum peak, the maximum of the wave envelope, or cross-correlation between the "input" and "output" waveforms [8,9]. In homogeneous media all these forms of velocity are expected to share the same value. However, for inhomogeneous media it has been shown that these velocities are not necessarily close [5,10].

From the above forms of velocity, the most common measurement in ultrasonics is the "pulse velocity". Considering that the material is homogeneous, pulse velocity is directly related to the elasticity modulus [11] and correlated to the strength of concrete through empirical relations [1,11-13]. Since it is measured by the first detectable disturbance of the waveform this measurement depends on the strength of the signal with respect to the noise, which could have environmental and equipment-induced components. In case the initial arrivals of the wave are weaker than or similar to the noise level, pulse velocity is underestimated. This could certainly be the case in actual structures, where propagation distances through damaged materials are usually long. Rayleigh waves are also excited in a concrete surface; they propagate within a penetration depth of approximately one wave length and carry more of the excitation energy [3,14]. Their velocity is also related to elasticity and Poisson's ratio. Measurement of Rayleigh velocity is usually conducted by a reference peak point, so it is not directly influenced by noise level [11]. However, for cases of severe damage or long propagation, the strong reference cycle used for the measurement is severely distorted making the selection of reference points troublesome [8,15], as will be seen later. Frequency domain techniques like phase difference calculation between signals recorded at specific distances may provide solution for velocity measurement revealing also the dependence of velocity on frequency [7,16].

In addition to wave velocity, attenuation has also been widely used for characterization of microstructural changes or damage existence [17,18]. It represents the reduction of the wave amplitude per unit of propagation length. Attenuation is more sensitive to damage or void content as has been revealed in several studies [5,8,18-20] and has been correlated to the size of the aggregates, as well as air void size and content in hardened and fresh cementitious materials [5,18]. The sensitivity of attenuation to the microstructure is such, that the content of "heterogeneity" is not the only dominating factor; the typical size and shape of the inclusions play an equivalently important role and therefore, Rayleigh wave attenuation has been related to parameters like aggregate size, and damage content [8,21-23]. This sensitivity to the microstructure may complicate assessment but on the other hand offers possibilities for more accurate characterization.

Accurate characterization would require determination of several damage parameters like the number (or equivalent damage content) of the cracks, their typical size, as well as their orientation. Though this is a nearly impossible task, especially in-situ, advanced features sensitive to the above damage parameters should be continuously sought for in order to improve the maintenance services in structures. The valuable but rough characterization based on pulse velocity can be improved by the addition of features from frequency domain as will be explained below.

Elastic waves are applied on the surface of mortar specimens with inhomogeneity. In order to simulate damage, small flakey inclusions were added in different contents and sizes, and parameters of the surface waves are calculated. The effect of "damage" content as well as the size of the inclusions on wave parameters is discussed showing that the typical size of inhomogeneity is equally important to the content. In similar studies, sphere is a first approximation for the damage shape, which produces quite reliable results [24]. However, for the case of actual cracks, which were simulated by thin, flakey, light inclusions, the random orientation and general shape complicates the anticipated results and increases experimental scatter. This study aims to supply more experimental data in the area of surface waves in media with random and randomly oriented inhomogenity, which has not been studied as widely as stratified media [11,25] and media with surface breaking cracks [26-29]. The results presented herein are realistic due to the shape of the light inclusions, while size and population of inclusions leave their fingerprint on the phase velocity and attenuation curves since large population of small inclusions impose stronger attenuation that small population of larger size. Additionally the application of other features from frequency domain, like the coherence function, is discussed as to their contribution in damage characterization.

Concerning aerospace materials, innovative NDE methods based on linear and nonlinear acoustics are of outmost importance for developing damage tolerance approaches by monitoring the accumulation of damage under cyclic loading.

High strength titanium alloys, as well as fiber reinforced metal matrix composite materials, are being considered for a number of applications because of their improved mechanical properties in high temperature applications. In applications where cyclic loading is expected and where life management is required, consideration must be given to the behavior of the material in the vicinity of stress risers such as notches and holes. It is in these regions that damage initiation and accumulations are expected. In the case of metal matrix composites for aircraft structural and engine components, several damage modes near stress risers have been identified [30]. One important damage mode under cyclic loading is the nucleation and growth of matrix cracks perpendicular to the fiber direction. In some composite systems, the matrix crack growth occurs without the corresponding failure of the fibers. This process results in the development of relatively large matrix cracks that are either fully or partially bridged by unbroken fibers. The presence of bridging fibers can significantly influence the fatigue crack growth behavior of the composite. To develop a life prediction methodology applicable to these composite systems, an understanding must be developed of both the matrix cracking behavior as well as the influence of the unbroken fibers on the crack driving force and the effect of interfacial degradation and damage on the eventual failure of the composite.

Paramount to understanding the influence of unbroken fibers is to identify the mechanisms which transfer the load from the matrix to the fiber. The mechanics of matrix cracking and fiber bridging in brittle matrix composites has been addressed [31,32]. The analysis is based on the shear lag model to describe the transfer of load from the fiber to the matrix. In the shear lag model, the transfer of load occurs through the frictional shear force (τ) between the fiber and the matrix. The analyses indicate that size of the region on the fiber over which t acts can have a significant effect on the influence of unbroken fibers on crack growth rate behavior. However, although some indirect ultrasonic experimental techniques have been developed to determine the extent of influence of τ [33-35], no direct nondestructive experimental techniques currently exist. Another important interfacial phenomenon is the degradation, fracture and/or failure of the interface resulting from crack initiation and growth [36,37].

This chapter has an objective to discuss the utility and versatility of surface waves application in cementitious materials as well as two state-of-the-art surface acoustic wave techniques, ultrasonic microscopy and nonlinear acoustics, for material behavior research of aerospace materials.

Ultrasonic microscopy utilizes high focused ultrasonic transducers to induce surface acoustic waves in the material which can be successfully used for local elastic property measurement, crack-size determination, as well as for interfacial damage evaluation in high temperature materials, such as metal matrix composites [38]. Nonlinear acoustics enables real-time monitoring of material degradation in aerospace structures [39]. When a sinusoidal ultrasonic wave of a given frequency and of sufficient amplitude is introduced into a non-harmonic solid the fundamental wave distorts as it propagates, and therefore the second and higher harmonics of the fundamental frequency are generated. Measurements of the amplitude of these harmonics provide information on the coefficient of second and higher order terms of the stress-strain relation for a nonlinear solid. As it is shown here, the material bulk nonlinear parameter for metallic alloy samples at different fatigue levels exhibits large changes compared to linear ultrasonic parameters, such as velocity and attenuation [40].

2. Damage characterization in concrete using surface waves

In order to characterize damage, several parameters are required. These include the quantification of: damage content, typical size of cracking or even distribution of sizes and possibly orientation preference. This task is nearly impossible in real structures or even in laboratory due to several variables that need to be determined. However, dealing with some of the aspects of characterization is extremely important bearing in mind that so far, the assessment of concrete condition is based on the rough and empirical correlations between pulse velocity and strength. In the next part, the simulated damage content and typical size of cracks in concrete are examined in terms of their influence on different parameters of the propagating wave studied complementary to the conventional pulse velocity.

2.1 Experimental process

Concerning the part of the study relevant to cementitious materials, the experimental specimens were made of cement mortar with water to cement ratio of 0.5 and sand to cement 2 by mass. The maximum sand grain size was 3 mm. Two minutes after the ingredients were mixed in a concrete mixer, vinyl inclusions were added and the mixing continued for two additional minutes. Then the mixture was cast in square metal forms of 150 mm side. The specimens were demolded one day later and cured in water for 28 days [41]. The vinyl inclusions were added in different volume contents (specifically 1%, 5% and 10%), while a mortar specimen was also cast without any inclusions. Vinyl inclusions were cut in small square coupons from sheets with thickness of 0.2 mm and 0.5 mm. The exact sizes of the vinyl coupons were 15x15x0.5 mm, 15x15x0.2, 30x30x0.5 and 30x30x0.2 mm. One specimen was cast for each inclusion size and content.

After fracturing one preliminary specimen, it was revealed that the dispersion of the fillers can be considered random, as would possibly be the case for actual cracks. Although the total volume content of the fillers was strictly measured to the specified value (i.e. 1%, 5% and 10%) in order to exclude local variations twenty measurements were taken on the surface and the results were averaged.

For the ultrasonic measurements, three broadband piezoelectric transducers, Fujiceramics 1045S, were placed on the specimens' surface with a distance of 20 mm, see Fig. 1. A pencil lead break was excited (HB 0.5) producing frequencies in the bandwidth of 0-300 kHz, in front of the first receiver. The signals received by the three sensors were preamplified by 40 dB and digitized with a sampling rate of 10 MHz in a Mistras system of Physical Acoustics Corporation. Silicone grease was used as couplant between the sensors and the specimen.

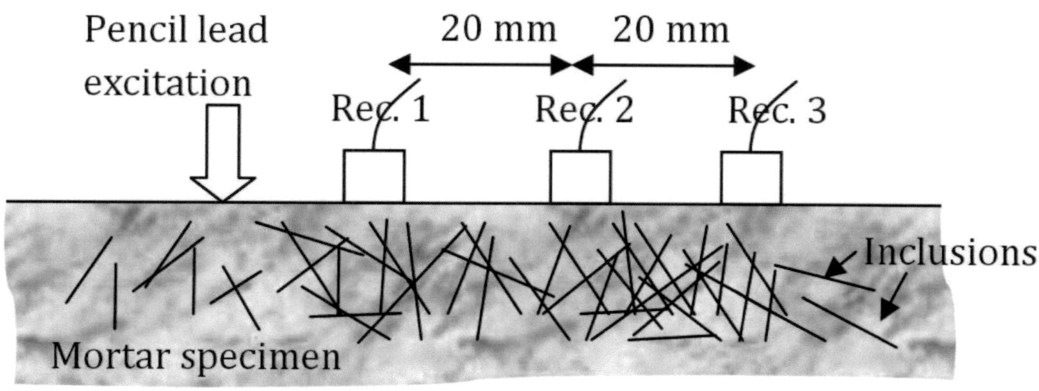

Fig. 1. The experimental setup.

2.2 Shape distortion and pulse velocity

Figure 2a shows typical waveforms as captured by the first and last receivers on a plain mortar specimen. In both receivers the strong Rayleigh cycle is observed, preceded by the weaker longitudinal mode. Measuring the delay between the strong negative peaks (Fig. 2a), the Rayleigh velocity is calculated at 2116 m/s. When simulated damage is included in a content of 1%, the Rayleigh cycle is slightly distorted (Fig. 2b). The Rayleigh velocity in this case is measured at the value of 2062 m/s. It is mentioned that the pulse velocity, dictated by the first detectable disturbance of the waveforms, is essentially similar for both cases measured at around 4140 m/s, meaning that Rayleigh waves are more sensitive to the slight amount of simulated damage than longitudinal. When the content of damage is 10% (see Fig. 2c) the Rayleigh cycle is severely distorted and a reference point cannot be easily selected for the measurement. In case the minimum point of the 2nd cycle is chosen (marked by an arrow), the Rayleigh velocity is calculated at 1770 m/s, being reduced by 17% relatively to the sound material, while pulse velocity is measured at 3773 m/s, with a corresponding decrease of 8.5%.

Figure 3 depicts the average longitudinal wave velocity for different types of inclusions and constant content of 5% normalized to the velocity of plain mortar. As observed, the different size has considerable effect on the velocity since the inclusions of 15x15x0.5 mm size reduce the velocity to approximately 96.5% of plain mortar while inclusions of 30x30x0.2 mm reduce the velocity even to 83%. This shows that the material cannot be treated as homogeneous considering only the volume fraction of inclusions (or cracks in an actual situation), since their shape and size is also important. The velocities presented come from the average of 20 individual measurements.

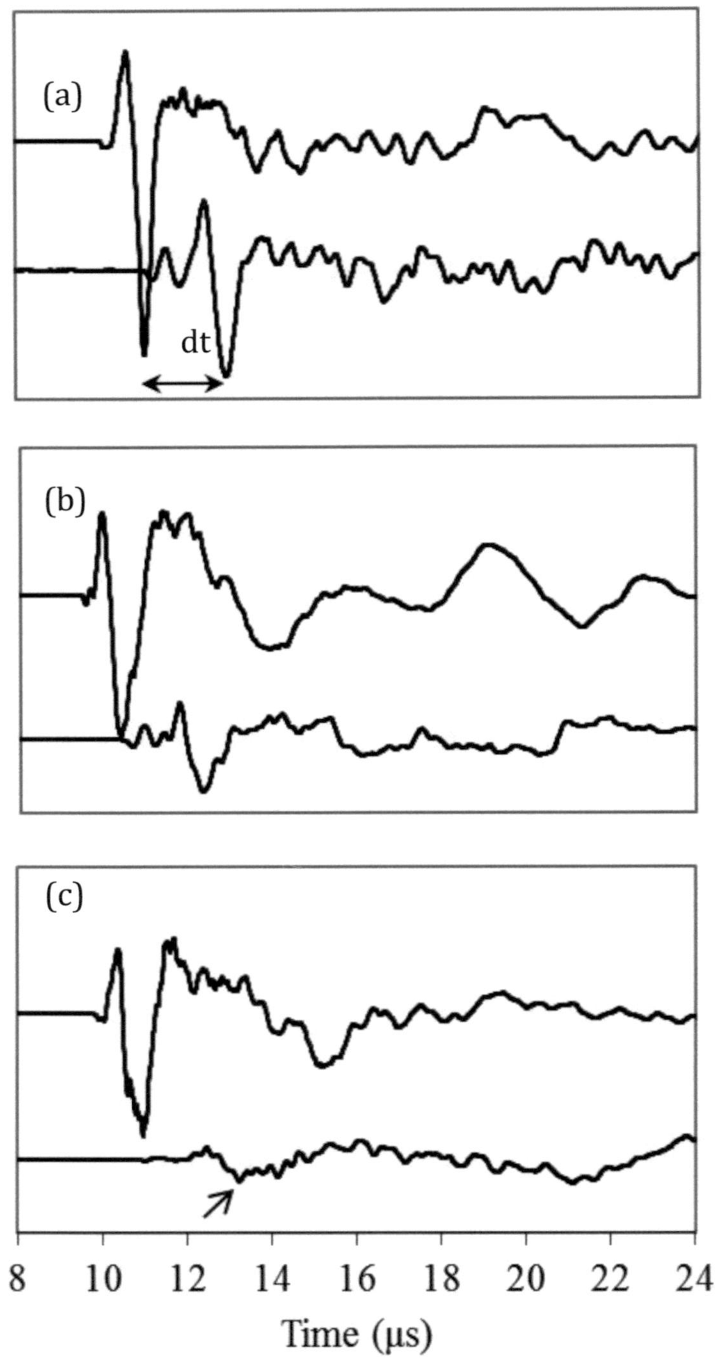

Fig. 2. Sample waveforms of 1st and 3rd receiver for mortar with inclusions contents 0% (a), 1% (b) and 10% (c) and inclusion size 30x30x0.5 mm.

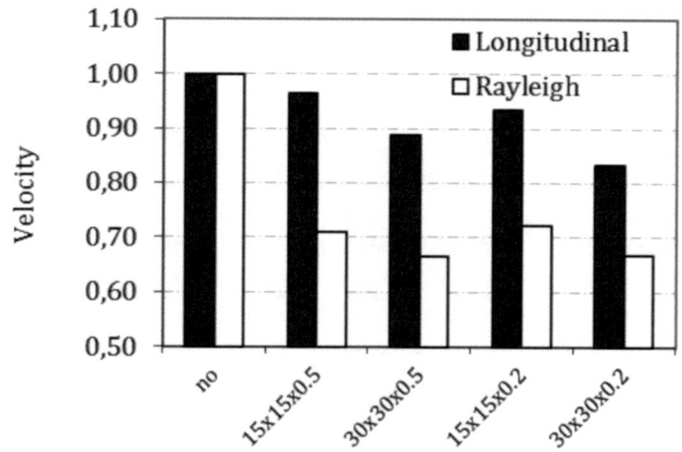

Fig. 3. Wave velocities for different inclusion sizes and constant content (5%).

As to the Rayleigh wave propagation, if a reference point can be identified, e.g. the strong negative peaks in Fig. 2(a), the Rayleigh velocity can be calculated easily. However, in case of strong distortion, case of Fig. 2(c) selection of a single point would not be safe. Therefore, an approach with cross-correlation was followed. The time lag resulting after cross correlation between the signals, over the separation distance between the first and third transducer (40 mm), gives a measure of the velocity with which the energy propagates [8]. For this task the first 100 μs of the waveforms were used where the Rayleigh contribution is certain to exist.

The results are shown again in Fig. 3 normalized to the plain mortar Rayleigh velocity of 2116m/s. The Rayleigh velocity follows similar diminishing trend with the longitudinal, being however more strongly reduced, to even 66% of the plain mortar, for the case of 30x30x0.2 inclusions. This again shows the importance of the scatterer size. Additionally, the differential influence of inclusion shape on longitudinal and Rayleigh wave velocities should be highlighted. This is another indication of the inhomogeneous nature of the material. The addition of lower elasticity inclusions reduces the "effective" modulus of elasticity of mortar. In case the material could be regarded as homogeneous, the influence on longitudinal and Rayleigh velocities should be the same, since they are both firmly connected to the elastic properties. However, for the material at hand, the Rayleigh is obstructed much more intensively, especially by the thin inclusions, showing that a traditional homogenized approach used for concrete is not adequate for cementitious material with inclusions. This should also be related to the propagation mechanism of Rayleigh waves, which includes displacement components in two directions (parallel and vertical to the direction of propagation) being therefore more sensitive to inhomogeneity.

2.3 Phase velocity

The above mentioned approach yields a measure of the velocity of the whole pulse. In a heterogeneous medium like the one studied herein, the velocity is expected to be strongly dependent on the frequency. Therefore, it is significant to calculate the dispersion curve (phase velocity vs. frequency). In the case of Rayleigh waves this is not trivial since there are always contributions from longitudinal and shear waves that are faster. Therefore, the Rayleigh cannot be isolated. However, since they generally carry more energy than the other types, concentrating on a time window where the Rayleigh are expected, can yield information about this wave with little influence from other types. In this case a window of 30 μs located around the major Rayleigh arrivals was isolated and the rest of the waveform was zero-padded as presented in similar cases [42]. Using Fast Fourier Transform, the phase of the waveform is calculated and unwrapped. Therefore, the difference of phase between waveforms collected at different distances from the excitation (i.e. the first and third receiver) leads to the calculation of phase velocity vs. frequency curve [7]. The results are depicted in Fig. 4 for materials with different inclusion type and volume content 5%. Each curve is the average of 20 individual curves in order to diminish variation effects.

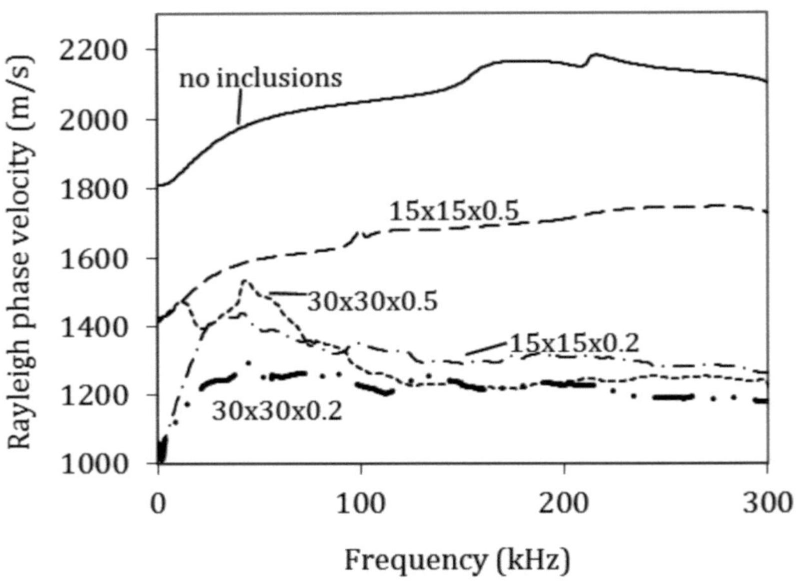

Fig. 4. Phase velocity vs. frequency curves for different size of inclusions and content 5% by vol.

It is seen that even plain mortar exhibits dispersive behavior with velocity increasing throughout the first 200 kHz. The dispersion curve for inclusions with dimensions 15x15x0.5 is translated to lower levels by about 400 m/s. However, other inclusion sizes seem to have much stronger influence, with the large and thin inclusions (30x30x0.2) lowering the curve by more than 800 m/s in average. Additionally, all the curves exhibit strong velocity increase for the band up to 50 kHz or 60 kHz. For higher frequencies each curve seems to converge. This is a behavior generally observed in composite systems. For low frequencies, the velocity may exhibit strong variations or resonance peaks but, as the frequency increases, the variations seem to diminish [43-44] as is the case for this study.

It is mentioned that the level of the dispersion curve is not necessarily close to the value of Rayleigh velocity measured by cross-correlation. For plain mortar, the Rayleigh phase velocity curve seen in Fig. 4, averages at 2073 m/s being close to the Rayleigh velocity of

2116 m/s measured by cross-correlation. However, as the inhomogeneity increases, considerable discrepancies arise. Indicatively, for the inclusions of 30x30x0.2 mm the phase velocity curve of Fig. 4 averages at 1217 m/s, while the cross-correlation leads to a velocity of 1426 m/s, see Fig. 3. This discrepancy has been observed in other cases of cementitious materials and increases with the level of inhomogeneity, i.e. from cement paste to mortar with sand grains or concrete with large aggregates [5], as well as other strongly scattering systems [10]. This trend is reasonable because in heterogeneous, dispersive materials the pulse shape changes during propagation and the expressions of velocity may well differ as they depend on the reference points selected for the calculation.

2.4 Spectral distortion

Apart from the velocity decrease, which is approximately of the order of 35% for Rayleigh waves, the signal suffers strong distortion which is evident from the shape of the time domain waveform. The distortion of the spectral content can be evaluated by the coherence function [45]. The frequency dependent coherence $\gamma_{xy}(f)$ between two time domain waveforms x(t) and y(t) is a measure of the similarity of their spectral content and can be described as the corresponding of cross-correlation in the frequency domain. It is given by:

$$\gamma_{xy}^2(f) = \frac{|G_{xy}(f)|^2}{G_{xx}(f)G_{yy}(f)}, \quad 0 \le \gamma_{xy}^2(f) \le 1 \tag{1}$$

where $G_{xy}(f)$ is the cross-spectral density function between time domain waveforms x(t) and y(t), while G_{xx} and G_{yy} are the auto-spectral density functions of x(t) and y(t).

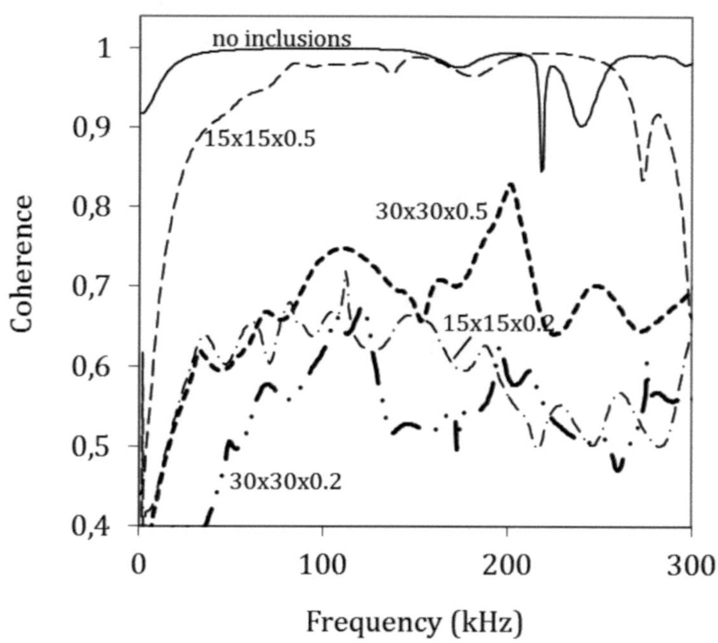

Fig. 5. Coherence function for different size of inclusions and content 5% by vol.

Figure 5 shows the coherence between signals collected at separation distance of 40 mm in mortar with different typical sizes of inhomogeneity. For no inclusions, the coherence is almost maximum (value close to 1), showing little distortion. Addition of inclusions in the content of 5% certainly diminishes the level of coherence but the result strongly depends on the size of the heterogeneity as well. Specifically for the size of 15x15x0.5 mm the average coherence is 0.92, while for 30x30x0.5 mm is 0.66. The lowest coherence is exhibited between signals recorded in specimens with 30x30x0.2 mm inclusions with an average value of 0.54. This shows the importance of the dominant size of inhomogeneity in spectral similarity and makes coherence a descriptor with strong characterizing power over the typical size of damage. It has also been used for characterization of concrete composition through ultrasonic signals and classification of acoustic emission signals [13,46].

2.5 Attenuation curves

The comparison of the peak amplitude of waveforms captured at different positions is a measure of attenuation. If the time domain signals are transformed into the frequency domain by fast Fourier transform (FFT) their response leads to the frequency dependent attenuation coefficient through the next equation:

$$a(f) = -\frac{20}{x}\log\left(\frac{A(f)}{B(f)}\right) \tag{2}$$

where a(f) is the attenuation coefficient with respect to frequency, A(f) and B(f) are the FFTs of the responses of the two sensors and x is the distance between the sensors (in this case 40 mm).

a(f) reveals the attenuative characteristics for specific frequency bands improving the characterization capacity, especially in controlled laboratory conditions, since the typical size of inhomogeneity affects specific wavelengths. Attenuation curves were calculated according to Eq. 2 for different inclusion contents and are seen in Fig. 6 for material with inclusion size of 30x30x0.5 mm. The curves exhibit strong fluctuations throughout the frequency band of 0 to 300 kHz. These fluctuations are difficult to be accurately evaluated and therefore, the curves are indicatively fitted by exponential functions, which although do not map the fluctuations, follow the general increasing trend. It is seen that plain mortar exhibits the lowest attenuation, while the attenuation curve of the 10% inclusions is the highest. The inclusions force an attenuation increase of approximately 200% - 300% compared to the attenuation of the plain material, while at the same time they were responsible only for a slight decrease of 35% in Rayleigh velocity, as was seen earlier. Additionally, the curve of 1% damage content is distinctly higher than plain material, showing the strong sensitivity of attenuation even to slight damage content.

As mentioned earlier, apart from the content, the size of the inhomogeneity plays an important role in the wave behavior and especially the attenuation. This is demonstrated in Fig. 7 where the attenuation curves for different sizes of inclusions are shown for the content of 5%. In general all "damaged" material curves exhibit an increase with frequency, as is expected for scattering media in a moderate frequency regime. The inclusions size 30x30x0.2 mm exhibits the highest attenuation in average (approximately 0.016 dB/mm), while 30x30x0.5 mm the lowest of all "damaged" specimens (0.013 dB/mm). The attenuation curve of sound material is much lower (average at 0.005 dB/mm). This attenuation behavior is a result of the combined effect of geometric spreading, damping and scattering. Geometric spreading has exactly the same effect on all curves, due to the same experimental conditions. Damping depends on the viscosity parameters of the constituents and therefore, for material with the same content of inclusions should not lead to strong changes. Therefore, the strong discrepancies between the curves within the graph, are attributed directly to scattering on the flakey inclusions. The different size of them imposes different scattering conditions and crucially affects the scattered wave field. For certain frequency bands differences of the order of 100% may arise depending on the size of the inclusions alone, even though the inclusion content is constant. This trend once again shows the complexity of wave propagation in damaged concrete and the need of multi-parameter approach for structural condition characterization. In general, it can be mentioned that 30x30x0.5 mm inclusion size exhibits the lowest attenuation curve. This is reasonable since for this shape, the volume of each particle is larger, and therefore, less individual inclusions are necessary to build the specified content. This leads to less scattering incidences as the wave propagates from the excitation point to the receivers, only moderately reducing the amplitude of the wave recorded.

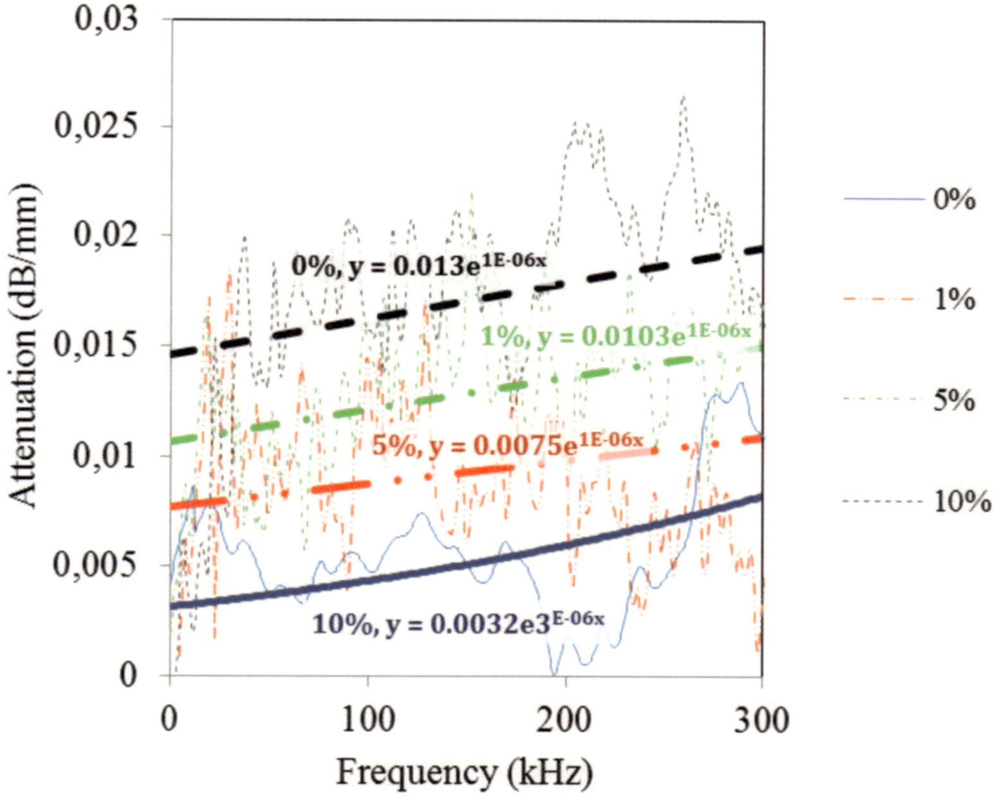

Fig. 6. Attenuation vs. frequency for mortar with different inclusion contents and inclusion size 30x30x0.5 mm.

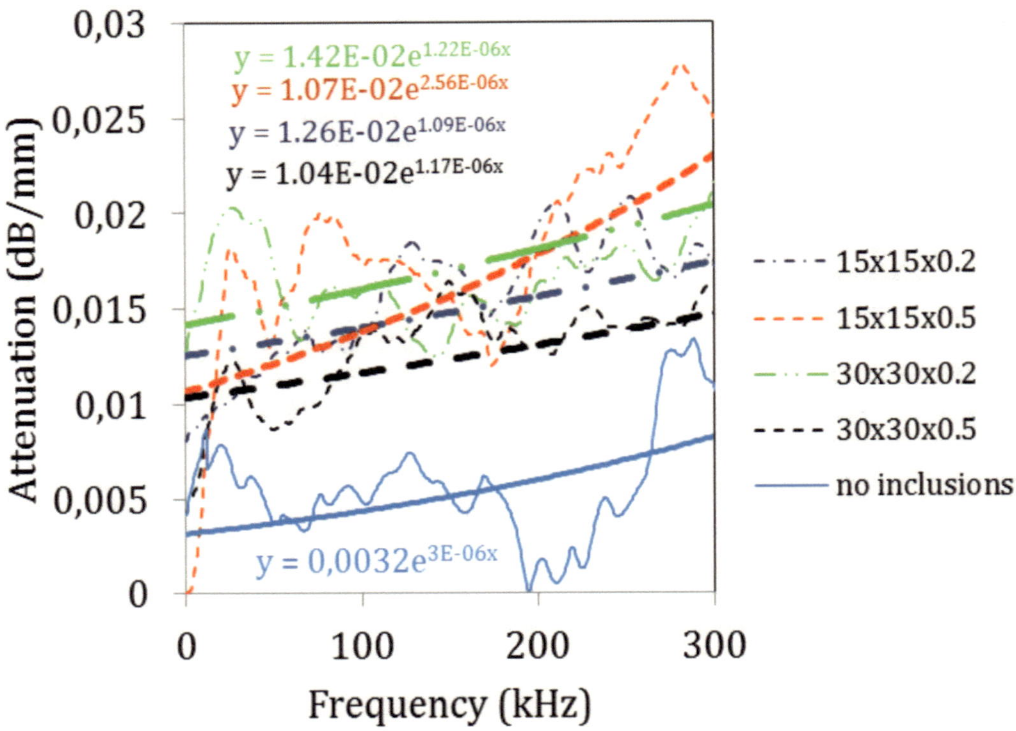

Fig. 7. Attenuation vs. frequency for mortar with different inclusion size and content 5%.

As a conclusion to the study on concrete it is worth to mention that surface wave propagation has not been previously studied for media with randomly distributed inhomogeneity. In this case surface wave features are used for accurate material characterization. Wave velocity of both longitudinal and Rayleigh waves exhibit close connection to the damage content as well as typical size. However, phase velocity vs. frequency curves prove more sensitive since they exhibit stronger decrease for the same damage content. Additionally spectral distortion can also be exploited by means of the coherence function and supply another descriptor sensitive to damage content and size. Finally, attenuation coefficient seems to be the strongest parameter since it suffers an increase of even 300 % compared to sound material. The merit from such a work would be the application of these parameters in real structures, since the complementary use of a combination of parameters will certainly enhance the characterization accomplished by pulse velocity alone.

3. Ultrasonic microscopy for characterization of damage in fiber-reinforced composites

The principle of operation of ultrasonic microscopy is based on the production and propagation of surface acoustic waves (SAW) as a direct result of a combination of the high curvature of the focusing lens of the transducer and the defocus of the transducer into the sample [47, 48]. The most important contrast phenomenon in this technique is based on the attenuation of Rayleigh waves which are leaking toward the transducer and are very sensitive to local mechanical properties of the materials being evaluated [36]. The generation and propagation of a leaky Rayleigh wave is modulated by the material's properties, thereby making it feasible to image even very subtle changes of the mechanical properties. The sensitivity of the SAW signals to surface and subsurface features depends on the degree of defocus and has been documented in the literature as the V(z) curves [49]. A V(z) curve is obtained when the transducer, kept over a single point, is moved toward the specimen. Then, the signal, rather than simply decreasing monotonically, can undergo a series of oscillations. The series of oscillations at a defocus distance can be associated with Rayleigh wave excitation and interaction of a SAW with the specular reflection received directly by the transducer. The Rayleigh wave velocity, v_R, can then be calculated using a simple relationship:

$$v_R = \frac{v_o}{\sqrt{1 - \frac{v_o}{2\omega\Delta z}}} \tag{3}$$

where, v_o is the sound velocity in the coupling medium, ω is the frequency of ultrasound, and Δz is the periodicity of the V(z) curve.

The defocus distance also has another important effect on the SAW signal obtained by the SAM transducer and dictates whether the SAW signal is separated in time from the specular reflection or interferes with it. Thus, depending on the defocus, the technique can be used either to map the interference phenomenon in the first layer of subsurface fibers, or to map the surface and subsurface features in the sample.

The conventional technique for measuring SAW velocity is based on a V(z) curve acquisition and analysis procedure utilizing a tone-burst system to interrogate the sample at a specific frequency using specially designed acoustic lenses. This technique requires calibration of the specific lens as well as of the response of the electronic circuit using a V(z) curve obtained from a lead sample, a material for which the Rayleigh wave has too low a velocity to be excited. This procedure requires specialized instrumentation, is time-consuming, and cannot be used for on-line measurements in interrupted testing mode. However, a novel ultrasonic microscopy method [37] overcomes the limitations of the conventional technique because it is based on automated SAW velocity determination via V(z) curve measurements using short-pulse ultrasound. The principle of ultrasonic microscopy is presented in Figure 8. A SAM transducer is schematically shown in Figure 1. The transducer used in ultrasonic microscopy has a piezoelectric-active element situated behind a delay line made of fused silica. The thickness of the active element is chosen to excite ultrasonic signals with a desired nominal frequency when an electrical spike voltage is delivered to the piezoelectric element. The silica delay has a highly focused spherical acoustical concave lens that is ground to an optical finish. A numerical aperture (NA; ratio of the diameter of the lens to the focal distance) of >1 (or F number—focal distance/diameter—of the lens <1) is essential for the ultrasonic microscopy technique to effectively generate and receive surface waves [50] in the sample being imaged.

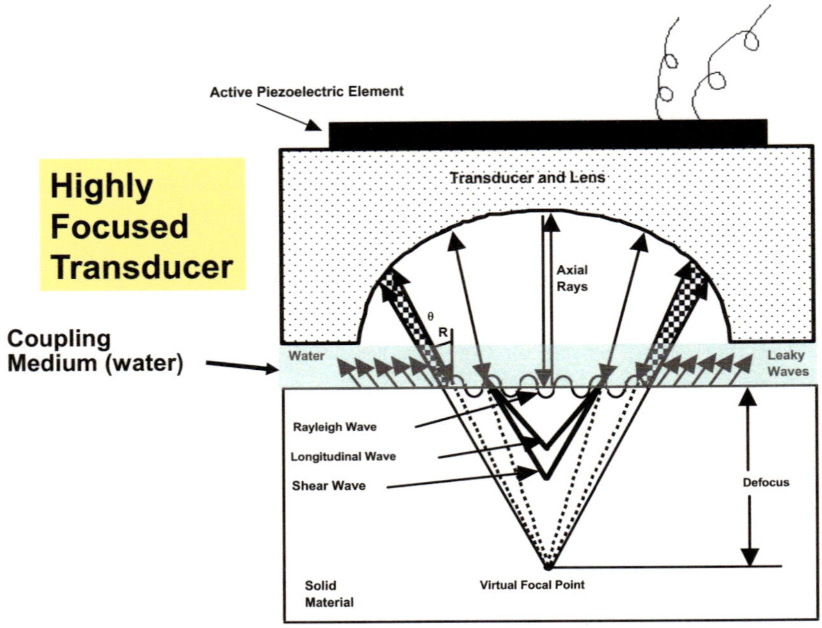

Fig. 8. Principle of ultrasonic microscopy.

The sensor used is a highly focused ultrasonic transducer with a central frequency of 50 MHz. The method employed here is self-calibrated, and is used to obtain Rayleigh velocity maps of the specimen through automated V(z) curve acquisition and analysis [37]. The resolution of the technique for characterizing individual fibers and determining interfacial properties strongly depends on the lens defocus from the surface of the sample. In addition, it should be underlined that the choice of the coupling medium is essential for resolving individual fibers in the composite, since, for a specific ultrasonic transducer with a fixed lens curvature, the generation of Rayleigh waves on the surface of the composite only depends on the sound velocities of the coupling medium and of the material under interrogation. Based on Snell's law, the curvature of a transducer's lens required to generate SAW in a material is given by the relationship,

$$\theta = \sin^{-1}\left(\frac{c_{coupling}}{c_{material}}\right) \tag{4}$$

where, θ is the half-arc of the lens, and $c_{coupling}$ and $c_{material}$ are the ultrasonic velocities of the coupling medium and the material, respectively.

Figure 9 shows ultrasonic microscopy imaging of a Ti-24Al-11Nb/SCS-6 composite subjected to thermo-mechanical fatigue (TMF) using a highly focused 50 MHz transducer, designed to generate SAW in metals such as titanium and steel with water as a coupling medium.

Due to environmental exposure, oxides were formed on the material's surface. This altered the sound velocity of the surface of the composite and, therefore, SAW could not be generated (Fig. 9a). The use of methanol as a coupling medium alleviated this difficulty (Fig. 9b).

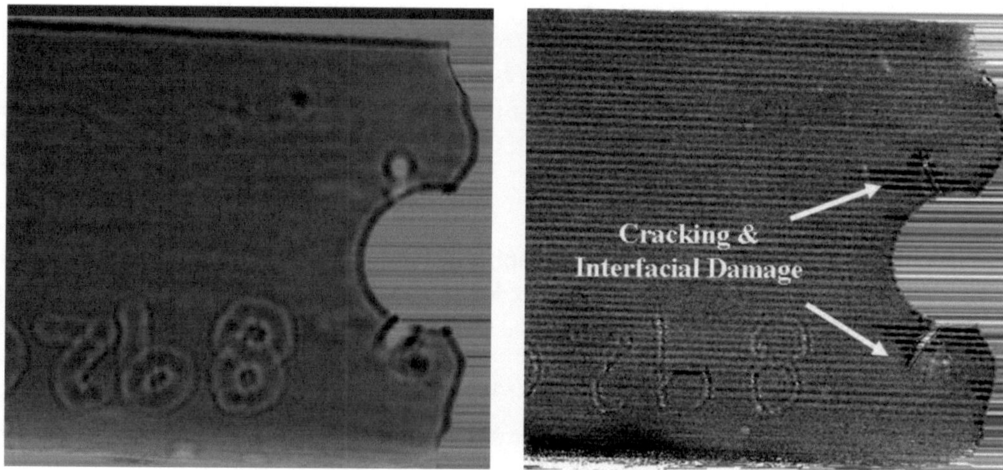

(a) (b)

Fig. 9. The role of coupling medium in ultrasonic microscopy; using (a) water as coupling medium for imaging damage in Ti-24Al-11Nb/SCS-6 composite subjected to TMF, (b) methanol as coupling medium for imaging the same specimen.

The capability of ultrasonic microscopy to determine cracks size and evaluate interfacial damage is depicted in Figure 10. This figure shows the first ply of titanium matrix composites with $[0/90]_S$ cross-ply and unidirectional lay-up of fibers subjected to isothermal mechanical fatigue [51, 52]. Matrix cracks and interfacial debonding are clearly observed in the figure. Crack bridging by unbroken fibers resulting to interface debonding dominate the fatigue crack growth life as evidenced by the characteristic decrease in crack growth rates as the crack length increased during fatigue cycling.

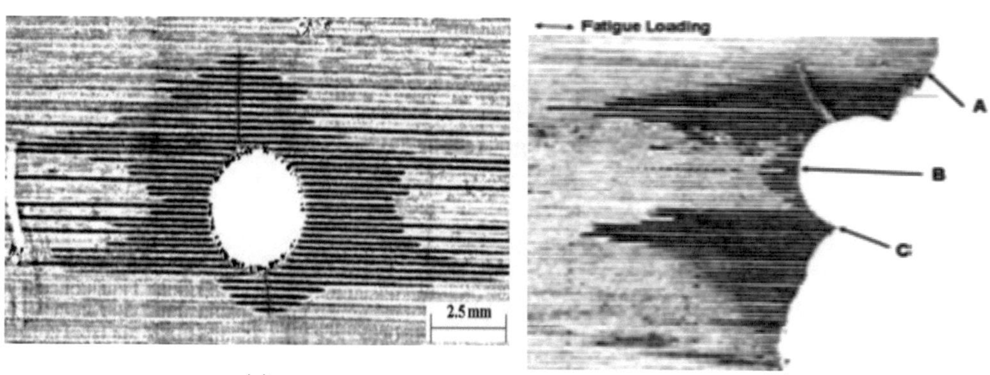

(a) (b)

Fig. 10. Ultrasonic microscopy micrographs of (a) Ti-15Mo-3Nb-3Al-0.2Si/SCS-6 composite with [0/90]S cross-ply lay-up of fibers subjected to 70 hours of isothermal (650°C) fatigue; (b) Ti-15Mo-3Nb-3Al-0.2Si/SCS-6 composite with unidirectional lay-up of fibers after 1.8×10^5 cycles of isothermal (650° C) fatigue: A. Point of accelerated crack growth to failure. B. Interfacial degradation due to compressive stresses. C. Interfacial degradation due to tensile stresses.

4. Real-time assessment of damage in aerospace materials based on nonlinear surface acoustic waves

This section presents an innovative NDE technique based on nonlinear surface-wave acoustics, which is sensitive to early stages of the fatigue process. A nonlinear parameter is derived and monitored to quantify the state of damage during cyclic loading of the material.

4.1 Background

A reliable inspection methodology for quantifying damage in space structures and for relating the level of damage to the remaining life of the material is essential for preventing catastrophic failures in aerospace systems. Many researchers have tried to develop techniques for fatigue damage characterization of aerospace materials based on linear acoustics, i.e. measurements of ultrasonic velocity and attenuation.

Studies of sound velocity as a function of number of fatigue cycles do not show appreciable changes. On the other hand, ultrasonic attenuation exhibits large changes however, relating these changes to the level of material damage is almost impossible since many other factors (experimental parameters, grain size, etc.) can affect the attenuation in a similar way. Studies [53] on nonlinear property of aluminium alloy and stainless steel have shown dramatic changes in the nonlinearity parameter by the time the material undergoes 30-40% of total fatigue life. However, this technique could not be applied, in real time, in test specimens during fatigue. Several dog-bone fatigue specimens were prepared, then fatigued to different number of cycles and the middle section of each one was cut off. This process is reliable as long as the test coupons are assumed to have the same microstructural characteristics before fatigue. However, due to normal statistical variability of test coupons this methodology may not provide meaningful information about the fatigue process. Moreover, the method is not applicable in real structures for health monitoring purposes. Other studies were carried out [38, 39], where two piezoelectric crystals were placed on the two opposite ends of a dog-bone titanium alloy specimen, and nonlinear measurements were performed in real time on the same specimen while undergoing cyclic loading. These studies utilized bulk acoustic waves and was a first step in the direction of developing a methodology for continuing monitoring of fatigue damage in the laboratory. The main drawback of the method for evaluating damage evolution in real aerospace structures is that it is based on bulk acoustic waves and requires access on the two sides of the structure that must also be perfectly parallel to each other, which is not the usual case. Based on the excellent results that the bulk-wave nonlinear method showed for evaluating the level of damage in aerospace materials, a whole new approach has been developed by performing second harmonic nonlinear measurements of surface acoustic waves, which enabled true health monitoring of damage evolution in space structures since the use of SAW overpasses the limitations of existing methods and do not require access to both sides of the structure.

In order to enhance the understanding and predict fatigue failure in critical components used in aerospace applications an innovative NDE technique based on nonlinear acoustics has been developed. This method is sensitive to early stages of fatigue damage accumulation. Failures of engine components, which often occur much earlier than predictions by initial design, increase the need for reliable NDE methods for early fatigue damage characterization. In order to characterize fatigue mechanisms using acoustic waves it is necessary to understand the physics of propagation of acoustic or elastic waves in solids and also the physics involved in the process of fatigue damage in materials. In this direction, the "vibrating string model of dislocation damping" developed in the 1950's [54] is the starting point for all the theories on acoustic wave interaction with dislocations.

It is well known that linearized relation between stress and strain "linearized Hooke's law" is sufficient to describe the mechanical properties of solids. The Hooke's law provides a way to relate stress to strain through the second order elastic constants or moduli of the solid. The linear approximation allows the properties of the material that can be measured experimentally to two properties namely, the velocity of sound (elastic modulus) and attenuation (damping) in the material. However, it has been shown [53] that these parameters are not robust enough to describe the fatigue mechanism. Generally a solid possesses nonlinear elastic behavior, but for practical engineering applications and for the purpose of simplification it is ignored and treated as a linear material. Thus it is necessary to understand acoustic wave propagation in nonlinear elastic material. Introduction of nonlinear terms into stress-strain relationship leads to inclusion of higher-order elastic constants.

4.2 Theoretical analysis

Derivation of analytical expressions for the nonlinear parameter enables using nonlinear acoustics techniques for real-time monitoring of fatigue damage accumulation in engineering materials.

4.2.1 Nonlinear bulk-wave propagation

In conventional wave propagation analysis the solid medium is considered as linear. Figure 11 depicts the difference in wave propagation between a linear and nonlinear medium.

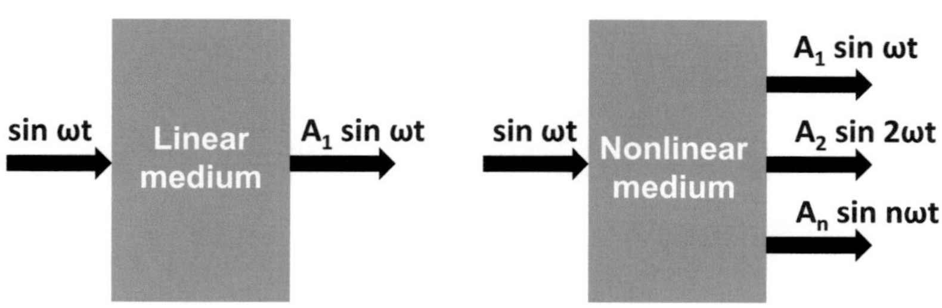

Fig. 11. Linear and nonlinear wave propagation in solid materials.

Engineering materials are nonlinear media. The fundamental wave (sin ωt) that propagates in such material will distort as it propagates, therefore the second-order (sin 2ωt) and higher-order (sin nωt) harmonics will be generated. The anharmonicity of the lattice and dislocation structures contribute in particular to the nonlinearity parameter, β, of the material. This parameter is a measure of the degree of material nonlinearity.

A longitudinal stress, σ, associated with an ultrasonic wave propagating in the material produces a longitudinal strain, ε, which is a combination of elastic, ε_{el}, and plastic, ε_{pl}, strains: $\varepsilon = \varepsilon_{el} + \varepsilon_{pl}$. The plastic strain component is associated with the motion of dislocation in the dipole configuration [53-55]. The relation between the stress and elastic strain can be written in the nonlinear form of Hooke's law (quadratic nonlinear approach),

$$\sigma = A_2^e \varepsilon_e + \frac{1}{2} A_3^e \varepsilon_e^2 + \ldots \text{ or } \varepsilon = \frac{1}{A_2^e}\sigma - \frac{1}{2}\frac{A_3^e}{\left(A_2^e\right)^3}\sigma^2 + \ldots \tag{5}$$

where A_2^e and A_3^e are the Huang coefficients.

By considering the dipolar forces one can easily obtain the relation between the stress and the plastic strain [53]. For edge dislocation pairs with opposite polarity, for example, these forces can be written in the following form:

$$F_x = -\frac{Gb^2}{2\pi(1-\nu)} \frac{x\left(x^2 - y^2\right)}{\left(x^2 + y^2\right)^2} \tag{6}$$

where, b is the Burgers vector, v the Poisson's ratio, G the shear modulus, and x and y the Cartesian coordinates of one dislocation in the pair in respect to the other. At equilibrium state, y equals the dipole height, h.

The relation between the plastic strain and the relative dislocation displacement $\xi = x - h$ is given by the expression $\varepsilon_{pl} = \Omega \Lambda_{dp} b \xi$, where Ω is a conversion factor and Λ_{dp} is the dipole density [53].

Using these relationships and an expansion of Eq. (6) in a power series in x with respect to h leads to the following equation

$$\sigma = A_2^e \left[\varepsilon - \frac{1}{2}\left(\frac{A_3^e}{\left(A_2^e\right)^3} + \frac{A_3^{dp}}{\left(A_2^{dp}\right)^3} \right) \right] \varepsilon^2 + \ldots \tag{7}$$

The wave equation with respect to the Lagrangian coordinate X is given by

$$\rho \frac{\partial^2 \varepsilon}{\partial t^2} = \frac{\partial^2 \sigma}{\partial X^2} \tag{8}$$

Replacing σ, given by Eq. (7), in Eq. (8) the following equation can be obtained:

$$\frac{\partial^2 \varepsilon}{\partial t^2} - c^2 \frac{\partial^2 \varepsilon}{\partial X^2} = -c^2 \beta \left[\varepsilon \frac{\partial^2 \varepsilon}{\partial X^2} + \left(\frac{\partial \varepsilon}{\partial X} \right)^2 \right] \tag{9}$$

where $c = \sqrt{\frac{A_2^e}{\rho}}$ and $\beta = \beta_e + \beta_{dp}$ with $\beta_e = -\frac{A_3^e}{A_2^e}$ and $\beta_{dp} = \frac{16\pi\Omega R^2 \Lambda_{dp} h^3 (1-\nu)^2 \left(A_2^e\right)^2}{G^2 b}$

The Huang coefficients can be expressed in terms of higher elastic constants. Specifically,

$A_1^e = C_1$ (C_1 equals the initial stress)

$A_2^e = C_1 + C_{11}$

$A_3^e = 3C_{11} + C_{111}$

Assuming $C_1 = 0$, the portion of β describing the nonlinear contribution from the elasticity of the lattice can be written in the form, $\beta_e = -\left(3 + \frac{C_{111}}{C_{11}}\right)$.

Assuming a purely sinusoidal input wave, $\varepsilon_o \sin(\omega t - kX)$, a solution to Eq. (9) is:

$$\varepsilon = \varepsilon_o \sin(\omega t - kX) - \frac{1}{4}\beta k \varepsilon_o^2 X \sin\left[2(\omega t - kX)\right] \tag{10}$$

Therefore β can be described by the following expression containing the amplitudes of the fundamental wave, A_1, and the second harmonic, A_2,

$$\beta = \frac{4k}{X}\frac{A_2}{A_1^2} \tag{11}$$

Eq. (11) permits the experimental determination of the nonlinear parameter β. This parameter depends on the amplitudes of the fundamental as well as the second harmonic, the frequency, wave velocity and propagation distance. To obtain experimentally the nonlinear parameter β of a material, the amplitude of the second harmonic needs to be determined experimentally using a specific frequency and propagation distance (38-40,53).

4.2.2 Nonlinear Rayleigh wave propagation

The analysis for obtaining β as described in Section 4.1.1 is only valid for longitudinal ultrasonic waves. The derivation of the nonlinear parameter β for the propagation of 2-D Rayleigh waves is more challenging. Surface acoustic waves are in general a superposition of bulk waves. Longitudinal waves are sensitive as it concerns the generation of higher harmonics, contrary to shear waves which are not considered prone to higher harmonic generation. In this respect, Rayleigh waves are expected to have a similar nonlinear wave propagation behavior as the longitudinal waves. In practice, it is essential to derive the nonlinear parameter β for surface acoustic waves as a function of the amplitudes of the fundamental frequency and the second order harmonic.

If one considers a Rayleigh wave propagating in the positive x direction, assuming the z axis into the material, the displacement potentials describing the longitudinal and shear ultrasonic waves,

$$\varphi = Ae^{-k\sqrt{1-\left(\frac{c}{c_L}\right)^2}z} e^{ik(x-ct)} \tag{12}$$

$\phi = Be^{-k\sqrt{1-\left(\frac{c}{c_S}\right)^2}z} e^{ik(x-ct)}$ can be re-written in the form:

$$\varphi = -i\frac{B_1}{k_R}e^{-\sqrt{k_R^2-k_l^2}\,z}e^{i(k_Rx-\omega t)} \tag{13}$$

$$\phi = -i\frac{C_1}{k_R}e^{-\sqrt{k_R^2-k_S^2}\,z}e^{i(k_Rx-\omega t)}$$

where, k_R, k_l, k_s are the wave-numbers for Rayleigh, longitudinal, and shear ultrasonic waves, respectively [56].

Considering the boundary conditions for a stress-free surface in Eq. (13), a relation between the constants B_1 and C_1 can be obtained:

$$B_1 = -i\frac{2k_R\sqrt{k_R^2-k_l^2}}{2k_R^2-k_S^2}C_1 \tag{14}$$

Taking into account that the surface acoustic waves are a superposition of longitudinal and shear waves propagating along a stress-free surface that have the same velocity, the displacement components can be decomposed into their longitudinal and shear components [56-60]:

$$u_x = B_1\left(e^{-\sqrt{k_R^2-k_l^2}\,z} - \frac{2\sqrt{\left(k_R^2-k_l^2\right)\left(k_R^2-k_S^2\right)}}{2k_R^2-k_S^2}e^{-\sqrt{k_R^2-k_S^2}\,z}\right)e^{i(k_Rx-\omega t)} \tag{15}$$

$$u_z = iB_1\frac{\sqrt{k_R^2-k_l^2}}{k_R}\left(e^{-\sqrt{k_R^2-k_l^2}\,z} - \frac{2k_R^2}{2k_R^2-k_S^2}e^{-\sqrt{k_R^2-k_S^2}\,z}\right)e^{i(k_Rx-\omega t)}$$

The first term in Eq. (15a) presents the pure longitudinal wave motion. Considering a material with a weak nonlinearity, the second order harmonic Rayleigh waves that propagate a large enough distance can be expressed in the form [56-60]:

$$u_x \approx B_2\left(e^{-2\sqrt{k_R^2-k_l^2}\,z} - \frac{2\sqrt{\left(k_R^2-k_l^2\right)\left(k_R^2-k_S^2\right)}}{2k_R^2-k_S^2}e^{-2\sqrt{k_R^2-k_S^2}\,z}\right)e^{i2(k_Rx-\omega t)} \tag{16}$$

$$u_z \approx iB_2\frac{\sqrt{k_R^2-k_l^2}}{k_R}\left(e^{-2\sqrt{k_R^2-k_l^2}\,z} - \frac{2k_R^2}{2k_R^2-k_S^2}e^{-2\sqrt{k_R^2-k_S^2}\,z}\right)e^{i2(k_Rx-\omega t)}$$

Taking into account the fact that the acoustic nonlinear behavior of the shear waves in an isotropic medium ceases to exist due to the symmetry of the third order elastic constants, it can be inferred that the longitudinal wave component is only one contributing to the higher harmonic generation in the case of the propagation of surface acoustic waves. Hence, the amplitudes of the in-plane displacement, u_x, of the fundamental and second harmonic of a near-surface Rayleigh wave can be related as those in the case of bulk longitudinal waves [56]:

$$B_2 = \frac{\beta k_1^2 x B_1^2}{8} \tag{17}$$

where β is the acoustic nonlinear parameter for the bulk longitudinal waves, and x is the propagation distance.

In case of experimental determination of β for Rayleigh waves using contact ultrasonic transducers of interferometry, the out-of-plane component on the surface (z=0) of the particle velocity, $\overline{u}_z$ or displacement, u_z, is detected.

From Eqs. (15b), (16b), and (17), the ratio of the amplitudes of the second harmonic to the fundamental is given by:

$$\frac{\overline{u}_z(2\omega)}{\overline{u}_z^2(2\omega)} = \frac{\beta k_1^2 x}{8i\left(\frac{\sqrt{k_R^2 - k_l^2}}{k_R}\right)\left(1 - \frac{2k_R^2}{2k_R^2 - k_S^2}\right)} \tag{18}$$

where $\overline{u}_z(\omega) = u_z(\omega; x, z=0)$. Therefore, the measured out-of-plane displacement components are related to the acoustic nonlinear parameter β for the bulk longitudinal waves [56],

$$\beta = \frac{8i\ \overline{u}_z(2\omega)}{k_l^2 x\ \overline{u}_z^2(\omega) k_R} \frac{\sqrt{k_R^2 - k_l^2}}{k_R}\left(1 - \frac{2k_R^2}{2k_R^2 - k_S^2}\right) \tag{19}$$

While the shear wave alone does not generate higher harmonics, shear wave components interact with longitudinal ones, as it becomes evident from the second term in the parenthesis of Eq. (19).

The harmonic ratio $\frac{\overline{u}_z(2\omega)}{\overline{u}_z^2(\omega)} \approx \frac{A_2}{A_1^2}$ can be used to determine material nonlinearity and to assess the state of damage in materials with different levels of fatigue. The dimensionless nonlinear parameter β in case of propagating Rayleigh surface waves in the material is also dependent on the amplitude ratio $\frac{A_2}{A_1^2}$ as in case of bulk longitudinal waves.

Although the nonlinear parameter β remains constant for different propagation distances, the amplitude of the second harmonic changes linearly with the propagation distance, as required for a quadratic nonlinearity [56].

4.3 Real time measurement of material nonlinearity during fatigue

In the measurement of nonlinearity parameter β under cyclic loading the change in β is more important than its absolute values. Hence, measurements of relative changes in the nonlinearity of the material from the virgin state to a fatigued state are discussed here. Therefore, the β parameter defined in Eq. (19) and normalized by the value β_o (nonlinear parameter of the material at the virgin state) is experimentally measured. In this sense, in order to assess the level of fatigue damage using nonlinear acoustics measurements a determination of the absolute value of the material's nonlinearity is not required, enabling real-time experiments [38-39].

Below it is shown an example of assessing in real-time the state of fatigue damage in titanium alloys undergone cyclic loading. The piezoelectric detection of second harmonic ultrasonic amplitude is based on propagating a pure single frequency f ultrasonic wave through the sample. As the elastic wave propagates through the medium, it is distorted as the result of the anharmonicity of the crystalline lattice and other microstructural disturbances, such as the grain boundaries and dislocations. During the fatigue process of Ti-6Al-4V, the lattice anharmonicity remains constant since the stress level applied to the specimen is far below the yield strength i.e., in the elastic region. However, the other factors like grain boundaries, dislocations, and other impurities change as a function of fatigue level. The distorted signal is composed of the combination of the harmonics and grows as it propagates until the attenuation factor stops its growth. The harmonic portion of the distorted ultrasonic signal is very sensitive to the changes in the strain energy density due to the changes of these factors. The second harmonic wave, of frequency 2f, is detected by a second piezoelectric transducer. The transducer was manufactured using 36° Y-cut $LiNbO_3$ crystals placed inside specially designed brass housing and plexiglass tubing. Lithium niobate single-crystals were used since they are linear materials and also exhibit higher electromechanical coupling compared to quartz crystals.

An important issue for correctly measuring material nonlinearity is to use linear instrumentation. The experimental data shown below are obtained using amplifiers and filters of linear response. For example, the fundamental and second harmonic signals were detected using high quality linear band-pass filters with a rejection ratio better than 60 dB. Two such low power (up to 100 W) filters were designed, one with a central frequency of 5 MHz, the other 10 MHz. Figures 12(a) and 12(b) show the 5 MHz and 10MHz low power filters' characteristics and simulation performance data, respectively.

Transducer holder and the grips for the fatigue load frame were designed to enable on-line monitoring of the material's nonlinearity parameter during the fatigue process, since the conventional grips are inadequate for attaching transducers to the specimen.

The experimental configuration for on-line piezoelectric detection of second harmonic signal during mechanical fatigue is based on a tone-burst generator and a power amplifier to launch longitudinal sound waves into the specimen at a frequency of 5 MHz. A linear high-power band-pass filter was placed between the power amplifier and the transducer to make sure that unwanted harmonic signals are filtered out. The same transducer was used to detect the fundamental signal reflected from the other end of the specimen. A 10 MHz transducer bonded to the other end of the specimen was used to receive the second harmonic signal. After the second harmonic signal is detected, it was fed to a linear narrow band amplifier through the 10 MHz band-pass filter. Both fundamental V_1 (mV) and second harmonic V_2 (mV) signals were sent to the A/D converter for digitization and the nonlinear parameter β was finally determined from the sampled signals.

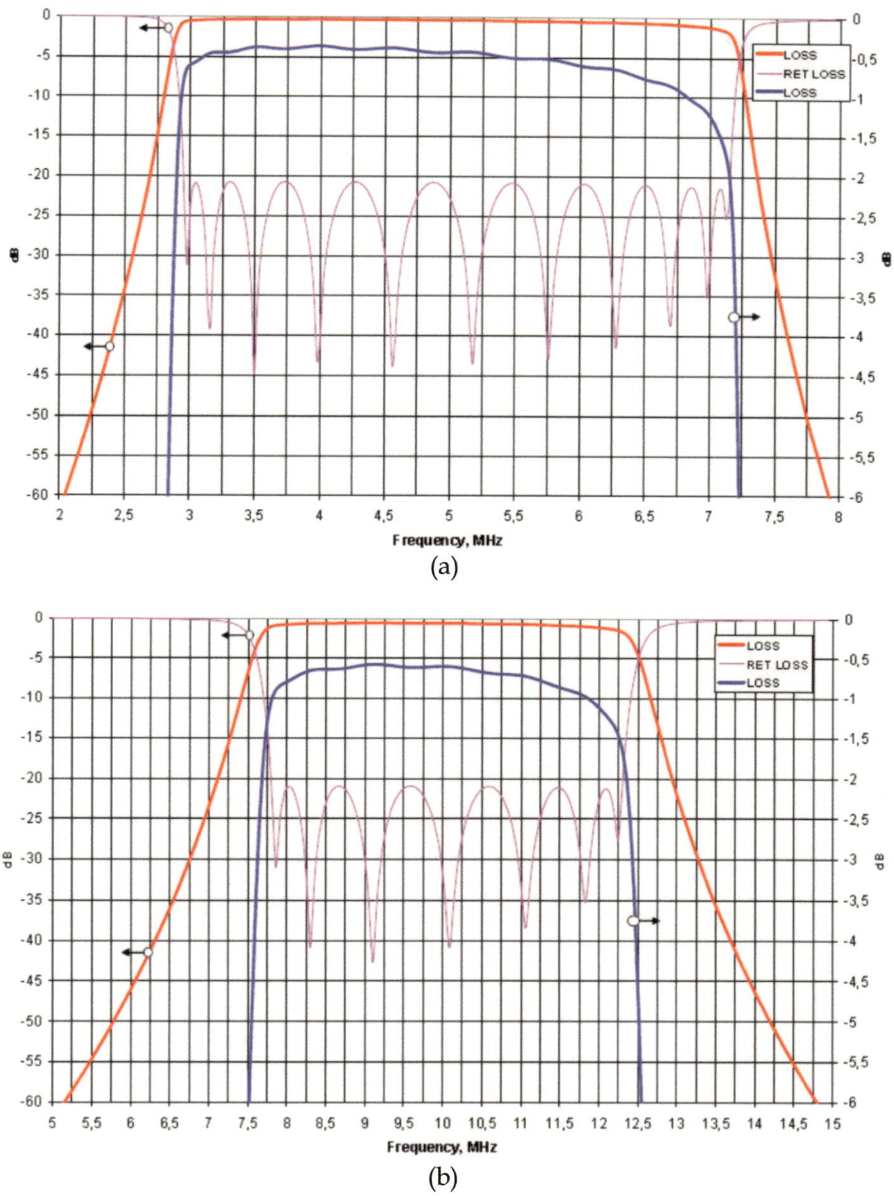

Fig. 12. Performance data of the (a) 5 MHz and (b) 10 MHz 100 W band-pass filters.

Since the nonlinear property of the specimen was measured, it was necessary to verify that the measurement setup itself was linear indeed. For checking the system's linearity a simple experiment was performed with an unfatigued Ti-6Al-4V sample at room temperature by changing the input voltage to the transmitting transducer. It was thus demonstrated that the slope of the curve of the amplitude of second harmonic vs. the fundamental using the linear filters was linear.

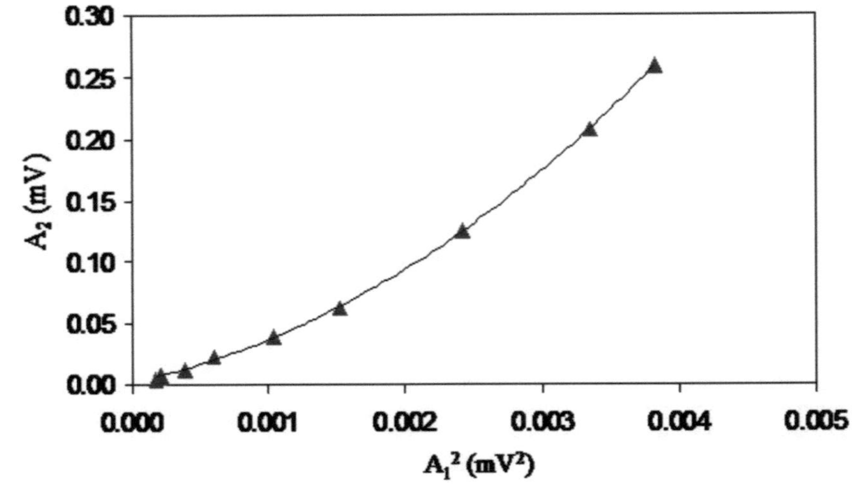

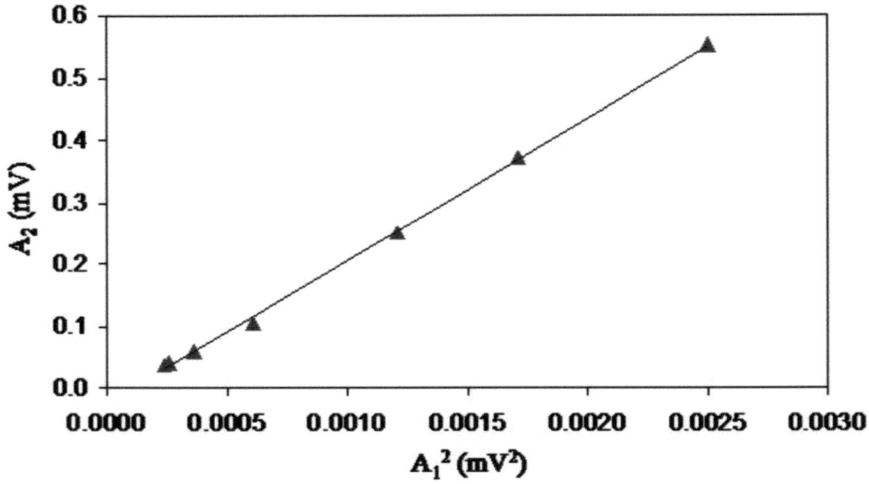

Fig. 13. Linearity check of the measurement setup; (a) without the linear band-pass filter for the second harmonic signal, and (b) with the linear band-pass filter for the second harmonic signal.

As it is demonstrated by comparing Figs. 13(a) and 13(b), using linear band-pass filters leads to a totally linear experimental setup. Thus, any determination of nonlinear behavior can be attributed to material nonlinearity and not to the measurement system.

Since the measurements require a relatively long time for experiments performed at a cyclic frequency of 1 Hz, it is necessary to check the stability of the measuring system over a period time. The amplitudes of the fundamental and second harmonic signals were monitored over a period of 24 hours in the laboratory and only small, almost negligible fluctuations, compared to the size of the measured values during the fatigue test, were observed. Figure 14 shows the amplitudes of the fundamental, A_1, and second harmonic, A_2, signals as a function of time for period of 24 hours in laboratory conditions.

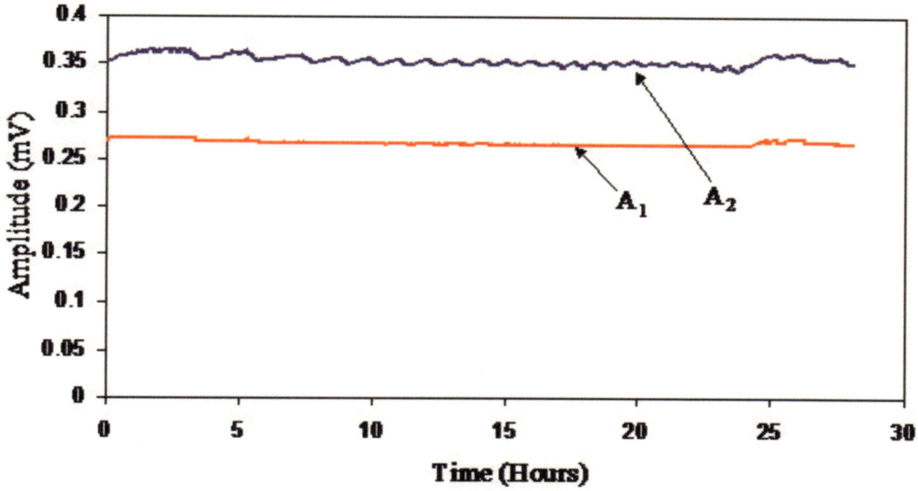

Fig. 14. Long term stability test of the measurement system at room temperature.

The experimental technique described above was used to measure the nonlinear acoustic properties of Ti-6Al-4V alloys and characterize in real-time their fatigue behavior. During the fatigue tests, the samples were subjected to cyclic loading at the frequency of 1 Hz under low cycle fatigue conditions (σ_{max} = 850 MPa, and R ratio = 0.1). The ultrasonic velocity and nonlinear property were measured at zero-load on the sample, at an interval of 100 cycles of fatigue.

Attenuation and velocity of longitudinal wave measurements were performed at a frequency of 5 MHz at various stages of fatigue. It was observed that the longitudinal sound velocity had a minute measurable change in the beginning of the fatigue process (Fig. 15). This can be explained by the small increase of the specimen's length that occurs during the fatigue process. For an accurate determination of the velocity of sound it is necessary to incorporate the changes in the specimen length.

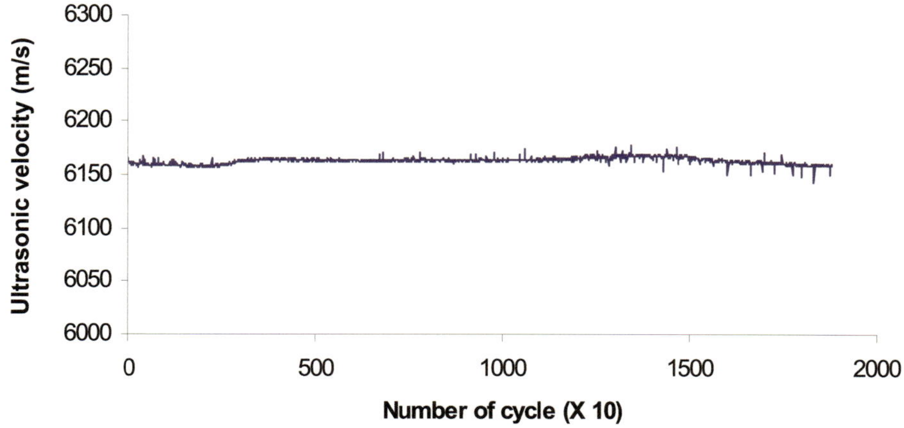

Fig. 15. Variation of longitudinal wave velocity with fatigue life

Additionally, it was observed that the attenuation increased significantly in the initial stages of fatigue (Fig. 16). The initial increase of attenuation of 50 % is quite significant; however, it is less sensitive to fatigue process beyond 20% of the fatigue life. The higher attenuation at the higher fatigue cycles may indicate an increase in the scattering of sound waves due to the increased dislocation dipole density from fatigue. As the increase in dislocation density saturates, the level of scattering of sound wave within the material become stable. It should be pointed out, however, that the general tendency of dislocation movement is known to migrate to the surface of the material. This could mean that the attenuation measurement in the bulk is less meaningful throughout the entire lifetime of the material.

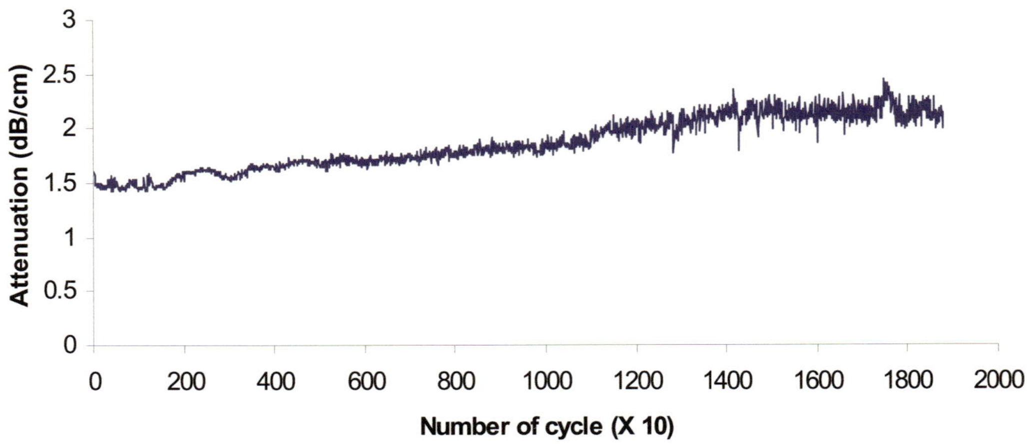

Fig. 16. Variation of ultrasonic attenuation with fatigue life

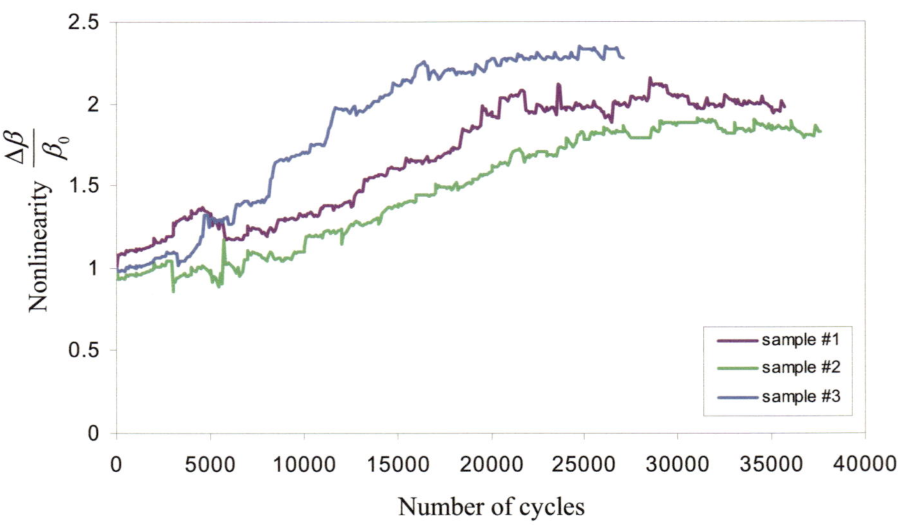

Fig. 17. Variation of the nonlinear parameter β with fatigue life in Ti-6Al-4V

Variation in amplitude of the second harmonic signal, as the amplitude of the fundamental signal is changed, was used for measurement of nonlinear acoustic behavior of the material.

As the material was undergoing fatigue the amplitude of the second harmonic signal increases to give a steeper slope. Figure 17 shows measurements of β as a function of fatigue cycles for three different Ti-6Al-4V samples. The trend of the variation of relative β as a function of fatigue life is similar for the three samples; however, the level of nonlinearity is different due to the fact that different samples are fatigue damaged in a dissimilar way.

Figure 18 depicts the normalized nonlinearity (shown with line plot) of Ti-6Al-4V samples as a function of the number of cycles and correlation with transmission electron microscopy (TEM) analysis of the dislocation density (shown with data points). It can be observed in Fig. 18 that acoustic nonlinearity of the material exhibits large changes during the fatigue process. This finding is in contrast with the measurements of attenuation and elastic behavior, where the majority of variation occurred before the 20% of fatigue lifetime. The second harmonic signal generated during the fatigue process is not only sensitive to the early stage of the process, but also to later stages of damage. This implies that the harmonic signal is very sensitive to the microstructural changes in the material. The variation of nonlinearity continues due to the generation of additional dislocation dipoles by the fatigue process and their interaction with the acoustic waves, as predicted by relevant models and experimental work [38-40, 53].

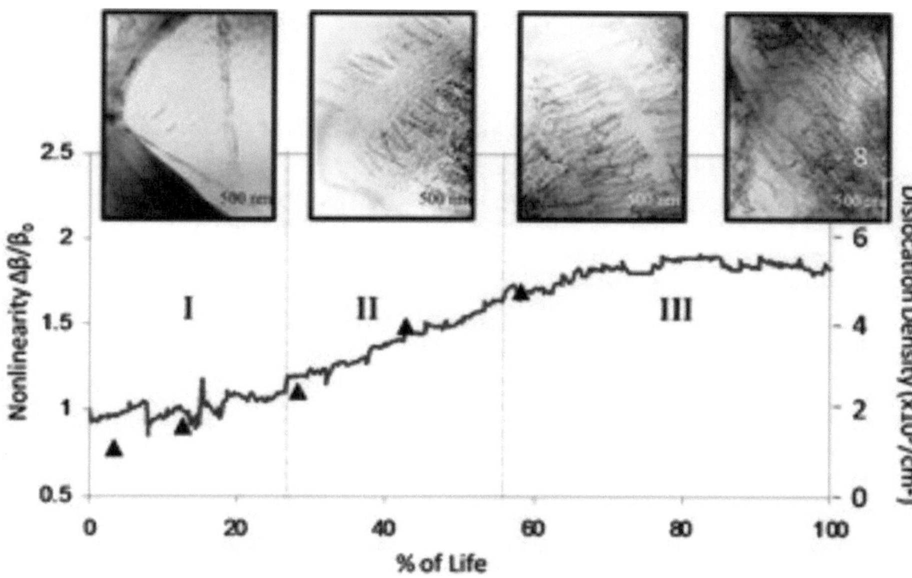

Fig. 18. Correlation of TEM imaging and dislocation density in Ti-6Al-4V with normalized nonlinearity as a function of fatigue level.

5. Conclusion

The present chapter gives an overview of contemporary techniques to assess the quality of materials using surface elastic waves. It is highlighted that different instrumentation and methodology facilitates non destructive evaluation in quite distinct material groups, like concrete and aerospace composites. In both kinds of material, damage is primarily initiated on the surface. Therefore, early assessment by surface waves is crucial for the estimation of remaining life or repair method. Concrete, including large scale of inhomogeneity is typically examined by comparatively long wavelengths by means of pulse velocity correlation with strength or damage. The present study complements the information of pulse velocity using features like the dispersion curve, the attenuation coefficient as a function of frequency, as well as the spectral distortion between pulses recorded at different points of the surface. It is seen that combined wave features can act complementary, while their acquisition does not require further upgrade on the equipment used for conventional pulse velocity measurements. Concerning alloys and metal matrix composites used in aerospace applications, due to their much shorter typical size of inhomogeneity, delicate equipment should be employed enabling the evaluation on a microscopic scale. In these components fatigue loading is of primary concern, which may result in different damage modes. Acoustic microscopy enables the scanning of the material for surface or subsurface defects providing a detailed point by point assessment on the elastic properties. Additionally, nonlinear surface waves are extremely sensitive to the evolution of micro-damage, including dislocation motion and micro-cracking. This is due to the development of higher harmonics and, therefore, is an excellent way to monitor in real time the damage as evolved during cyclic loading from early stages of fatigue to final failure of the structure. The derived nonlinear parameter enables the prediction of remaining life in critical components.

6. References

[1] Kaplan, M. F. (1959). The Effects of Age and Water/Cement Ratio Upon the Relation between Ultrasonic Pulse Velocity and Compressive Strength. *Mag. Concrete Res.,* Vol. 11, No. 32, pp. 85-92

[2] Popovics, S., Rose, J. L. and Popovics, J. S. (1990). The Behavior of Ultrasonic Pulses in Concrete. *Cement Concrete Res.,* Vol. 20, pp. 259–270

[3] Qixian, L. and Bungey, J. H. (1996). Using Compression Wave Ultrasonic Transducers to Measure the Velocity of Surface Waves and Hence Determine Dynamic Modulus of Elasticity for Concrete. *Constr Build Mater,* Vol. 4, pp. 237–242

[4] Graff, K. F. (1975). Wave Motion in Elastic Solids. *New York: Dover Publications,* Vol., No.

[5] Philippidis, T. P. and Aggelis, D. G. (2005). Experimental Study of Wave Dispersion and Attenuation in Concrete. Ultrasonics. 43, No. 7, pp. 584-595

[6] Kinra, V. K. and Rousseau, C. (1987). Acoustical and Optical Branches of Wave Propagation. *J.Wave-Mater. Interact.,* Vol. 2, pp. 141–152

[7] Sachse, W. and Pao, Y.-H. (1978). On the Determination of Phase and Group Velocities of Dispersive Waves in Solids. *J. Appl. Phys.,* Vol. 49, No. 8, pp. 4320–4327

[8] Aggelis, D. G. and Shiotani, T. (2007). Experimental Study of Surface Wave Propagation in Strongly Heterogeneous Media. *Journal of the Acoustical Society of America,* Vol. 122, No. 5, pp. EL 151-157

[9] Washer, G. A., Green, R. E. and Jr, R. B. P. (2002). Velocity Constants for Ultrasonic Stress Measurement in Prestressing Tendons. *Res. Nondestr. Eval.,* Vol. 14, pp. 81-94

[10] Cowan, M. L., Beaty, K., Page, J. H., Zhengyou, L. and Sheng, P. (1998). Group Velocity of Acoustic Waves in Strongly Scattering Media: Dependence on the Volume Fraction of Scatterers. *Phys. Rev. E,* Vol. 58, pp. 6626–6636

[11] Sansalone, M. and Carino, N. J. (2004). Stress Wave Propagation Methods. *CRC Handbook on Nondestructive Testing of Concrete, Malhotra V. M., Carino N. J., eds. CRC Press, Boca Raton FL,* pp. 275-304

[12] Naik, T. R., Malhotra, V. M. and Popovics, J. S. (2004). The Ultrasonic Pulse Velocity Method. *Malhotra VM, Carino NJ. editors. CRC Handbook on nondestructive testing of concrete. Boca Raton: CRC Press*

[13] Philippidis, T. P. and Aggelis, D. G. (2003). An Acousto-Ultrasonic Approach for the Determination of Water-to-Cement Ratio in Concrete. *Cement and Concrete Research,* Vol. 33, No. 4, pp. 525-538

[14] Gudra, T. and Stawinski, B. (2000). Non-Destructive Characterization of Concrete Using Surface Waves. *NDT&E INT,* Vol. 33, pp. 1-6

[15] Aggelis, D. G., Kordatos, E. Z., Strantza, M., Soulioti, D. V. and Matikas, T. E. (2011). NDT Approach for Characterization of Subsurface Cracks in Concrete. *Construction and Building Materials,* Vol. 25, pp. 3089-3097

[16] Aggelis, D. G. and Shiotani, T. (2008). Surface Wave Dispersion in Cement-Based Media: Inclusion Size Effect. *NDT&E INT,* Vol. 41, pp. 319-325

[17] Landis, E. N. and Shah, S. P. (1995). Frequency-Dependent Stress Wave Attenuation in Cement-Based Materials. *J. Eng. Mech.-ASCE,* Vol. 121, No. 6, pp. 737-743

[18] Punurai, W., Jarzynski, J., Qu, J., Kurtis, K. E. and Jacobs, L. J. (2006). Characterization of Entrained Air Voids in Cement Paste with Scattered Ultrasound. *NDT&E International,* Vol. 39, pp. 514–524

[19] Shah, S. P., Popovics, J. S., Subramanian, K. V. and Aldea, C. M. (2000). New Directions in Concrete Health Monitoring Technology. *J. Eng. Mech.-ASCE,* Vol. 126, No. 7, pp. 754-760

[20] Selleck, S. F., Landis, E. N., Peterson, M. L., Shah, S. P. and Achenbach, J. D. (1998). Ultrasonic Investigation of Concrete with Distributed Damage. *ACI Materials Journal* Vol. 95, No. 1, pp. 27-36

[21] Jacobs, L. J. and Owino, J. O. (2000). Effect of Aggregate Size on Attenuation of Rayleigh Surface Waves in Cement-Based Materials. *J. Eng. Mech.-ASCE,* Vol. 126, No. 11, pp. 1124-1130

[22] Owino, J. O. and Jacobs, L. J. (1999). Attenuation Measurements in Cement-Based Materials Using Laser Ultrasonics. *J. Eng. Mech.-ASCE,* Vol. 125, No. 6, pp. 637-647

[23] Aggelis, D. G. (2010). Damage Characterization of Inhomogeneous Materials: Experiments and Numerical Simulatios of Wave Propagation. *Strain,* in press (10.1111/j.1475-1305.2009.00721.x)

[24] Chaix, J. F., Garnie, r. V. and Corneloup, G. (2006). Ultrasonic Wave Propagation in Hetero- Geneous Solid Media: Theoretical Analysis and Experimental Validation. *Ultrasonics,* Vol. 44, pp. 200-210

[25] Kim, D. S., Seo, W. S. and Lee, K. M. (2006). IE–SASW Method for Nondestructive Evaluation of Concrete Structure. *NDT&E International,* Vol. 39, pp. 143-154

[26] Hevin, G., Abraham, O., Pedersen, H. A. and Campillo, M. (1998). Characterisation of Surface Cracks with Rayleigh Waves: A Numerical Model. *NDT&E Int* Vol. 31, No. 4, pp. 289–297

[27] Aggelis, D. G., Shiotani, T. and Polyzos, D. (2009). Characterization of Surface Crack Depth and Repair Evaluation Using Rayleigh Waves. *Cement and Concrete Composites,* Vol. 31, No. 1, pp. 77-83

[28] vanWijk, K., Komatitsch, D., Scales, J. A. and Tromp, J. (2004). Analysis of Strong Scattering at the Micro-Scale. *J. Acoust. Soc. Am.,* Vol. 115, pp. 1006–1011

[29] Pecorari, C. (1998). Rayleigh Wave Dispersion Due to a Distribution of Semi-Elliptical Surface-Breaking Cracks. *J Acoust Soc Am,* Vol. 103, No. 3, pp. 1383-1387

[30] Harmon, D. and Saff, C. (1988). Metal Matrix Composites: Testing, Analysis, and Failure Modes. *Johnson EWS, editor. Damage Initiation and Growth in Fiber Reinforced Metal Matrix Composites. Philadelphia, PA,* Vol. ASTM STP 1032, pp. 237-250

[31] Marshall, D. B., Cox, B. N. and Evans, A. G. (1985). The Mechanics of Matrix Cracking in Brittle-Matrix Fiber Composites. *Acta Metallurgica,* Vol. 33, No. 11, pp. 2013-2021

[32] McCartney, L. N. (1987). Mechanics of Matrix Cracking in Brittle-Matrix Fibre-Reinforced Composites. *Proc. R. Soc. Lond.,* pp. A409:329-350

[33] Karpur, P., Matikas, T. E., Krishnamurthy, S. and Ashbaugh, N. (1992). Ultrasound for Fiber Fragmentation Size Determination to Characterize Load Transfer Behavior of Matrix-Fiber Interface in Metal Matrix Composites. *Thompson DO, Chimenti DE, editors. Review of Progress in Quantitative NDE. La Jolla CA,* Vol. 12B, pp. 1507-1513

[34] Matikas, T. E. (2008). High Temperature Fiber Fragmentation Characteristics of SiC Single-Fiber Composite with Titanium Matrices. *Advanced Composite Materials,* Vol. 17, No. 1, pp. 75-87

[35] Matikas, T. E. and Karpur, P. (1992). Matrix-Fiber Interface Characterization in Metal Matrix Composites Using Ultrasonic Shear-Wave Back-Reflection Coefficient Technique. *Thompson DO, Chimenti DE, editors. Review of Progress in Quantitative NDE. La Jolla CA,* Vol. 12B, pp. 1515-1522

[36] Matikas, T. E., Rousseau, M. and Gatignol, P. (1993). Theoretical Analysis for the Reflection of a Focused Ultrasonic Beam from a Fluid-Solid Interface. *Journal of the Acoustical Society of America,* Vol. 93, No. 3, pp. 1407-1416

[37] Matikas, T. E. (2000). Quantitative Short-Pulse Acoustic Microscopy and Application to Materials Characterization. *Microsc. Microanal.,* Vol. 6, pp. 59–67

[38] Frouin, J., Sathish, S., Matikas, T. E. and Na, J. K. (1999). Ultrasonic Linear and Nonlinear Behavior of Fatigued Ti-6Al-4V. *Journal of Materials Research,* Vol. 14, No. 4, pp. 1295-1298

[39] Frouin, J., Maurer, J., Sathish, S., Eylon, D., Na, J. K. and Matikas, T. E. (2000). Real-Time Monitoring of Acoustic Linear and Nonlinear Behavior of Titanium Alloys During Cyclic Loading. *Nondestructive Methods for Materials Characterization, Materials Research Society,* Vol. 591, pp. 79-84

[40] Matikas, T. E. (2010). Damage Characterization and Real-Time Health Monitoring of Aerospace Materials Using Innovative NDE Tools. *Journal of Materials Engineering and Performance,* Vol. 19, No. 5, pp. 751-760

[41] Aggelis, D. G. and Shiotani, T. (2008). Effect of Inhomogeneity Parameters on the Wave Propagation in Cementitious Materials. *American Concrete Institute Materials Journal,* Vol. 105, No. 2, pp. 187-193

[42] Dokun, O. D., Jacobs, L. J. and Haj-Ali, R. M. (2000). Ultrasonic Monitoring of Material Degradation in FRP Composites. *J. Eng. Mech.,* Vol. 126, pp. 704–710

[43] Aggelis, D. G., Tsinopoulos, S. V. and Polyzos, D. (2004). An Iterative Effective Medium Approximation (Iema) for Wave Dispersion and Attenuation Predictions in Particulate Composites, Suspensions and Emulsions *Journal of the Acoustical Society of America,* Vol. 116, pp. 3443-3452

[44] Mobley, J., Waters, K. R., Hall, C. H., Marsh, J. N., Hughes, M. S., Brandenburger, G. H. and Miller, J. G. (1999). Measurements and Predictions of Phase Velocity and Attenuation Coefficient in Suspensions of Elastic Microspheres. *J. Acoust. Soc. Am.,* Vol. 106, pp. 652-659

[45] Bendat, J. S. and Piersol, A. G. (1993). Engineering Applications of Correlation and Spectral Analysis. *2nd ed., Wiley, New York*

[46] Grosse, C., Reinhardt, H. and Dahm, T. (1997). Localization and Classification of Fracture Types in Concrete with Quantitative Acoustic Emission Measurement Techniques. *NDT&E INT,* Vol. 30, No. 4, pp. 223-230

[47] Quate, C. F., Atalar, A. and Wickramasinghe, H. K. (1979). Acoustic Microscopy with Mechanical Scanning - a Review. *Proceedings of the IEEE,* Vol. 67, pp. 1092-1114

[48] Bertoni, H. L. (1985). Rayleigh Waves in Scanning Acoustic Microscopy. *Ash EA, Paige EGS, editors. Rayleigh-Wave Theory and Application. The Royal Institution, London,* Vol. 2, pp. 274-290

[49] Liang, K. K., Kino, G. S. and Khuri-Yakub, B. T. (1985). Material Characterization by the Inversion of V(Z). *IEEE Transactions on Sonics and Ultrasonics,* Vol. SU-32, No. 2, pp. 213-224

[50] Matikas, T. E., Rousseau, M. and Gatignol, P. (1992). Experimental Study of Focused Ultrasonic Beams Reflected at a Fluid-Solid Interface in the Neighborhood of the Rayleigh Angle. *IEEE Transactions on Ultrasonics Ferroelectrics and Frequency Control,* Vol. 39, No. 6, pp. 737-744

[51] Blatt, D., Karpur, P., Matikas, T. E., Blodgett, M. P. and Stubbs, D. A. (1993). Elevated Temperature Degradation and Damage Mechanisms of Titanium Based Metal Matrix Composites with SCS-6 Fibers. *Scipta Metallurgica et Materialia,* Vol. 29, pp. 851-856

[52] Waterbury, M. C., Karpur, P., Matikas, T. E., Krishnamurthy, S. and Miracle, D. B. (1994). In Situ Observation of the Single-Fiber Fragmentation Process in Metal-Matrix Composites by Ultrasonic Imaging. *Composites Science and Technology,* Vol. 52, No. 2, pp. 261-266

[53] Cantrell, J. H. and Yost, W. T. (1994). Acoustic Harmonic Generation from Fatigue-Induced Dislocation Dipoles. *Philosophical magazine A,* Vol. 69, No. 2, pp. 315-326

[54] Granato, A. and Lücke, K. (1956). Theory of Mechanical Damping Due to Dislocations. *Journal of Applied Physics,* Vol. 27, pp. 583

[55] Planat, M. (1985). Multiple Scale Analysis of the Nonlinear Surface Acoustic Wave Propagation in Anisotropic Crystals. *J. Appl. Phys.,* Vol. 57, pp. 4911–4915

[56] Herrmann, J., Kim, J. Y., Jacobs, L. J., Qu, J., Littles, J. W. and Savage, M. F. (2006). Assessment of Material Damage in a Nickel-Base Superalloy Using Nonlinear Rayleigh Surface Waves, *Journal of Applied Physics,* Vol. 99, No. 12, pp. 124913

[57] Kim, J.-Y., Jacobs, L. J., Qu, J. and Littles, J. W. (2006). Experimental Characterization of Fatigue Damage in a Nickel-Base Superalloy Using Nonlinear Ultrasonic Waves. *J Acoust Soc Am,* Vol. 120, No. 3, pp. 1266-1273

[58] Shui, G., Kim, J. Y., Qu, J., Wang, Y.-S. and Jacobs, L. J. (2008). A New Technique for Measuring the Acoustic Nonlinearity of Materials Using Rayleigh Waves. *NDT&E International,* Vol. 41, pp. 326-329

[59] Shull, D. J., Hamilton, M. F., Ilinsky, Y. A. and Zabolotskaya, E. A. (1993). Harmonic Generation in Plane and Cylindrical Nonlinear Rayleigh Waves. *J. Acoust. Soc. Am.*, Vol. 94, pp. 418-427

[60] Zabolotskaya, E. A. (1992). Nonlinear Propagation of Plane and Circular Rayleigh Waves in Isotropic Solids. *J. Acoust. Soc. Am.*, Vol. 91, No. 5, pp. 2569-2575